C000078196

Prevention of Accidents Through Experience Feedback

Prevention of Accidents Through Experience Feedback

Urban Kjellén

London and New York

First published 2000 by Taylor & Francis
11 New Fetter Lane, London EC4P 4EE

Simultaneously published in the USA and Canada
by Taylor & Francis
29 West 35th Street, New York NY10001

Taylor & Francis is an imprint of the Taylor & Francis Group

© 2000 Urban Kjellén

Typeset in 10/12 pt Sabon by Graphicraft Limited, Hong Kong
Printed and bound in Great Britain by Biddles Ltd, Guildford
and King's Lynn

Every effort has been made to ensure that the advice and
information in this book is true and accurate at the time of going to
press. However, neither the publisher nor the authors can accept
any legal responsibility or liability for any errors or omissions that
may be made. In the case of drug administration, any medical
procedure or the use of technical equipment mentioned within
this book, you are strongly advised to consult the manufacturer's
guidelines.

British Library Cataloguing in Publication Data
A catalogue record for this book is available
from the British Library

Library of Congress Cataloging in Publication Data

Kjellén, Urban, 1948–
    Prevention of accidents through experience feedback / Urban Kjellén.
        p.   cm.
    Includes bibliographical references and index.
    ISBN 0–7484–0925–4 (hb : alk. paper)
        1. Industrial safety.   2. Accidents–Prevention.   I. Title.
    T55 .K523 2000
    363.11′63—dc21   99–087075

Contents

List of figures   xiii
List of tables   xviii
Preface   xxi
Acknowledgements   xxv

PART I
Introduction   1

1   Introducing the concept of SHE information systems   3

    1.1   Model of a SHE information system 4
    1.2   Human information-processing analogy 6
    1.3   What does research tell us about the effects of SHE
          information systems? 8
    1.4   Developing the model further 9

2   Boundary conditions   11

    2.1   Conditions inside the company 11
    2.1.1 Size, type of technology and resources 11
    2.1.2 The organisational context 12
    2.2   The outer context 14
    2.2.1 The employer's responsibilities 14
    2.2.2 Regulations on record keeping and on the
           reporting of injuries and incidents to the
           authorities 16
    2.2.3 Workers' compensation systems 17
    2.2.4 International standards and guidelines 18
    2.2.5 Other non-governmental organisations 19

3 **Alternative accident-prevention approaches** 20

    3.1 Barriers against hazards 20
    3.2 Administrative system for feedback control 21
    3.3 Arenas for organisational learning 22
    3.4 Risk homeostasis 24

4 **Case study: Reducing emissions to the air from a fertiliser plant** 25

PART II
Theoretical foundation 29

5 **Accident models** 31

    5.1 On the need for accident models 31
    5.2 Causal-sequence models 32
    5.3 Process models 36
    5.4 Energy model 39
    5.5 Logical tree models 43
    5.6 Human information-processing models 44
    5.7 Moving the perspective to the organisational context 45
    5.7.1 SHE management models 45
    5.7.2 The SHE culture 51

6 **Framework for accident analysis** 53

    6.1 Characteristics of the accident sequence 53
    6.2 Consequences of accidents 58
    6.2.1 Types of consequences 58
    6.2.2 Consequence measures 59
    6.2.3 Economic consequences of accidents 61
    6.2.4 Actual versus potential losses 63
    6.3 Incident (uncontrolled energy flow) 65
    6.4 Deviations 67
    6.4.1 Heinrich's classical man–environment taxonomy 67
    6.4.2 Ergonomics and industrial-engineering systems views 69
    6.5 Contributing factors and root causes 70
    6.5.1 Contributing factors at the functional department and work-system levels 72
    6.5.2 Root causes at the general and SHE-management-systems levels 76
    6.5.3 Problems in identifying causal factors 77

7   Accident counter-measures   82

    7.1   Barriers against losses 82
    7.1.1   Prevention of occupational accidents 82
    7.1.2   Prevention of major accidents due to fires and
        explosions 83
    7.2   Active and passive barriers 86
    7.3   Different time frames in the implementation and
        maintenance of barriers 86
    7.4   The role of experience transfer 87
    7.5   Designing for safety of machinery 89
    7.6   Safety measures in operation 92
    7.6.1   The permit-to-work system 93

8   The human element in accident control   95

    8.1   Human information processing 95
    8.2   Human errors 100
    8.2.1   Definition 100
    8.2.2   Human-error taxonomies 101
    8.2.3   Error recovery 102
    8.2.4   The influence of emotion 104
    8.2.5   Preventing human errors and promoting
        error recovery 105
    8.3   The role of the operators in major-accident prevention 107
    8.3.1   Unscheduled manual interventions 107
    8.3.2   Fallacy of the defences-in-depth philosophy 109
    8.3.3   High-reliability organisations 110

9   The occurrence of accidents over time   111

10   Feedback and use of experiences in decision-making   114

    10.1   Overview of feedback mechanisms 114
    10.2   Uses of SHE-related information in decision-making 115
    10.3   The diagnostic process 117
    10.3.1   Effects of limitations in human information-processing
        capacity 119
    10.3.2   Hale's problem-solving cycle 120
    10.3.3   Deming's circle 122
    10.4   Persistent feedback control 123
    10.5   Ashby's law of requisite variety 124
    10.6   Van Court Hare's hierarchy of order of feedback 126

10.7   *Obstacles to an efficient learning from experience 129*
10.7.1   *Organisational defences 129*
10.7.2   *Local information and the SHE information system 130*
10.7.3   *Culpability and liability 131*
10.8   *A balanced approach 132*

11   Requirements for a SHE information system   **134**

11.1   *Requirements for SHE performance indicators 135*
11.2   *Requirements for the SHE information system
        as a whole 136*
11.2.1   *Data collection 137*
11.2.2   *Distribution and presentation of information 137*
11.2.3   *The SHE information system as a whole 139*

PART III
Learning from incidents and deviations   **141**

12   Sources of data on accident risks   **143**

12.1   *The ideal scope of different data-collection methods 143*
12.2   *Filters and barriers in data collection 144*

13   Accident and near-accident reporting and investigation   **146**

13.1   *Why report and investigate accidents and
        near accidents? 146*
13.2   *Investigations at three levels 147*
13.3   *Reporting 149*
13.3.1   *Reporting to the authorities 150*
13.3.2   *Problems of under-reporting 151*
13.3.3   *Near-accident reporting 154*
13.4   *Immediate investigation and follow-up 160*
13.4.1   *Quality of the supervisor's first report 160*
13.4.2   *Use of checklists and reporting forms 161*
13.4.3   *Displaying the sequence of events 163*
13.4.4   *Computer-supported accident investigations 167*
13.4.5   *Registration of accident costs 168*
13.5   *Group problem-solving 168*
13.6   *In-depth accident and near-accident investigations 173*
13.6.1   *The steps in an in-depth investigation 174*
13.6.2   *Applying SMORT in in-depth investigations 179*
13.6.3   *Legal aspects of the commission's report 185*

13.7　*Computer-supported distribution of the*
　　　investigation report 186
13.8　*A procedure for accident and near-accident*
　　　reporting and investigation 187

14　SHE inspections and audits　　　　　　　　　189

14.1　*Inspections 189*
14.1.1　*Workplace inspections 190*
14.1.2　*Inspecting and testing barrier integrity 193*
14.2　*SHE audits 194*
14.2.1　*Application of SMORT in audits 196*

15　Accumulated accident experience　　　　　　　198

15.1　*Database on accidents and near accidents 199*
15.1.1　*Database definition 199*
15.1.2　*Accessing the database 200*
15.1.3　*Coding of accident and near-accident data 205*
15.2　*Analysis of accident and near-accident data 209*
15.2.1　*Finding accident repeaters 209*
15.2.2　*Uni- and bi-variate distribution analyses 210*
15.2.3　*Accident-concentration analysis 211*
15.2.4　*Analysis of accident causes 215*
15.2.5　*Severity-distribution analysis 216*
15.2.6　*Extreme-value projection 218*
15.3　*Experience carriers 221*

PART IV
Monitoring of SHE performance　　　　　　　　225

16　Overview of SHE performance indicators　　　227

17　Loss-based SHE performance indicators　　　　228

17.1　*The lost-time injury frequency rate 228*
17.1.1　*The control chart 228*
17.1.2　*The problems of SHE performance measurement 233*
17.1.3　*Zero-goal mindset 236*
17.2　*Other loss-based SHE performance indicators 237*
17.2.1　*Measures of risk 237*
17.2.2　*Standard loss-based SHE performance indicators 238*
17.2.3　*Untraditional SHE performance indicators 239*

18   Process-based SHE performance indicators   242

   18.1   SHE performance indicators based on
   near-accident reporting 242
   18.2   Behavioural sampling 243

19   Causal factor-based SHE performance indicators   248

   19.1   Rating the elements of a company's SHE
   management system 248
   19.1.1   International Safety Rating System (ISRS) 249
   19.1.2   Self-rating as a means of improving SHE
   management 251
   19.1.3   Tripod Delta 254
   19.2   Measurement of safety climate 255
   19.3   Measuring the degree of learning from incidents 256

20   Selecting key SHE performance indicators   258

   20.1   Combinations of SHE performance indicators 258
   20.2   Indicators of barrier availability 260

PART V
Risk analysis   263

21   The risk-analysis process   265

   21.1   What is risk analysis? 265
   21.2   Acceptance criteria for the risk of losses due
   to accidents 266
   21.3   Methods of risk analysis 267

22   Coarse or energy analysis   271

   22.1   Planning 272
   22.2   Execution and documentation 273
   22.2.1   Identification of hazards and causes 274
   22.2.2   Risk estimation 275
   22.2.3   Development of safety measures 277
   22.2.4   Documentation and follow-up of results 277
   22.3   Establishing a database on potential accidents 278

23   Detailed job-safety analysis   280

   23.1   Analysis object 280
   23.2   Resource needs and scheduling 280
   23.3   Description of the steps of the job 281

23.4 *Subsequent steps 282*

23.5 *Accidental exposure to chemicals 282*

23.6 *Systematic mapping of hazards within an organisation 284*

24 Risk assessments of machinery   285

24.1 *Requirements as to risk assessments 285*

24.2 *Method for risk assessment 286*

24.2.1 *Determination of the limits of the machinery (Step 1) 287*

24.2.2 *Coarse risk assessment (Step 2) 289*

24.2.3 *Detailed risk assessment of the machinery (Step 3) 293*

25 Comparison risk analysis   294

25.1 *Acceptance criteria for the risk of occupational accidents 294*

25.2 *Risk-assessment model 295*

25.2.1 *Assumptions 296*

25.3 *The steps of the analysis 297*

26 CRIOP   302

PART VI

Putting the pieces together   309

27 The oil and gas industry   311

27.1 *Accidents in offshore oil and gas production 311*

27.2 *The Ymer Platform 311*

27.2.1 *Design 311*

27.2.2 *Organisation and manning 312*

27.3 *Prevention of accidents in design 313*

27.3.1 *The phase model for offshore field exploration and development 313*

27.3.2 *SHE management principles 316*

27.3.3 *Prevention of major accidents 322*

27.3.4 *Prevention of occupational accidents 327*

27.4 *Construction-site safety 330*

27.4.1 *SHE management principles 331*

27.4.2 *Step 1: Pre-qualification 331*

27.4.3 *Step 2: Tender evaluation and clarification, contract award 332*

27.4.4 *Step 3: Evaluation of the SHE programmes 333*

27.4.5   Step 4: Follow-up during construction 333
27.5   Safety during plant operation 335
27.5.1   SHE management principles 336
27.5.2   Policy and goals 337
27.5.3   Implementation 338
27.5.4   Control and verification 339

28   The trucking industry   344

28.1   Accidents in road transportation 344
28.1.1   Measures of the risk of traffic accidents 345
28.2   The man–vehicle–road–environment model 346
28.2.1   The driver 348
28.2.2   The vehicle 349
28.2.3   The traffic environment 351
28.3   Sources of information on traffic-accident risks 352
28.4   Feedback mechanisms 356
28.4.1   The trucking company 357
28.4.2   The truck manufacturer 359
28.4.3   The roads administration 360

PART VII
Improving the corporate SHE information system   363

29   The improvement process   365

29.1   Evaluation of existing conditions 366
29.2   Establishing goals and defining user needs 367
29.3   Developing solutions and following up results 369

30   Design of the system   371

30.1   Database definition 371
30.2   Organisation and routines 372
30.3   Personnel 372
30.4   Instruments and tools 373

31   Epilogue   374

Appendix I: Definitions   376
Appendix II: SMORT questionnaire   379
Bibliography   409
Name index   417
Subject index   419

Figures

1.1	Flow of information in a SHE information system	5
1.2	The SHE information system seen in relation to its effects and the resources required	9
3.1	The relationship between hazard, victim and barriers	20
3.2	System controlled through negative feedback	21
3.3	Developments of the LTI-rate after the introduction of a SHE management system	22
3.4	Different learning mechanisms through experience exchange	23
4.1	Integrated process control system	25
4.2	Emissions to the air during a four-year period	26
5.1	'Domino theory' according to Heinrich	33
5.2	The ILCI model	34
5.3	The TRIPOD model of accident causation	35
5.4	Analysis of an accident at a construction site by means of the OARU model	38
5.5	Relations between the phases of different models of accidents	39
5.6	Haddon's ten accident prevention strategies	40
5.7	Statistics on fatalities in Norway during 1984–93	42
5.8	Accident types and their share of the reported accidents for selected branches of industry and for all industry in Norway in 1993	42
5.9	Example of a typical combination of deviations likely to lead to a traffic accident	44
5.10	Accident sequence model showing failures in the various human information-processing stages leading to an accident	44
5.11	The top checkpoints of MORT	46
5.12	The levels of a SMORT analysis	49
5.13	E&P Forum's SHE management model	50
6.1	Accident analysis framework	53
6.2	Framework for the analysis of an accident sequence	56
6.3	Distribution of accidents by actual and potential losses	64

6.4   Hierarchical relation between contributing factors and
       root causes   72
6.5   Deviations and contributing factors of an accident during
       transportation of a 12-ton pressure pipe in a tunnel   75
6.6   Accident-investigation stairs   78
6.7   Two dimensions in the attribution of causes, illustrated
       by an example showing different causes of a crane
       accident   80
7.1   Illustration of when the different barriers according to
       Haddon intervene in the accident sequence   83
7.2   Measures to prevent exposure to hazardous chemicals   84
7.3   The prevention of fires and explosions in a plant for
       the processing of hydrocarbons   85
8.1   Surry's Decision Model of the accident process   96
8.2   Wilde's risk homeostasis model of the relation between
       operator behaviour and accident risk   97
8.3   Hale and Glendon's model of behaviour in the face
       of danger   99
8.4   Error recovery may occur at different stages in human
       information processing   103
8.5   Unscheduled manual interventions (UMIs) following
       a process disturbance   108
9.1   Recorded number of accidents per year   111
9.2   Graphical illustration of the frequency function for
       a Poisson distribution where $(c \times t)$ equals 4   113
10.1   Examples of uses of SHE-related information in
       decision-making   116
10.2   Mintzberg's model of the decision-making process   118
10.3   The marginal subjective value and cost of additional
       information   120
10.4   The problem-solving cycle according to Hale   121
10.5   Deming's circle and the quality standards necessary to
       prevent fall-back   122
10.6   Feedback cycle for the control of anything   123
10.7   Percentage of the accident reports at seven yards
       containing information on safety measures by type
       of measure   125
10.8   Number of actions per inspection at a construction site,
       by type of action   127
10.9   Single-loop and double-loop learning according to Argyris   129
10.10 Balanced approach in the prevention of accidents   132
12.1   Overview of different means of collecting data on
       accident risks and their ideal scope   143
13.1   Comprehensive approach for accident investigations
       at three levels   148

13.2   Relation between reporting reliability and the average
        severity of accidents   152
13.3   Iceberg theory according to Heinrich   155
13.4   The development in the frequency of accidents and
        near accidents after the introduction of new near-accident
        reporting routines   156
13.5   What different participants in a human-error reporting
        system give and receive in return   157
13.6   The Finnish accident-analysis model   165
13.7   Basic outline of a cause–effect diagram   166
13.8   Results of an experiment involving training of
        supervisors and safety representatives at three
        construction sites in the use of the OARU
        accident-investigation method   170
13.9   The victim's position at the time of the accident   184
13.10   The results of the SMORT investigation displayed
        in a diagnostic diagram   186
15.1   Overall data model for the proposed 'smallest efficient
        set' of data elements   202
15.2   Simplified database environment   202
15.3   Illustration of errors relating to degree of retrieval and
        degree of precision   203
15.4   Tendency of coding reliability to decrease as the analysis
        moves backwards in the 'causal chain' of Table 15.2   208
15.5   Distribution of lost-time and first-aid injuries at two
        offshore installations by type of event and department   212
15.6   Results of an accident-concentration analysis of
        construction work in an offshore project   213
15.7   Severity-distribution analysis of the accidents at a
        steel mill   217
15.8   Extreme-value projection Type I of the accidents
        at the steel mill   220
15.9   Extreme-value projection Type II of the accidents
        at the steel mill   221
15.10   Examples of experience carriers used to transfer
        experiences from an accident database to existing and
        new production systems   222
16.1   Overview of different SHE performance indicators   227
17.1   Control chart showing the development of the lost-time
        injury frequency rate for a steel mill for ten consecutive
        periods   229
17.2   Chart displaying a trend in the development of the
        LTI-rate   232
17.3   Percentage of the lost-time accidents at seven yards that
        were of a less severe type as a function of LTI-rate   234

17.4   Relation between the minimum period length in months
        and the number of employees in order to produce
        meaningful control charts   235
17.5   Chart showing the parallel developments of the LTI-rate,
        the accumulated LTI-rate from the first period and the
        moving 3-months average of the LTI-rate   235
17.6   Total recordable injury rates and their components for
        five plants producing car components   240
17.7   Nelson Aalen plot of accidents in a large construction
        project   240
18.1   Application of behavioural sampling in controlling
        performance   244
18.2   Development of the so-called housekeeping index at
        a shipyard   245
19.1   Example of a failure-state profile   255
19.2   Distribution of actions taken after accidents by level
        of feedback   257
19.3   Weighted average level of feedback in accident reports
        from three offshore projects   257
20.1   Control chart for the high-potential incident rate
        at a refinery   260
21.1   Relation between acceptance criterion and goal for
        the risk of accidents   266
21.2   The scope of different risk-analysis methods   269
22.1   Typical analysis team   272
22.2   Distribution of all identified potential accidents in the
        electrolysis departments by type of energy involved   279
24.1   The main steps of the total risk assessment   287
24.2   Coarse and detailed risk assessments   290
25.1   The steps of a Comparison risk analysis   298
26.1   The steps of the CRIOP scenario analysis   304
26.2   First parts of a STEP diagram of the scenario involving
        collision between a supply boat and a floating offshore
        installation   306
27.1   The Ymer platform   312
27.2   Phase model for offshore field exploration and
        development   314
27.3   The role of different control and verification activities
        in meeting goal-oriented and detailed requirements
        to design   317
27.4   Acceptance criteria for loss of evacuation possibilities
        due to major accidents   318
27.5   Experience flow from operations to project to support
        the design process   319
27.6   Use of different risk analysis methods in design   328

27.7	Norskoil's principles for follow-up of contractor's construction site safety	332
27.8	The lost-time injury frequency rate from start of construction	335
27.9	The SHE control loop	336
27.10	Registration and follow-up of accidents and near accidents	339
27.11	Control chart for the Ymer platform, showing the upper and lower control limits	340
28.1	A model of the driver–vehicle–environment system	346
28.2	The driver-error iceberg	353
28.3	Borderline between serious and minor conflicts according to the Swedish traffic-conflicts technique	356
28.4	Different feedback mechanisms in safety in road transportation	357

Tables

5.1	Relationship between classes of deviations and different systems of production planning and control	37
5.2	Examples of safety measures relating to Haddon's ten strategies	41
5.3	Hazards generated by machinery	43
6.1	Application of the accident-analysis framework to illustrate the development of different types of losses	58
6.2	Type of injury and part of the body injured	59
6.3	Overview of measures of the severity of losses due to injury or damage	60
6.4	Common consequence measures for different types of losses	61
6.5	Company cost elements of an accident	62
6.6	Two examples of accident type classifications	66
6.7	Injury agency classification	66
6.8	Overview of different schemes for the classification of deviations	68
6.9	Checklist of deviations	69
6.10	Examples of accident models applying a causal hierarchy	71
6.11	Extracts from Swain's checklist on human-performance-shaping factors	73
6.12	Checklist of contributing factors at the workplace and department levels	74
6.13	The different elements of a root-cause analysis	76
7.1	Examples of long-term measures for the implementation and maintenance of barriers against accidents	88
7.2	Strategy for selecting safety measures according to EN 292	89
7.3	Examples of safety measures according to EN 292 and their relation to Haddon's accident-prevention strategies	90
7.4	Summary of requirements as to the involvement of an authorised institution in the documentation of the safety of machinery	91
7.5	A checklist of measures for use in evaluating applications of hot-work permits	94

8.1	Overview of human-error taxonomies	101
10.1	Hierarchy of feedback systems arranged by order of feedback	126
11.1	Requirements for SHE performance indicators	135
11.2	Requirements for a SHE information system	136
13.1	Accident-reporting requirements in Norway	151
13.2	Example of an internal form for an immediate investigation	164
13.3	Form for registration of accident costs	169
13.4	SMORT checklist levels 1 and 2	180
13.5	SMORT checklist levels 3 and 4	181
14.1	Example of a checklist on inspection themes	191
14.2	Example of a theme-specific checklist	191
15.1	Proposed 'smallest efficient set' of data elements in accident and near-accident records for computer storage	201
15.2	Overview of a coding schedule for accidents and near accidents	207
15.3a	Distribution of injuries by job experience	215
15.3b	Distribution of injuries at a hamburger chain by job experience	216
15.4	Severity-distribution analysis of accidents at a steel mill	217
15.5	Data for extreme-value projection of the consequences of occupational accidents within a steel mill	219
17.1	Loss-related SHE performance indicators	238
17.2	Example of untraditional SHE performance indicators	239
18.1	Example of process-related SHE performance indicators	243
19.1	Measurement of safety performance according to ISRS	250
19.2	Scale applied in evaluation of commitment to SHE through leadership	251
19.3	Safety element method	252–3
21.1	Overview of different risk-analysis methods	268–9
22.1	Checklist of typical jobs	274
22.2	Checklist of hazards	274
22.3	Record sheet for documentation of the results of a Coarse analysis	275
22.4	Matrix for risk estimation for a workplace with about 100 employees	276
22.5	Matrix for estimation of the individual risk	276
23.1	Record sheet for documentation of a job-safety analysis	281
23.2	Matrix for risk estimation for exposure to chemicals	283
24.1	Record sheet for documentation of the Coarse risk assessment	291
24.2	Record sheet for documentation of the detailed risk assessment	292
25.1	Decision rules in risk estimation	297

25.2	Checklist of factors that affect the risk of accidents	299
25.3	Extracts of results from the comparison analysis of kitchen work	300
25.4	Results of a comparison analysis	300
26.1	Record sheet for documentation of the CRIOP scenario analysis	307
26.2	Checklist of phases in human-information processing	308
26.3	Checklist of performance-shaping factors	308
27.1	Examples of results from the coarse REPA for the two development alternatives for Ymer	324
27.2	Overview of different barriers and method for control of barrier availability	342
28.1	Some common measures of the risk of traffic accidents	345
28.2	Performance indicators for different traffic-safety areas	361
29.1	Checklist of uses of a SHE information system	368

Preface

Industrial accidents are unwanted events that cause losses to the individual, the company and the society as a whole. They are also symptoms of underlying weaknesses within the company. When accidents occur, they represent an invitation to improve by learning about these weaknesses. Near accidents, production disturbances and unsafe conditions are other examples of such symptoms.

There is no simple answer to the question of how to use the experiences from unwanted events and conditions in order to prevent accidents. Accident risks usually cannot be engineered away altogether. Neither is it sufficient to train and motivate people to avoid the risks or to establish and enforce strict work procedures. Only a combination of well-balanced measures will do the job. Each company must find its own unique combination of measures, and the organisation must feel ownership of the measures and trust their efficiency. The measures also need to be adapted to the changing conditions inside and outside the company.

Experience feedback plays a central role in any management systems for the prevention of accidents. This is accomplished through different channels. There are formal channels for experience feedback such as accident and near-accident reporting and workplace inspection. These channels do not necessarily tap the abundant collective experiences of the personnel from their day-to-day encounters with near accidents and production disturbances. This can only be accomplished by involving the personnel in the problem-solving itself. Each company must find its own solutions on how to design and manage efficient experience feedback and learning processes.

Students of the management of safety will usually observe that companies' systems for feedback control of accidents are far from ideal. There are many examples in the literature of the shortcomings in, for example, companies' accident and near-accident reporting systems. A number of technical, organisational and motivational obstacles have to be overcome in order to accomplish efficient feedback.

Many normative models of how such ideal systems should be designed are described in the literature. The problem is often that they do not work in accordance with the pre-conditions in practice. In order to improve, the decision-makers inside a company must not only have an idea of what they want to accomplish, but also about the constraints and obstacles that have to be overcome. Often, the change process is at least as significant as the design of the solution, when it comes to the efficiency of the results.

This book is addressed to current and future facilitators of the processes inside companies aimed at improving experience transfer and learning in the area of safety. It will help the reader in finding answers to the question of how to develop and manage corporate systems for control of accidents through experience feedback. It does not provide the reader with any ready-made answer. Rather, it presents a variety of tools and methods, both for the change process and for the final outcome, and provides a basis for the selection and adjustment of the right tools for the actual circumstances. The book also raises a number of critical questions that have to be addressed in the design and implementation process.

Experience feedback in the area of safety is a complex phenomenon. The book analyses it from a number of different perspectives rooted in the technical, social and management sciences. The reader with expertise within one field of science may find the treatment too superficial from his or her particular perspective. It has been the author's ambition to present the totality rather than to go into depth within one particular area. This ambition usually coincides with the practicalities of problem-solving in industry. Where relevant, the reader is referred to more specialised literature for further reading.

In the literature, accidents are subdivided into different categories such as occupational, major and traffic accidents. Among the major accidents, so-called process accidents are an important special subset involving loss of containment of hazardous chemicals. Safety practitioners tend to specialise in the prevention of one single category of accidents. Each community of practitioners emphasises a specific set of remedial actions and has developed extensive skills in mastering them. Accidents are complex phenomena needing a great variety of measures to be adequately controlled. It is the author's ambition to build bridges between students and professionals specialising in the different fields of accident prevention. The book presents a common framework for the analysis of all kinds of accidents and for the design of adequate experience feedback loops and remedial actions. This framework will serve as a communication tool in experience exchange on the generalities and uniqueness of each area of application. Numerous examples are presented in the areas of major, occupational and traffic accidents that reflect different applications of this framework.

The book has its origin in the author's lecturing in safety management at the Norwegian University of Science and Technology in Trondheim. It combines theories and principles from safety science with the author's professional experience from safety practice in industry. It has been written with a mixed readership in mind including safety professionals within industry, insurance companies and the authorities, safety researchers and students in the area of safety management. When used as a textbook in undergraduate and post-graduate courses, the objective is to give the students an essential knowledge and understanding to be able to function effectively in their jobs as high-level SHE (safety, health and environment) experts in industry and within the authorities. The goals are to provide the reader with:

- Knowledge about principles of accident control through experience feedback;
- An understanding of the challenges within SHE management in industry today and of modern principles for problem solving;
- Knowledge about various methods and tools for use in SHE practice and an ability to develop skills to put them into use; and
- Proficiency in assessing the potentials and limitations of various existing methods and new trends in SHE management.

In approaching these different goals, the author has made a careful balance between the need to give a comprehensive presentation of alternative approaches from the safety-research literature and the need to give precise recommendations. Whereas the experienced SHE professional may be most benefited by the former, students with minimal practical experience in the field may in the short term be best served by a cookbook approach. The author has tried to solve this dilemma by presenting alternative solutions and at the same time giving recommendations as to the preferred alternative.

Part I introduces the concept of the SHE information system and presents four complementary perspectives on accident prevention. It also presents a case and explains the outline of the book in the light of this case.

Part II reviews some of the basic theories and principles in the design of SHE information systems. Thereafter, an accident model is introduced that represents a synthesis of many existing models. It provides a framework for a presentation of the different types of methods and tools in Parts III to V. Part II also introduces a number of theoretical concepts and principles including accident-barrier theory, theories on human performance and errors, and feedback mechanisms in decision-making. This part ends with a set of criteria for the evaluation of SHE information systems.

Part III leads the reader through a number of different methods for the collection, analysis and use of data on accident risks. The focus is on how to prevent accidents through learning from individual events and conditions and from accumulated accident experience from many incidents. The advantages and disadvantages of the different methods are discussed.

Part IV is dedicated to the control of accidents through performance feedback. It reviews the different measures of SHE performance used in the monitoring and control of the accident risk level and discusses their merits and shortcomings.

Part V is a descriptive section. It presents different methods of risk analysis. The selected methods fall within the scope of this book, since the employee's experience of incidents and accidents is a basic data source in the analysis. Seen from this perspective, risk analysis provides an arena for exchange and documentation of tacit knowledge on accident risks. This knowledge will then become available to formal decision-making.

Part VI brings the different pieces presented in Parts II–V together in case descriptions of SHE information systems. Three of the cases involve different phases of the lifecycle of an offshore installation: design, construction and operation. A separate chapter is dedicated to traffic safety. Although this theme falls outside the main scope of the book, i.e. the prevention of industrial accidents, the chapter gives a demonstration of how the same basic principles of experience feedback are applied in a new setting.

Part VII applies an output-driven perspective to the development of a company's computerised SHE information system and presents a scheme for how this may be accomplished.

Acknowledgements

Many people have been involved in the development of this book, including many of my colleagues at Norsk Hydro ASA and NTNU. I want especially to mention Jan Hovden at NTNU, who has inspired me in completing this work and John Monsen at Norsk Hydro, who has been an interesting discussion partner. I have also received constructive comments and support from Jorma Saari at the Finnish Institute for Occupational Health and from two former students at NTNU, Karsten Boe and Roy Hallås. Marianne Döös at the Swedish Institute of Working Life has helped me in quality assurance of the sections on organisational learning. Similarly, Ragnar Rosness and Trygve Steiro at Sintef in Trondheim have reviewed the chapter on human errors, Arne Tiltnes at Norsk Hydro the chapter on risk assessment of accidental exposure to chemicals, and Thomas Lekander at the Swedish National Road Administration the chapter on traffic safety.

I have also been dependent on many people for source input to the book. I have used examples from the project work of my students at NTNU. Here I especially want to mention Gro Blindheim, Børge Godhavn, Bjørnar André Haug, Jørn Lindtvedt, Vidar Tuven and Elisabeth Wendel. Karsten Boe at Norsk Hydro has been my discussion partner in the area of risk acceptance criteria. Bengt Elf at ASG Road Transport, Lennart Svensson at Volvo Lastvagnar AB and Thomas Lekander at the Swedish National Road Administration have helped me with input to the chapter on traffic safety.

Rykkinn, October 1999
Urban Kjellén

Acknowledgements

Figure 5.3 Reprinted from 'Too little and too late: A commentary on accident and incident reporting systems' Reason, 1991 In: 'Near Miss Reporting as a Safety Tool' Van der Schaaf, T.W., Lucas, D.A. and Hale, A.R. (eds) by permission of Butterworth Heinemann Publishers, a division of Reed Educational & Professional Publishing Ltd

Figure 8.2 Reprinted from 'Risk homeostasis and risk assessment' Stanton, N. and Glendon, I., 1996, *Safety Science*, 22: 1–13, page 4 with permission from Elsevier Science

Figure 8.3 Reprinted from 'Individual Behaviour in the Control of Danger' Hale, A.R. and Glendon, A.I., 1987, page 40 with permission from Elsevier Science

Figure 18.2 Reprinted from 'The effect of positive feedback on industrial housekeeping and accidents; a long term study at a shipyard' Saari, J. and Näsänen, M., 1989, *International Journal of Industrial Ergonomics*, 4: 201–211, page 207 with permission from Elsevier Science

Section 25.2 Reprinted from 'Integrating analyses of the risk of accidents into the design process – Part II: Method for prediction of the LTI-rate' Kjellén, U., 1995, *Safety Science*, 19: 3–18, pages 4–7 with permission from Elsevier Science

Figure 28.2 Reprinted from 'Accident liability – the human perspective' Maycock, G., 1997 In: Rothengatter, T. and Vaya, E.C. (eds), Traffic and Transport Psychology, Pergamon, Amsterdam, page 5 with permission from Elsevier Science

Part I

Introduction

In Chapter 1 we introduce the concept of SHE (safety, health and environment) information systems. It will provide a frame of reference in our subsequent analysis of the different tools and methods used in accident control through experience feedback. We make a comparison with the human information processes and identify basic similarities and differences. Chapter 2 gives an overview of different boundary conditions of a SHE information system, both inside and outside a company. Chapter 3 introduces four different approaches in safety practice and describes how these will contribute in subsequent parts of the book to our understanding of how to prevent accidents. In Chapter 4 we will look into a case from the environmental field. It demonstrates a successful application of basic principles of experience feedback in the reduction of emissions from a fertiliser plant. We use this example to present some of the issues dealt with in later parts of the book and demonstrate how they form a coherent whole.

1 Introducing the concept of SHE information systems

In pre-industrial society, human muscles accounted for much of the energy used in production of goods and services. The individual and the work group had hands-on control of the hazards associated with these energies. At the same time, humans were at the mercy of nature and its hazards, as represented by weather extremes, wild animals, etc. Accidents were viewed as a phenomenon beyond human control. They were determined by fate or were a punishment for sin.

Today, electrical motors and other non-human energy sources keep most of production going. Mechanisation and automation, together with new principles for management and organisation, contribute to the separation of individuals from the direct control of the energies in production. At the same time, the energy flow has increased and so also has the potential for severe accidents. It has become necessary for industry to control accident risks by the same management principles that are applied to the control of production.

In connection with industrialisation, safety was identified as a specific activity area needing dedicated attention. Accident prevention was already at an early stage, accomplished through the development and enforcement of safety regulations, standards and rules. These were based on the accumulated experiences from investigations into accident cases. This 'trial and error' approach is still valid today, but has become insufficient to master the rapidly changing technology and means of organisation of modern industrial society. Today, the management of risks is accomplished by means similar to those of management in general. It involves the establishment, implementation and follow-up of corporate policies, acceptance criteria and goals related to safety, health and the environment (SHE). The systematic feedback of experiences on accident risks is still a cornerstone in any management system for the prevention of accidental losses. Compared to the early 'trial and error' approach, the types of feedback loops are so much more diverse and include reactive as well as pro-active means.

Accidents are infrequent events in modern industry, and accident prevention is usually not at the top of the agenda of management, supervisors and operators. Accidents belong, however, to a set of unplanned and unwanted events that in all are relatively common. Production stoppages, substandard product quality and material damages are more common consequences of these events than, for example, injury to personnel. Management and operators are used to handling unplanned events on a daily basis and to taking action to prevent recurrence. This is accomplished through improvements in the planning and control of production and maintenance. A reduced risk of accidents will follow, although this is not the primary aim. The question then arises whether SHE management really is necessary as a distinguishable set of activities. Our answer to this question is yes. There are synergetic effects between the management of SHE and the management of production and maintenance (Kjellén *et al.*, 1997). It follows that the best SHE results can only be achieved when there are adequate production and maintenance planning and control systems and an adequate SHE management system. Such combinations also have positive effects on the economic results through improved production quality and productivity.

1.1 Model of a SHE information system

We are here concerned with the prevention of accidents through formal systems for experience feedback. In analysing the necessary processes to accomplish this, we will apply the concept of **a SHE (safety, health and environment) information system**. Such a system provides the information needed for decisions and signalling related to safety, health and environmental matters. It is a vital element in a company's SHE management system. A SHE information system provides decision-makers with support in several areas such as:

- Prevention of accidents that may result in injury to personnel, accidental emissions of pollution to the environment and damage to material assets;
- Prevention of occupational diseases;
- Reductions in emissions to the air and discharges to the sea of harmful substances and land contamination;
- Waste handling;
- Keeping various physical parameters such as noise, illumination and concentrations of toxic gases in the working environment within accepted limits; and
- Follow-up of the psychosocial working environment.

We will here focus on the first application. In our analysis of the feedback mechanisms in accident prevention, we will use a model of the SHE information system and its different subsystems according to Figure 1.1.

Our underlying assumption is that accidents are preventable through systematic experience feedback. This feedback goes from the workplaces to the decision-makers at different levels of the organisation. An important part of this feedback takes place in informal settings in the day-to-day contacts between people in the organisation. We will focus on how the experiences are made explicit through a documentation of the results of

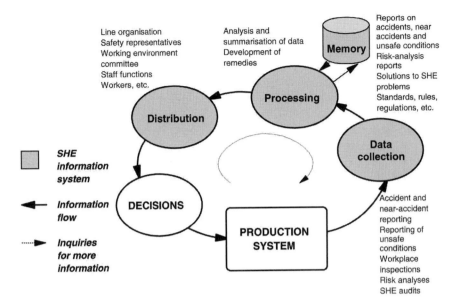

Line organisation
Safety representatives
Working environment
committee
Staff functions
Workers, etc.

Analysis and
summarisation of data
Development of
remedies

Memory

Reports on
accidents, near
accidents and
unsafe conditions
Risk-analysis
reports
Solutions to SHE
problems
Standards, rules,
regulations, etc.

Processing

Distribution

☐ **SHE information system**

◀ **Information flow**

Data collection

DECISIONS

Accident and
near-accident
reporting
Reporting of
unsafe
conditions
Workplace
inspections
Risk analyses
SHE audits

PRODUCTION SYSTEM

⋯▶ **Inquiries for more information**

Figure 1.1 Flow of information in a SHE (safety, health and environment) information system. Inquiries for more information go in the opposite direction.
Source: Adapted from Kjellén, 1983.

investigations into accidents and near-accidents, workplace inspections, SHE
audits, risk analyses, etc. Positive effects on safety are only achieved when
the *loop is closed*, i.e. when the results of the decisions are implemented
in a way that affects the work places (existing or future).

A SHE information system provides the following accident prevention
functions:

1 *Reporting and collection of data* on accident risks by means of
   accident and near-accident investigations, workplace inspections,
   SHE audits and risk analyses. Methods of data collection include
   observation, interviews, self-reporting, group discussions, electronic
   registration, etc.
2 *Storing* of data in a memory (paper file, electronically, etc.) and
   retrieval of data from it.
3 *Information processing*, i.e. retrieval and analysis of data, compilation into meaningful information, development of remedial
   actions, etc.
4 *Distribution* of the information to decision-makers inside the organisation, e.g. line management, staff officers, working environment
   committee.

The effectiveness of a SHE information system has traditionally been determined by its ability to provide the necessary basis for decisions on remedial actions. In order to close the loop, i.e. to go from experience to action, the efficiency of each subsystem or function is critical. A weak link such as an inadequate distribution of information will break up this loop. This is irrespective of the quality of other functions. The end result, the prevention of accidents, is ultimately determined by the decision-makers' ability to ask for and use the available information.

More specifically, a SHE information system must support such different SHE management activities as:

- Prioritising SHE measures, developing them and evaluating their effect;
- Establishing SHE goals and monitoring the development in SHE performance for comparison with these goals and other companies' performance;
- Providing source data as input to risk analyses;
- Evaluating the efficiency of SHE programme elements and the SHE management system as a whole.

The rapid developments in information technology have made it possible to solve many of the traditional problems associated with SHE information systems. Computer networks provide efficient support in the collection, processing and distribution of information and 'unlimited' storage space. The focus in the development of SHE information systems has thus shifted from the 'mechanical aspects' of providing adequate computer hardware and software. We are now more concerned about how a SHE information system may support the shaping of knowledge and commitment to SHE among decision-makers.

1.2 Human information-processing analogy

Let us compare our model of a SHE information system with the way we as human beings collect, process and use information in order to manage our environment. We use our eyes, ears and other sense organs to collect data. We process the data in our brain and relate it to pre-stored information in our memory. We make decisions and execute these through talking, writing, moving, etc.

The analogy also brings us beyond this simple model of human information processing. Research has explored how learning is accomplished through an interaction between the individual and his/her environment. In so-called *experiential learning*, the individual does not merely study a phenomenon but also does something with it and observes the results (Kolb, 1984). Experiential learning resembles the feedback loop that we recognise from our model of the SHE-information system. It takes place in four different ways:

1 Through reflective observations of the environment,
2 Through assimilation of the observations and conceptualisation of them,
3 Through transformation of the experiences into actions to affect the environment (experimentation), and
4 Through concrete experience of the results of the actions.

Knowledge is achieved as a result of a combination between grasping experience and transforming it. In a positive learning process, the individual will gradually improve in his/her understanding of real-world phenomena (know-why) and his/her ability to act effectively (know-how). This is called *qualifying experience*.

We are here not primarily concerned with individual learning but with the industrial organisation's ability to learn from experience. We will focus on how the organisation learns from unwanted occurrences and errors, and thereby develops a shared understanding and know-how among its members. This learning is used to improve the organisation's ability to control and prevent such occurrences. SHE information systems provide different possibilities for the organisation to acquire such skills.

We will also be concerned with constraints and obstacles in efficient organisational learning and how to circumvent them. We know, for example, that direct experience through trial and error is an efficient way of learning. A problem in a large, hierarchical organisation is that decision-makers often do not get in direct contact with the consequences of their decisions. Another concern is that information on accident risks is often collected by a different part of the organisation than that responsible for using the information. In human information processing, observation is not a passive reception of information but involves active search for missing information. Decision-makers who are distant from the information source rarely have the opportunity of filling in missing information about the circumstances at the accident site and the conditions under which the information was collected. They run the risk of misinterpreting and distorting the information.

Our human information-processing analogy takes us even further. Natural information systems such as our own have developed through an evolutionary process. Species and individuals equipped with effective information systems have a competitive advantage. This evolutionary process is *output-driven*. Success has been accomplished through adequate decisions and actions in relation to the demands from the environment. Our information system is adapted to meet the actual needs.

Company SHE information systems, on the other hand, are man-made artefacts that have not been exposed to the rigorous laws of natural selection. In developing SHE information systems, experts have a tendency to ask questions such as: 'What information can we get?' 'What do we do about it?' These questions are typical for an *input-driven* development process, where the end result often is inefficient and even inadequate. History is full of examples of data graveyards, where elaborate databases on accidents have been created without finding any users that find it worthwhile to access them for information for use in accident prevention.

In the approach presented here, we will primarily be concerned with meeting the organisation's information needs. We thus strive to avoid mistakes made in designing many of today's SHE information systems.

1.3 What does research tell us about the effects of SHE information systems?

W.W. Heinrich pioneered the research into industrial accident prevention in the 1920s. He described the accomplishment of safety through five separate steps: accident investigation; analysis; selection of remedy; implementation; monitoring of results (Heinrich, 1959). The safety information system concept was introduced in the 1970s (Johnson, 1980). It represents a more recent approach that parallels developments in information and decision-support systems in general. At that time, safety management focused on the prevention of accidents resulting in injury to personnel and major economic losses. We now deal with all types of accidental losses in the areas of safety, health and environment in our SHE management systems; hence we use the term SHE information system here.

Researchers have studied whether the qualities of the various subsystems of a SHE information system affect the accident risk. Two different research methods have been applied. In so-called *ex-post facto studies*, scientists explore whether there are differences in the SHE information systems between companies with high and low accident rates. They have found that companies with low accident rates have better injury record-keeping systems (Simonds and Shafai-Sahrai, 1977), better formal routines for workplace inspections (Smith *et al.*, 1978), and are more inclined to inquire into minor injuries and near accidents (Cohen *et al.*, 1975).

In *action research*, scientists participate in the introduction of changes in the SHE information system of a company. They evaluate how these changes affect the behaviour of its organisation and the company's accident rates. Many studies report reductions in accident rates following the introduction of such measures as improved accident investigation and workplace inspection routines and introduction of near-accident reporting (Adams *et al.*, 1981; Menckel, 1990; Komaki *et al.*, 1978).

In *evaluation research*, scientists study effects of programme changes made inside companies, without being involved in the changes themselves. An aluminium plant had introduced a documented SHE information system, involving improved routines for accident and near-accident reporting, workplace inspections and follow-up of SHE results. An evaluation study showed that the introduction of internal control, in combination with other measures such as improved maintenance, contributed to an improved control of production. This resulted not only in substantially reduced losses from accidents but also in reduced overall operational expenditures (Kjellén *et al.*, 1997).

Such results help us in arguing about the importance of well-functioning SHE information systems. They also give some hints about important qualities of such a system.

1.4 Developing the model further

Let us look at the immediate effects or output from a SHE information system, Figure 1.2. There are two types of effects (Nonaka and Takeuchi, 1995). *Explicit effects* at the organisational level have to do with what we normally associate with results of decision-making in the area of safety, health and environment. It is expressed in formal documentation of procedures and work methods, changes in design and organisation at the workplaces, safety training programmes, etc. Explicit effects also involve improvements in the personnel's understanding of the SHE problem area in a way that they are able to articulate. A SHE information system will also affect the unconscious habits, routines and action patterns in the organisation and the members' unconscious mental models, skills and value systems. This we call *tacit output*. The aims of a SHE information system are to produce explicit and tacit output that will contribute to a reduction in the risk of accidents.

> There are certain resources needed to provide a SHE information system. These resources need to be well tuned together and include:
>
> - *Personnel* with certain authorities and responsibilities and their knowledge, skills and attitudes;
> - *Organisation and administrative procedures* and routines for accident reporting, investigation, distribution of information, etc.;
> - *Instruments and tools*, i.e. forms, checklists, analytic principles, computer software, etc. for the collection, processing, storing and distribution of information on accident risks, etc.

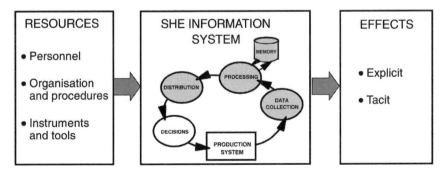

Figure 1.2 The SHE information system seen in relation to its effects and the types of resources required.

The different resources needed to establish and maintain a SHE information system have to be well tuned together and adapted to the environment in which the SHE information system exists. We will find different solutions, dependent on the size of the company, its production, traditions, management philosophy, influences from other companies, etc. There is no single solution that suits the needs of every company. Information-technology support is not a prerequisite, but will in most cases facilitate the establishment of efficient solutions.

2   Boundary conditions

When we design and manage a company's SHE information system, we have to consider a number of *boundary conditions*. These are conditions both inside and outside the company that we do not have immediate control of. They represent opportunities and threats to our ambitions in accomplishing efficient solutions.

2.1 Conditions inside the company

2.1.1 Size, type of technology and resources

Let us start by looking at three important structural variables describing a company, i.e. its size, type of technology and resources. Size affects the communication patterns within the company. In a small company, the organisational members know each other, and the communication links are often direct, informal and spontaneous. When the company grows, there is an increased need for communication and reporting through formal channels. These principles apply to communication in general, but also to communication on SHE issues. If, for example, an accident happens in a small company, the manager who shares his/her time between customer contacts and supervision will soon know about it. In a large company, management will be dependent on formal accident-reporting routines to get the necessary information.

There is a critical middle range between 50 and 500 employees. Here, the company is too big to have a well-functioning informal exchange of information on SHE issues. At the same time, the company is too small for a formal accident-reporting system to function well. The design of efficient SHE information systems for these companies represents a challenge that we have to address.

Technology is another important structural variable. High-risk industries run production processes involving substantial amounts of hazardous energy such as flammable or toxic substances, nuclear materials, etc. The potential consequences of an accident are catastrophic and may threaten the existence of the company. Often these industries take up new and complex technologies. It becomes increasingly difficult for management and workers to understand and anticipate the behaviour of the production processes. The company will be dependent on advanced SHE information systems to avoid unwanted events. Not only must all sorts of accidents, incidents and deviations be reported and followed up, risk-analysis methods have to be applied to identify and assess potentially catastrophic events before they happen. Companies in branches of industry where the hazards are obvious and the consequences limited, on the other hand, do not face these advanced needs. They may rely on simple accident-reporting systems that meet the minimum regulatory requirements.

Resources and economy in general play a decisive role. Resourceful companies will be able to afford solutions that pay off in the long run. They will be able to acquire the necessary expertise for the design and management of efficient SHE information systems. Resources are also needed in order to close the loop from the reporting of accident risks to the design and implementation of remedial actions. When adequate resources are marshalled to solve SHE problems, the employees will experience this as a sign of a positive safety culture. They will in turn be more inclined to report accident risks and to participate in problem-solving activities. If the opposite situation is the case, even a well-designed SHE information system may not be allowed to develop into an efficient tool. Or, in periods of economic crisis, cost-cuts may affect the resources allocated to SHE activities in a way that has detrimental effects on the commitment of management and workers.

These different boundary conditions are visible in practice. Typically, we find well-functioning formal SHE information systems in large and resourceful companies in high-risk industries (Skaar, 1994). Here, the company's SHE policy and goals are often integrated as part of the company's overall criteria for success. At the other end of the spectrum we find small companies, and companies in the service sector in particular. Here, the formal SHE information systems are barely visible.

2.1.2 The organisational context

The stakeholders inside a company and their different interests represent another set of boundary conditions. Below are some examples:

- Top management's commitment and concerns about SHE.
- The SHE staff's need for professional tools.
- Line management's need for feedback information to secure an efficient and reliable production process.
- The employees' need to secure compensation in case of accidents and to have of a channel to bring safety concerns to management's attention.

On a more general level, we have to look into the organisational context within which the SHE information system operates. In doing so, we will here touch upon Bolman and Deal's four different organisational perspectives (Bolman and Deal, 1984; Hale and Hovden in Feyer and Williamson, 1998). Each perspective represents a way of looking at an organisation and what goes on inside it. They emphasise different aspects of a SHE information system and pre-conditions for it to work well.

The *structural perspective* emphasises goal accomplishment. This is achieved through feedback of SHE results to management for follow-up, and through rationality in the decision-making. The formal organisation, its distribution of responsibility and authority, and behaviour based on formal rules and procedures are also typical aspects associated with the structural perspective. The classical bureaucratic approach to safety management according to Heinrich fits well into this perspective (Heinrich, 1959). Most of the SHE management systems applied in industry today have been designed on the basis of this perspective. We have many examples of how management's decision to implement such a system has served as a springboard for the development of the company's SHE information system as well. It becomes a management tool that provides feedback data on actual performance for comparison with the pre-established SHE goals.

The *human resource perspective* takes the human needs as the point of departure and focuses on the interaction between people, empowerment through delegation of authority to the employees and their participation in the decision-making. Development of informal group norms, management caring and concerns are emphasised. The socio-technical research tradition in the Scandinavian countries fits well into this perspective. It is highly relevant to our understanding of how people in the organisation will respond to a formal SHE management system when it is put into use. Managers that are influenced by this perspective will seek to facilitate the solving of safety problems as close to the source as possible. We will here find organisational solutions, where arenas are established for worker participation in the decision-making. These arenas are part of the SHE information system and will facilitate the informal exchange of information and experiences on safety problems.

The *political perspective* emphasises the fact that decisions in an organisation are made in order to allocate scarce resources. Conflicts of interest will arise between different interest groups (management, workers, SHE experts, etc.), on what constitutes an acceptable solution to a safety problem. They will be settled through bargaining processes with the participation of the different interest groups. A SHE information system has to provide for organisational solutions that allow bargaining to take place. The political situation inside the company and the degree of mutual trust between management and labour is an important boundary condition. A reasonable degree of trust and good relations are preconditions for a SHE information system to be effective. A well-functioning SHE information system will also support processes inside the company that improve management–worker relations.

The *symbolic perspective* emphasises the symbolic value of promoting SHE for the organisation and its stakeholders. The 'cultural' element in SHE management and how people in the organisation interpret and attribute a meaning to accidental events are here important aspects. We cannot understand how a SHE information system will function in practice unless we take this perspective into account. We see a clear trend among leading companies to bring SHE as an important issue to the top management level. Management recognises the value of displaying the company's social responsibility by developing and promoting SHE policies and objectives. To maintain credibility both inside and outside the company, top management's commitment has to permeate all levels of the organisation. A SHE information system will serve as an important tool in this respect. It will have adequate impact only when it supports openness and debates, where different individual experiences and interpretations of why accidents occur and how to prevent them are shared. The SHE information system must not be a tool for transfer of secret and/or distorted data or for unilateral top management control.

2.2 The outer context

2.2.1 *The employer's responsibilities*

The employer is the primary body responsible for accident prevention at the workplaces. Although the legislation varies from country to country, typical duties of the employer are to ensure that:

- Design and organisation of the workplaces meet the requirements defined in laws and regulations;
- Hazards are identified and evaluated and that the necessary remedial actions are implemented;
- SHE conditions are monitored regularly;
- SHE work is carried out in co-operation with the employees; and
- Employees are informed about the hazards at the workplaces and about safe work practices.

These different requirements have clear implications for the design of the companies' SHE information systems. The employers have to establish and maintain administrative systems for the collection and use of data relating to hazards in the workplaces. They carry out their duties with a varying degree of enthusiasm. Compliance is basically ensured through the use of the stick, i.e. employers not meeting the requirements are penalised. Authorities also bring the employer or the responsible manager to court in cases where violations of regulatory requirements have caused severe accidents. The intention is that such actions shall have deterrent effects and counteract complacency. However, they may, on the contrary, result in self-protective actions, by which the employer hides sensitive information on accident causes.

Following a number of major accidents involving dangerous substances in the 1970s and 1980s, authorities in Europe and USA enacted legislation to prevent such accidents. This legislation goes further than the general working environment and environmental care legislation and makes explicit what the employers have to do to manage the risk adequately. The so-called Seveso directives in Europe, and the Clean Air Act Amendments in the USA, specify SHE management system elements that high-hazardous industries have to implement (European Council, 1996; EPA, 1990). Central to these regulations are the requirements to perform formal risk assessments. The employer must also carry out investigation of incidents and accidents to find causes and must implement remedial actions to prevent recurrence.

The principle of enacting regulations that specify requirements to the companies' SHE management systems have in some countries been extended to apply to SHE in general, including ordinary occupational accidents. The so-called internal control regulations in Norway and Sweden require companies to develop and implement adequate management systems to ensure that their activities are planned, organised and maintained in accordance with the SHE legislation (Arbetarskyddsstyrelsen, 1996; Arbeidstilsynet, 1996).

In the United Kingdom, the Management of Health and Safety at Work Regulations of 1992 have a similar objective. They require employers to carry out risk assessments, to implement measures to remedy the risks, to appoint qualified personnel and to provide for adequate information and training of the employees. The Health and Safety Executive has developed guidelines for health and safety management to support the implementation of this legislation (HSE, 1997a).

In the USA, the proposed safety and health programme rule requires employers to set up safety and health programmes to manage workplace safety and health (OSHA, 1989). The aim is to reduce injuries, illnesses and fatalities by systematically achieving compliance with the occupational safety and health regulations.

Let us look closer at the Norwegian internal control regulations, which are a typical example of the types of requirements that the employers have to meet. According to this legislation, there must be routines inside the company to:

- Keep an overview of the SHE regulations;
- Ensure that the employees have adequate knowledge and skills in the area of SHE work;
- Ensure that the employees participate in the SHE work in order to utilise their collective experiences;
- Establish goals in the area of SHE;
- Keep an overview of the company's organisation and distribution of responsibilities in the area of SHE;

- Systematically identify and assess hazards and implement action plans to control them;
- Identify nonconformities in relation to the regulatory requirements; and
- Perform audits and reviews to monitor the effectiveness of the company's SHE management system.

The SHE information will be an important element of the companies' SHE management system to meet these requirements.

2.2.2 Regulations on record keeping and on the reporting of injuries and incidents to the authorities

The legislation usually requires the employers to investigate incidents and accidents for the planning of preventive measures and for reporting to the authorities. In the USA, employers must keep records of work-related fatalities, illnesses and injuries (OSHA, 1971/97). According to the minimum requirements within the European Union, the employer must keep a list of occupational accidents resulting in a worker being unfit for work for more than three working days (European Council, 1989). Similar requirements exist in other countries. The aim is to ensure that the employers keep an overview of the accidents that have occurred, both for preventive purposes and to ensure that the information is available for possible later claims for compensation.

The requirements to report accidents to the authorities vary between different countries. The differences mainly concern the classes of employees to be included in the reporting and the severity thresholds for reporting (e.g. number of days of absence). There are usually two levels of reporting:

- Immediate reporting of severe accidents. It is common to require an immediate reporting of severe accidents involving fatalities, serious injury, fires and explosions, etc. to the enforcing authority. In the USA, accidents resulting in fatalities and multiple hospitalisation must be reported within eight hours to the Occupational Safety and Health Administration (OSHA). In Great Britain, the enforcing authority must be notified immediately of accidents resulting in death or major injury and of dangerous occurrences. The notification shall be followed by a report within ten days. Similar requirements exist in the Scandinavian countries.
- The national laws vary concerning the requirements to report less severe accidents to the responsible authorities. In the USA, so-called OSHA recordable accidents involving absence from work, medical treatment, loss of consciousness, restriction of work or motion or transfer to another job are reported to the authorities on a sampling basis.

In Great Britain, accidents resulting in over three days of absence must also be reported within ten days to the enforcing authorities (HSE, 1996). In

Sweden and Norway, the reporting threshold is absence from work beyond the day of injury.

The reporting of accidents to the authorities fulfils different aims. One is to supply the authorities responsible for the development of regulations with the necessary knowledge for prevention. A second aim is to provide input to the programmes for supervision of compliance with the SHE regulations at the workplaces.

2.2.3 Workers' compensation systems

Most countries have compensation systems for work-related disability (injury or illness) or premature death (Stellman, 1998). They usually cover loss of income, medical payments and rehabilitation expenses. There are different means of organising the workers' compensation system:

1   The employer has to provide insurance coverage through an insurance company. Such companies are subject to regulations and supervision by a governmental agency.
2   Workmen's compensation is part of the social insurance operated by government.
3   The system is operated by a special government agency, a so-called workers' compensation board.

It is usually the duty of the employer to report a disability or fatality to the insurer. The worker is, however, required to notify the employer about an injury or illness where this is possible. Compensation is payable for disabilities and fatalities that result from events and circumstances of employment. It is usually irrelevant whether there is any fault on the part of the victim, the employer or other party.

The worker or the employer has to provide a report on facts about the circumstances related to the disability or fatality for compensation to be paid. In the case of accidents, the question of whether the event has resulted from employment is usually not problematic.

The employer usually pays the costs of workers' compensation as a percentage of the payroll. It is common that the premiums are adjusted so that expenditures and revenues are equalised over time, but partial subsidy by government also occurs. The premium is usually established for each class of industrial activity based on historical data on expenditures for compensation. Experience rating is also practised, meaning that the premium to be paid by an employer is adjusted, based on previous experience with that employer. Some insurance companies also give premium incentives for improvements in the SHE standard and in the SHE management system made by the employer.

To sum up, accident reporting to the insurer provides a basis for compensation to the victim and for determination of insurance rates.

2.2.4 International standards and guidelines

In today's competitive world, managers tend to look at their colleagues in other companies in order to learn from their best practices. Successful companies in the area of SHE become models for many other companies.

Managers need to demonstrate their commitment to safety, health and environmental issues to interested parties such as stakeholders, customers, the community, voluntary organisations and the authorities. One way of doing so is to implement voluntary standards on SHE management and to receive proof of compliance from an independent certifying body. In order to receive such a certificate, the company's SHE management system has to be audited by the body to demonstrate that it is operating effectively in conformance with the standard.

The most important example is the ISO 14000 standard family on environmental management (ISO, 1996). These standards do not specify levels of environmental performance but provide a framework for the companies' approach to issues concerning environmental policy, plans and actions. They apply an environmental management system model similar to the quality management model of ISO 9000. The model includes the following elements:

* Commitment and policy, i.e. the company has to define its environmental policy and ensure commitment;
* Planning to fulfil its environmental policy;
* Implementation through development of capabilities and support systems;
* Measurement, monitoring and evaluation of environmental performance; and
* Review and continually improve the environmental management system.

The standards specify feedback mechanisms for the handling of incidents and deviations, including requirements on incident reporting, inspections, and follow-up action.

At present, there does not exist any internationally recognised standard for the total SHE area. There are examples of such standards that have been issued by certifying agencies and industrial associations. British Standards Institution has developed guidelines for occupational health and safety management systems based on the similar management principles of ISO 14000 as well as on British regulations and guidelines (British Standards Institution, 1996). This institution has also co-operated with Det Norske Veritas, Lloyds Register and other certifying bodies in developing an occupational health and safety management systems specification (British Standards Institution, 1999). The so-called OHSAS 18001 standard is intended for use in assessing and certifying companies' occupational health and safety management systems and is compatible with ISO 9001: 1994 and ISO 14001: 1996 (ISO, 1994; ISO, 1996).

The Oil Industry International Exploration and Production Forum has issued voluntary guidelines for the development and use of SHE management systems (E&P Forum, 1994). These guidelines apply a similar SHE management model to that of the ISO 14000 family.

In the USA, the Chemical Manufacturers Association has developed the 'Employee Health and Safety Code' covering management practices in six areas. They include the prevention of pollution, accident prevention, worker protection, distribution of responsibility, community relationships and product stewardship.

2.2.5 Other non-governmental organisations

There are numerous organisations on the national and international level that are active in the area of SHE such as:

- Associations of professionals in the area of SHE
- Employers' and workers' organisations
- Joint labour–management councils
- Associations of producers, manufacturers and operators
- Voluntary organisations.

They influence the corporate SHE practices in different ways. Some of the organisations provide professional support and services by arranging conference and training programmes for SHE professionals and by publishing educational materials, professional journals and textbooks. Other associations represent arenas for experience exchange and the development of common policies and practices. They represent, for example, a particular branch of industry such as chemical manufacturers, oil companies or primary aluminium producers. There are also interest groups taking care of the members' interests in the area of SHE such as worker and management organisations. Voluntary organisations, especially in the area of environmental care, are concerned with the effects of industrial production on the community.

3 Alternative accident-prevention approaches

There is no simple and uniform way to achieve a high safety standard. Throughout the book, we will see how different approaches contribute to our understanding of why accidents occur and how to prevent them. The intention is not to search for a unifying synthesis but rather to show the distinct characteristics of each approach and to analyse the contributions from each of them. To demonstrate the variety and to help the reader in recognising the differences, we will here look into four different approaches, each convincing from its own standpoint. We will later develop some of them further and present alternative ones to get a more comprehensive understanding.

3.1 Barriers against hazards

The prevention of accidents through *barriers* is an engineering approach and is a main principle behind safety in design. Injury and damage is caused by influences from the environment through a transfer of energy. We accomplish accident prevention primarily though the elimination of hazards, i.e. energy sources in the environment. If this is not feasible, we try to establish barriers between the hazards and the potential victim, Figure 3.1.

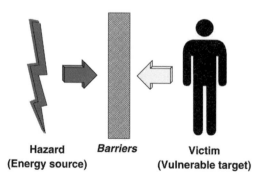

Hazard **Barriers** Victim
(Energy source) (Vulnerable target)

Figure 3.1 The relationship between hazard, victim and barriers.
Source: Haddon, 1980.

There is no need for a SHE information system in an ideal production system where all hazards have been eliminated or surrounded by durable and efficient barriers. In practice, this situation never occurs. Energy is needed for production purposes and SHE practice is an act of balancing between different SHE and economic goals. Production systems change and new hazards are introduced. Barriers are full of 'holes' or even lacking (Reason, 1997).

The energy-barrier approach is described more in detail in Chapter 5. We will make extensive use of this approach in our modelling of the accident sequence in Chapter 6. It also provides us with essential criteria for the identification and assessment of hazards and the selection of remedial actions (see Parts III and V).

3.2 Administrative system for feedback control

The process by which accident risks are controlled through feedback of experiences is necessary in dealing with hazards in a changing environment. This feedback may be accomplished through formal routines. The administrative or *bureaucratic control-system* approach is represented by the administrative systems for feedback control and structured decision-making within a company. This approach falls well into the structural perspective of organisations according to Bolman and Deal that we touched upon in Chapter 2. Traditional SHE management relies heavily on this approach. Negative feedback control is illustrated by the principles of quality control (Juran, 1989). The performance of the control subject (e.g. an industrial operation) is measured and compared to a norm, Figure 3.2. The difference between actual performance and norm is used as input to

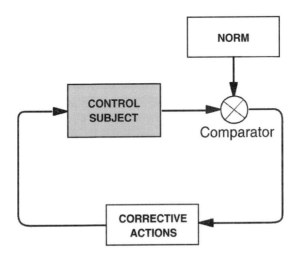

Figure 3.2 System controlled through negative feedback.

decisions about control actions. These actions are introduced in order to affect the performance of the control subject in a direction that approaches the standard.

SHE is an aspect of quality. In SHE management, we establish SHE norms for follow-up. Let us look at a typical example, where we use an upper limit for the number of accidents per year as a norm. Written procedures define the activities by which the organisation:

1 registers the accidents that occur,
2 calculates the actual number of accidents periodically for comparison with the goal (norm), and
3 takes action in case of deviations (i.e. the actual number of accidents exceeds the norm).

The structured decision-making process also belongs to the bureaucratic approach. This is a normative model of decision-making that prescribes how to make decisions to avoid irrationality.

3.3 Arenas for organisational learning

We are not only concerned with this bureaucratic feedback-control-system approach. Feedback also involves learning and the accumulation of qualified experience. This will result in gradual improvements in the organisation's performance. We will here be concerned with features of an organisation that shape behaviour of importance to SHE. Figure 3.3 shows

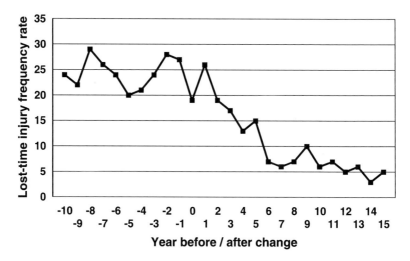

Figure 3.3 Developments of the lost-time injury frequency rate at an aluminium plant after the introduction of a formal SHE management system. The lost-time injury frequency rate (LTI-rate) is defined as the number of injuries resulting in absence from work per one million hours of work.

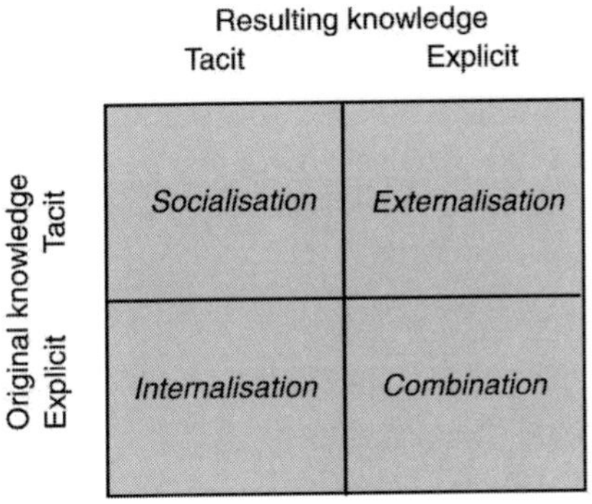

Resulting knowledge
Tacit   Explicit

	Tacit	Explicit
Original knowledge Tacit	*Socialisation*	*Externalisation*
Original knowledge Explicit	*Internalisation*	*Combination*

Figure 3.4 Different learning mechanisms during experience exchange between the members of the industrial organisation.
Source: Nonaka and Takeuchi, 1995.

a typical example of how SHE performance may develop after the introduction of a change to improve the company's SHE management system. Initially, there is a rapid rate of improvement. The improvement rate decreases and the SHE performance stabilises at a higher level than before the change. Such improvement patterns cannot be explained by simple feedback mechanisms.

Obviously, learning at the organisational level has been involved in the improvement process. In the *organisational learning approach*, we speak about different learning mechanisms involved in the exchange of tacit and explicit experience between the members of the industrial organisation, Figure 3.4.

A SHE information system provides several arenas for such learning. In accident investigations, for example, individuals' tacit knowledge and experience are embedded in the experiences they share, and this experience is *externalised* through documentation and made available to the organisation as a whole. We will also be concerned with group problem-solving in connection with accident investigations, SHE audits and in risk analyses. A team structure supports learning through *socialisation* by exchange of knowledge between the group members. In risk analyses, for example, the team members may *internalise* the accident models applied in the analysis, and this learning process will affect their unconscious mental models and skills in handling similar situations to those analysed.

The bureaucratic control-system approach and the organisational learning approach are complementary. In our analysis of central principles behind SHE information systems, we will be concerned with the handling and feedback of explicit information and the use of this information in decision-making. Feedback control is an essential component in many SHE management tools such as workplace inspections and SHE audits (Chapter 14), and in the monitoring of SHE performance (Chapters 16–20). The structured decision-making process plays a vital role in the classical models of accident investigations and SHE problem solving (Chapter 13).

We will also be concerned with the significance of the organisational learning approach in the prevention of accidents. Although daily work provides the most important arena for exchange of experiences and learning, the arenas offered by the SHE information system will support similar learning processes (Parts III and V). We will also see how informal processes may be counter-productive from an organisational perspective by disturbing experience exchange and the acquisition of qualified knowledge (Chapter 10).

3.4 Risk homeostasis

The *risk-homeostasis theory* focuses on the degree of caution exercised by individuals in the handling of accident risks. Accident rates are, according to the risk homeostasis theory, determined mainly by the accident risks that people on average are willing to take (Wilde, 1982). It is a rather pessimistic perspective as to the effects of non-motivational accident counter-measures. Elimination or reduction of hazards in the workplace will result in decreased caution on the part of the individual. This behavioural change will by and large offset the effects of the measures.

This theory has primarily been developed to explain traffic-safety data. In the traffic arena, risk-taking is to a large extent voluntary. In industry, on the other hand, behaviour including risk-taking is subject to management supervision and control. We must, however, expect that the same processes are present in industry but to a lower degree, since they represent basic human traits.

The immediate impression is that the risk-homeostasis theory may be in conflict with the basic theories and principles of SHE information systems. We tend to evaluate SHE information systems by looking into the safety measures that have been implemented as a result. They are often of the non-motivational type that according to the risk-homeostasis theory will have marginal effect. In general, the SHE-information-system approach does not involve any particular preferences as to the selection of remedial actions. They may be motivationally oriented as well as non-motivational.

We will return to the risk-homeostasis theory in Chapter 8 on the human element in accident control. It is also relevant to Part IV on SHE performance measures. Here, we will review how management decisions on an acceptable risk level are followed up through monitoring of actual performance. We will discuss to what extent such decisions actually affect the employees' risk-taking behaviour and result in genuine safety improvements. Together with the organisational learning approach, the risk-homeostasis theory conveys an important message. We can not solely rely on engineering measures in the prevention of accidents.

4 Case study: reducing emissions to the air from a fertiliser plant

We will here present a case from a fertiliser plant. It involves the successful introduction of a new system for feedback control of accidental emission to the environment. The system illustrates some basic qualities in accident prevention through experience feedback.

An integrated process control system was introduced at the plant. The goal was to reduce emissions of nitrogen compounds to the air from the plant. The system provided the control-room operators with more timely information on process deviations with the potential to cause environmental pollution, Figure 4.1.

Before the system was introduced, emissions were registered continuously by measuring the nitrogen content in the exhaust air in the pipe. This information was stored and summarised in internal reports and reports to the authorities. In case of major accidental emissions, the causes were analysed

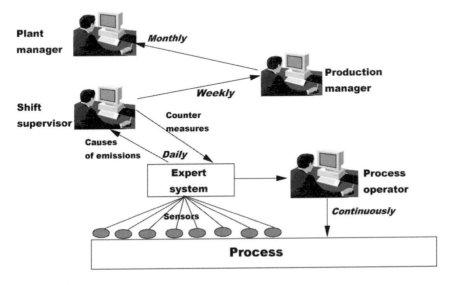

Figure 4.1 Integrated process control system.

Kg NH₄-N/h

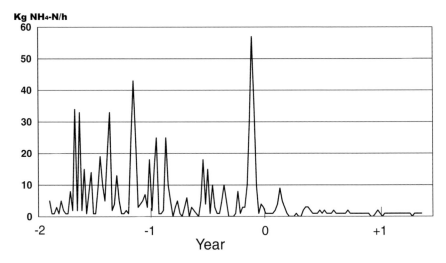

Figure 4.2 Emissions to the air from the fertiliser plant during a four-year period. The integrated process control and expert system was introduced at time 0.

and remedial actions taken. There was, however, no closed feedback loop operating on a day-to-day basis. Figure 4.2 shows typical output from the plant during a two-year period prior to the introduction of the new system.

Work with the new system started by analysing causes of typical emissions. These were traced back to process upsets in the form of deviating values for different process parameters such as pressure, temperature, flow, etc. A new system was designed that continuously measures these parameters and the results are fed into an expert system. Operators are furnished with the necessary information and recommendations on remedial actions from the expert system. They are then able to control the process before upsets result in increased emissions. In Figure 4.2, we see the effects of the new system through the reduced emissions after it was introduced.

The new information system also has some other interesting qualities. It provides management at different levels with filtered and summarised information according to their needs, Figure 4.1. The shift supervisor, for example, receives daily reports on causes of emissions and is responsible for measures to improve the information system and the production process in general.

Let us look into the contents of the remaining parts of the book in light of this example. They help us in answering such questions as:

- *How to analyse the causes of accidental events involving releases of pollution?* Chapter 5 gives an overview of accident models that have been developed to support analyses of this type. Chapter 6 presents one particular analysis framework that suits this purpose. Chapter 8 focuses on the role of the operators at the sharp end, who are responsible for controlling the process. Chapter 13 presents practical methods for use in investigations of accidental releases in order to find causes. Chapter 15 presents methods for use in the analysis of accident statistics with the same purpose.

- *How to prevent accidental releases from occurring?* Chapter 5 introduces the concept of barriers against accidents and this concept is further developed and exemplified in Chapter 7. We will there review different types of barriers from a generic perspective. Chapter 7 will help us in identifying the different possibilities of preventing accidental releases of pollution.

- *What principles should be applied in the distribution of information on accidental releases and underlying factors to the different levels of an organisation for use in decision-making?* Chapter 10 brings in a number of aspects of relevance to this issue. Feedback control and diagnosis are two central themes of this chapter. Typically, the process operators apply feedback control in their daily operation. Diagnosis is a central concept in problem-solving. It is applicable in two respects. One is in the analysis of causes of accidental releases. The second application is illustrated by the example above, where diagnosis was a part of the work to develop the process-control system.

- *When to react to increases or decreases in the frequency of accidental releases?* Accidents are a random phenomenon. Chapter 9 gives a short review of the basic statistical theory necessary for us to be able to analyse the occurrence of accidents in consecutive time periods. We will use the formulas that come out of this presentation in the presentation of SHE performance measures in Part IV of the book.

- *What types of sensors should be used in monitoring the process?* This is a specific question on the design of measurement devices for process control. A general presentation of different methods for the collection of data on accident risks is given in Chapters 12–14.

- *How to prevent the occurrence of accidental releases without first waiting for the accidents to happen?* Part V presents different methods of risk analysis. They are applicable already at the design stage and will help to identify the need for barriers against accidental releases. Use of risk analysis during the operations phase will speed up experience feedback and learning.

- *How to examine the process-control system in order to determine whether it functions as planned and is effective and suitable?* Chapter 14 presents the basic principles of quality audits. Audits may be used as a tool to answer these questions. Examples of questions for use in audits are also presented.

- *What can we learn from other application of SHE information systems?* Part VI presents different cases showing how the design and management of SHE information systems is handled in practice.
- *What specifications should we use when designing a process control system?* Chapter 11 presents a list of criteria or requirements to SHE information systems. These criteria are intended for use as a support in answering this question.
- *How to manage a project for the development of a process control system?* This is the focus of Part VII. It integrates the different principles and methods presented in previous parts of the book.

Part II
Theoretical foundation

This Part lays the theoretical foundation for the presentation of various methods in Parts III to V. Chapter 5 gives a presentation of different accident models. We need such models, for example, in order to ask the right questions during an accident investigation. Chapter 6 summarises many important aspects of the accident theories and models presented in Chapter 5 in an accident analysis framework. We use this framework when we review the different types of information on accident risks. Chapter 7 presents various accident counter-measures based on a barrier philosophy. Applications of this philosophy in the prevention of ordinary occupational accidents as well as major accidents are presented. Chapter 8 brings us into the important area of human errors. Chapter 9 summarises some basic statistical theory on Poisson processes. (We will use these basic statistics in Part IV on the monitoring of SHE performance.) Chapter 10 reviews some basic principles for experience feedback. We will introduce two important concepts, diagnosis and feedback control. We will also discuss some important obstacles to efficient learning from experience. Chapter 11 ends Part II with the presentation of a set of criteria for the evaluation of SHE information systems.

5 Accident models

5.1 On the need for accident models

Accident models play a vital role in the design of SHE information systems and in the training of users. They are simplified representations of the accidents occurring in real life. Each accident model has its own characteristics as to the types of 'causal factors' that it highlights. In an accident investigation, for example, accident models support the investigators in:

- Creating a mental picture of the accident sequence
- Asking the 'right' questions and defining the types of data to collect
- Establishing stop rules, i.e. rules for when to terminate the search for new causes further away from the accidental event
- Checking that all relevant data has been collected
- Evaluating the data and structuring and summarising the data into meaningful information
- Analysing relations between pieces of information and seeing interrelations
- Identifying and assessing remedial actions
- Communication between people by providing a common frame of references.

An important aim of introducing an accident model is to establish a shared understanding within the organisation of how and why accidents happen. It is especially important that those parts of the organisation responsible for the collection of information on accident risks and those responsible for using the information in decision-making use a similar frame of reference. Accident models will thus have direct influence on SHE practice, both consciously and unconsciously. A concern is whether the different users of the SHE information system actually interpret and internalise the models in a similar way and in accordance with the intentions of the systems designer.

Old accident 'models' or perceptions survive in parallel with modern models based on scientific research. Fate is still a major explanatory factor to many people (Hovden and Larsson, 1987). It follows from such an accident perception that there is not much you can do to prevent accidents.

Accident research during the early part of this century studied the relation between accidents and personal traits of the victim. 'Accident-proneness' was a commonly accepted theory at that time (McKennan, 1983). It stated that certain individuals due to personal traits are more susceptible to accidents than others. It follows from this theory that accident risk may be reduced substantially by removing 'accident-prone' persons from hazardous jobs. Today, 'accident-proneness' is considered to account for only a small part of accidents. It follows that a preventive strategy based on this theory in most cases will have only minor effects. We do not find accident models in use in industry today that focus solely on personal factors.

Another early approach that is still referred to in SHE practice is to use the 'human factor' as an overall explanatory factor. Early statistics showed that about 88 per cent of the accidents were caused primarily by dangerous acts on the part of the individual worker (Heinrich, 1959). The remaining 12 per cent were either caused by technical failures or were 'Acts of God'. We will later discuss the shortcomings of such statistics as an explanation of accidents.

In this Chapter, we shall present a selection of accident models from the research literature of relevance to the design of modern SHE information systems. This presentation will help us understand the framework for accident analysis in Chapter 6. This in turn forms the basis for our presentation of SHE management methods and tools in Parts III–V. The intentions with Chapters 5 and 6 are twofold. One is to help the reader develop an understanding of the nature of accidents. The second is to build a pedagogic structure with the aim of helping the reader organise his/her memory in relation to the distinct characteristics of the different methods and tools. We will also include models of historical significance to help the reader understand developments in the general understanding of the accident phenomenon.

We will see that there are many different models, representing the needs of individual researchers to highlight different aspects. Fortunately, there are converging trends in accident research in that different subsets of models have important aspects in common. We will focus on these aspects rather than on the particularities of each model. We will also see how different 'schools' contribute to an overall understanding of the accident phenomenon.

5.2 Causal-sequence models

The simplest types of accident models for use in the design of SHE Information Systems are the causal-sequence models. An early and historically very important example is the **Chain of Multiple Events** or 'Domino theory', Figure 5.1. In this model, an accident is described as a chain of conditions and events that culminate in an injury. A link in this chain is an unsafe act or unsafe condition at the workplace. It is suggested that accidents be prevented through the reduction of unsafe acts and conditions.

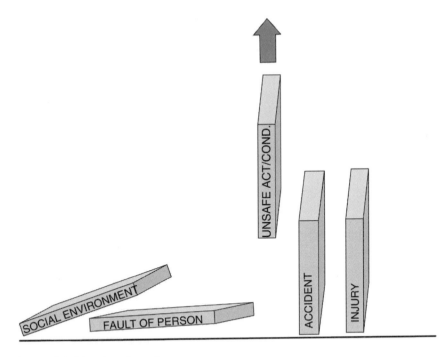

Figure 5.1 'Domino theory'.
Source: According to Heinrich (1959).

This model has had a very large influence on the development of schemes for the classification of accidents applied in many countries. Classification is used to standardise the collection of data on accidents and to reduce the complexity of the data to a manageable level for statistical purposes. An early and important example is the American National Standards Institute's system for the classification of accidents, ANSI Z16.2. In this system, the following facts are recorded about the accident sequence:

1 *Injury*
  a *Nature of injury*: cut, fracture, burns, etc.
  b *Part of body affected*, i.e. head, neck, upper extremities, etc.
2 *Accident*
  a *Accident type*: strike against something, fall, slip, etc.
  b *Agency of accident*: machine, vehicle, hand tools, etc.
3 *Hazardous condition*: unguarded, defective tools, unsafe design, etc.
4 *Unsafe act*: failure to secure, operating at unsafe speed, etc.

We recognise these different factors in many of the standard forms for use in accident reporting in Europe also. Statistics, based on the classification of accidents in accordance with these categories, usually show that a majority

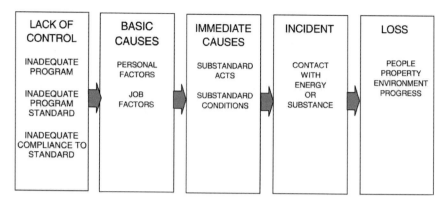

Figure 5.2 ILCI model.
Source: Adapted from Bird & Germain, 1985. Copyright by Det Norske Veritas and reproduced by permission.

of the accidents are 'caused' by unsafe acts. These types of analyses do not take into account the multiple causality of accidents.

The 'Domino theory' is the predecessor of many current accident models. In these models, the early part of the causal chain (fault of person, social environment) has been replaced by management factors. The aim is to evaluate how these affect the likelihood of unsafe acts and conditions. Figure 5.2 shows how the International Loss Control Institute has modified the original 'Domino theory' in their so-called **ILCI model** to include these types of factors. The three last 'events' in the sequence are similar to those of the 'Domino theory'. We note, however, that the 'injury' event has been replaced by the more general category of 'loss', i.e. injury or fatality to people, property or environmental damage, and process interruptions. The so-called root causes, i.e. the 'lack of control', include failures in SHE management factors similar to those found in quality assurance programmes.

The ILCI model has had a large influence on SHE practice in many countries through the International Safety Rating System (ISRS), a SHE management auditing and rating system. We will come back to this issue in Part IV. It also serves as a basis for the categorisation of information on accidents in accident and near-accident reporting systems in use by many companies.

A problem with causal-sequence models such as the Domino and ILCI models is that no clear distinction is made between the observable facts about the accident sequence on the one hand and the more uncertain 'causal' relationships at the personal, organisational and management levels on the other hand. The user is thus led into believing that information on, for example, personal factors such as mental stress, has the same objective status and is as unambiguous as information on observable facts about the sequence of events. There is thus a risk of misunderstanding and false

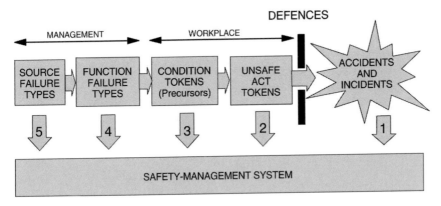

Figure 5.3 The TRIPOD model of accident causation.
Source: Reason, 1991. Copyright 1991 by Butterworth Heinemann Publishers, a division of Reed Educational & Professional Publishing Ltd, and reproduced by permission.

interpretations, especially at the higher management levels, where detailed knowledge about the accident occurrences is lacking.

A problem with the early 'Domino theory' was that the accident sequence was regarded as a simple chain of events. It has been criticised because it does not account for multiple causality.

The **TRIPOD model** presents an accident model of causal sequences rather similar to the logic principles of the ILCI model, Figure 5.3 (Reason, 1991; Reason, 1997). It has had a large influence on current thinking, because it models how 'erroneous' decisions at different management levels lead up to the circumstances of which the accident is a result. We will apply aspects from the TRIPOD model in Chapter 6. We will illustrate how accident occurrence is affected both by operational decisions at the work system level immediately before the accident, and by higher management decisions.

In the TRIPOD model, a distinction is made between different classes of human failure, mainly coinciding with the levels of organisational hierarchy:

1   *Unsafe act tokens* are specific failures made by individual operators at the sharp end, i.e. the work-system level.
2   *Failure types* are general classes of organisational and management failures. There are two failure types:
   a   *Function failure types* are latent failures resulting from decisions made by line management, designers, planners, etc.
   b   *Source failure types* are fallible top management decisions at the strategic level.

The '*defences*' in Figure 5.3 are similar to the safety barriers according to Haddon, see the Energy model in section 5.4 below.

The TRIPOD model also helps us in understanding the relationship between different feedback channels in a SHE information system. In Figure 5.3, we identify five different 'channels' for feedback of information on accident risks:

1 Accident and near-miss reporting will ideally cover both the accident sequence and underlying erroneous decisions.
2 Unsafe-act reporting will cover both the actual acts and their results such as poor housekeeping.
3 Reporting on unsafe-condition tokens will uncover poor workplace design, high workload, inadequate training, etc.
4 Reporting on function-failure types at the department level will identify faulty design, maintenance, procedures, housekeeping, communication, training, etc.
5 Reporting on source-failure types will uncover lack of top-level commitment and competence and a poor safety culture.

5.3 Process models

A generalised process model constitutes the backbone of our accident-analysis framework in Chapter 6. Process models help us in understanding how a production system gradually deteriorates from a normal state into a state where an accident occurs. Time is thus a basic factor. In contrast to causal-sequence models, process models make a clear distinction between the accident sequence on the one hand and the underlying causal or contributing factors on the other hand.

Modern process models distinguish between different phases of the accident sequence. A typical example is the **OARU** (Occupational Accident Research Unit) **model** (Kjellén and Larsson, 1981). The accident sequence is here divided into three phases: the initial phase; the concluding phase; and the injury phase. There are thus four transitions between different phases:

1 Transition from normal conditions to a state of lack of control;
2 Transition from lack of control to loss of control;
3 The target (the body in case of personal injury) starts to absorb energy; and
4 Energy absorption ceases.

The state of lack of control is characterised by the presence of *deviations* in the system. These are events or conditions that depart from the *norm* for the faultless or planned processes of the system. In the accident research literature, we find different terms used to illustrate the concept of deviation such as 'critical incident', 'unsafe act', 'unsafe condition', 'disturbance'. Various types of norms appear in the literature:

Table 5.1 Relationship between classes of deviations and different systems of production planning and control

Class of deviation	System of production planning and control
Flow of material	Material control
Personnel	Personnel control
Information	Supervision
Technical	Technical control, inspection and maintenance
Human action	First-line supervision
Intersecting/parallel activities	Activity planning
Environment	Health and safety services
Guards	Health and safety services
Personal protective equipment	Health and safety services

Source: Kjellén, 1984.

- Standard, code, rule or regulation
- Adequate or acceptable
- Normal or usual
- Expected, planned, intended.

A checklist of deviations has been developed to support accident investigations. It is based on an industrial engineering systems view and underlines the relationship between accident control and production control, Table 5.1.

The OARU model uses the term *determining factors* rather than causal factors. These are technical, organisational and social properties of the man–machine system and department that affect the accident sequence, but change only slowly in comparison with it. Figure 5.4 shows an example of an accident analysed by means of the OARU model. The distinction between the accident sequence and the underlying, determining factors is here made clear.

The significance of deviations as a risk-increasing factor is supported by empirical evidence. The risk of accidents increases when the production system develops into a state of lack of control, where there are production disturbances, defective equipment, non-ordinary personnel, etc. (Kjellén, 1983). These deviations represent symptoms of a bad 'control climate' at the plant, i.e. deficiencies in the management system for production control (Grimaldi, 1970). A bad control climate will not only result in an increased risk of occupational accidents but also in increased risk of property damages, unscheduled production stops, accidental contamination of the environment and production outside quality norms.

This does not mean, however, that all types of deviations will increase the risk of accidents. In some cases, deviations from, for example, work instructions represent a means used by the production personnel to adapt their behaviour to better suit the actual conditions at the workplace. Rather than changing the behaviour, it may here be a question of changing the

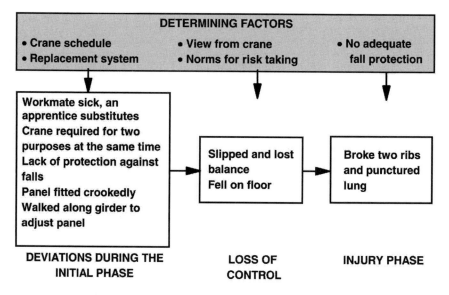

Figure 5.4 Analysis of an accident at a construction site by means of the OARU
model.
Source: Kjellén and Larsson, 1981.

norm. Each type of deviation has to be analysed with respect to its significance to the risk of accidents.

There are also other accident models that include a 'process component'.
Figure 5.5 shows that many of them distinguish between different phases in
a way similar to the OARU model.

Haddon's phase model was developed for the purpose of studying traffic
accidents, and the phases were denoted 'pre-crash', 'crash' and 'post-crash'
respectively. It was later made general in order to apply to studies of other
types of accidents as well (Haddon, 1968). The focus is on the 'second
phase' and the energy flows involved (see next Section).

The **Multi-linear Events Chartering Method** is based on the view that an
accident is a process involving interacting events and conditions (Benner,
1975). An accident begins when a system is transformed by a disturbance
and ends with the last injurious or damaging event. An event is described
in terms of an actor (animate or inanimate) and an action. The sequence
of actions of each actor and the interrelations between the actions are
displayed in a diagram. Also, conditions that influence the different actions
are displayed. This method is a central part of the so-called STEP method
for accident investigations that we will come back to in Part III.

The **ISA-model** applied by the Swedish information system on occupational
accidents was developed for the purpose of coding occupational accidents
that had been reported to the Swedish authorities (Andersson and Lagerlöf,

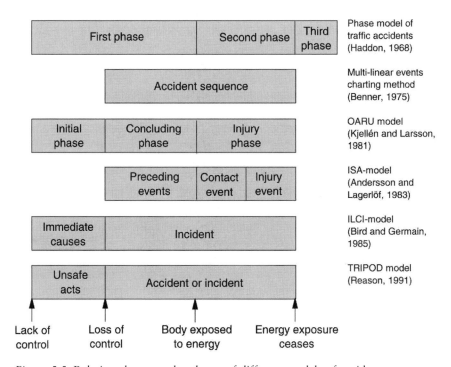

Figure 5.5 Relations between the phases of different models of accidents. *Source*: Kjellén and Hovden, 1993.

1983). It focuses on the later part of the event sequence, starting with the loss of control. This has to do with the fact that the accident reports usually lacked evidence about the early phase.

5.4 Energy model

The **energy model** is rooted in epidemiology. It represents an effort by the medical discipline to systematise the analysis of accident causes in a way similar to that of analysing causes of diseases. Gibson pioneered this development. He based his model on the fact that a transfer of energy in excess of body injury thresholds causes injury to a person (Gibson, 1961). The injury agent is energy exchange, which is mechanical, chemical, thermal, electrical, etc. In a development of this model, Haddon systematised known principles of accident prevention into ten strategies. These are related to different points of intervention according to Figure 5.6 (Haddon, 1980).

The energy model and Haddon's strategies have had a significant influence on the European legislation and standardisation work related to machinery safety and risk analysis (CEN, 1991; CEN, 1996). It is rarely used directly in SHE information systems but as a central component in more comprehensive

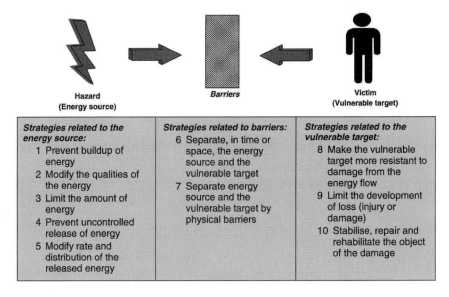

Strategies related to the energy source:	Strategies related to barriers:	Strategies related to the vulnerable target:
1 Prevent buildup of energy 2 Modify the qualities of the energy 3 Limit the amount of energy 4 Prevent uncontrolled release of energy 5 Modify rate and distribution of the released energy	6 Separate, in time or space, the energy source and the vulnerable target 7 Separate energy source and the vulnerable target by physical barriers	8 Make the vulnerable target more resistant to damage from the energy flow 9 Limit the development of loss (injury or damage) 10 Stabilise, repair and rehabilitate the object of the damage

Figure 5.6 Haddon's ten accident prevention strategies.
Source: Adapted from Haddon, 1980.

models such as the OARU, MORT, SMORT and TRIPOD models presented in this chapter.

Table 5.2 shows examples from different areas where safety measures have been classified in relation to Haddon's strategies.

The energy model has three distinct merits. One lies in the support it offers in checking that all possible preventive measures have been identified. In applying the energy model, different priorities have been given to the three main types of strategies. Hence, the primary strategies are those related to the energy source. If it is not possible to eliminate or reduce the hazard to an acceptable level, barriers such as fixed guards are introduced. Personal protective equipment is introduced as a last resort. These priorities have been implemented in the European legislation on machinery safety.

It should be remembered, however, that the energy model puts the focus on certain types of measures, whereas others are de-emphasised. Strategy no. 4, prevention of uncontrolled release of energy, is a primary strategy in many cases and includes a large array of measures. The energy model is not detailed enough to yield support in identifying and evaluating the specific types of measures that apply in a given situation.

A second merit of the energy model is the support it offers in anticipating the consequences of accidents. After the uncontrolled release of energy, the accident sequence basically follows the laws of physics. The consequences are to a large extent determined by the amount of energy involved. Accident statistics of the Norwegian Directorate of Labour Inspection shows that

Table 5.2 Examples of safety measures relating to Haddon's ten strategies

Type of strategy	Examples of hazards and safety measures		
	Rotating machinery	*Toxic chemical*	*Motor vehicle*
	(Circular saw)	*(Oil vapour and mist from drilling mud in shale-shaker)*	*(Car traffic)*
Prevent build-up	Eliminate use of circular saw by ordering pre-cut pieces of wood	Eliminate oil in mud by using water-based mud	Avoid car driving
Modify the qualities	Modified saw blade teeth	Use of low-toxic oil	Softening of hard objects in the cabin
Limit the amount	Limit rotational speed	Smaller evaporation area	Speed limits
Prevent release	Design of start button which prevents accidental start	Not applicable	Sanding and salting of roads
Modify rate and spatial distribution	Emergency stop	Ventilation	Cars with shock absorbing zones, safety belt
Separate in time or space	Automatic sawing machine	Remote control	Separate lanes for meeting traffic
Separate by barriers	Machine guarding	Air curtain	Cars with safety cage
Make the victim more resistant	Eye protection	Respirators	Helmet
Counter damage	First aid	Not applicable	First aid
Rehabilitation	Depends on type of injury	Depends on type of illness	Depends on type of injury

certain types of events with a 'high energy content' are common among the fatalities, Figure 5.7. We see that motor vehicle accidents (linear kinetic energy), falls to a lower level (potential energy that is transferred to kinetic energy), falling objects (potential energy that is transferred to kinetic energy), and being hit by a moving object (kinetic energy) account for about two-thirds of the fatalities. In Part III, we will use the relationship between the energy contents of accidental events and consequences in a method to assess the potential consequences of near-accidents.

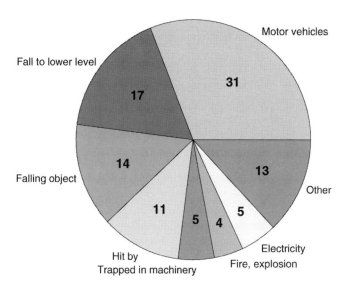

Figure 5.7 Statistics on fatalities in Norway during 1984–93.

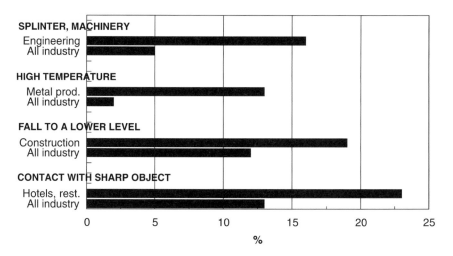

Figure 5.8 Accident types and their share of the reported accidents for selected branches of industry and for all industry in Norway in 1993.

A third important merit is the support the energy model offers in identifying hazards. Each type of industry is associated with certain types of hazards or 'energies', which show up in the accident records. Figure 5.8 shows an example from the accident statistics for Norway in 1993.

Checklists of 'energies' such as the example below have been developed for the purpose of identifying hazards.

Table 5.3 Hazards generated by machinery

---

1 Mechanical hazards
   a   crushing
   b   shearing
   c   cutting or severing
   d   entanglement
   e   drawing-in or trapping
   f   impact
   g   stabbing or puncture
   h   friction or abrasion
   i   high pressure fluid ejection
2 Electrical hazards
3 Thermal hazards
4 Hazards generated by noise or vibration
5 Hazards generated by radiation
6 Hazards generated by materials and substances
   a   toxic
   b   corrosive or irritant
   c   flammable or explosive
   d   viral or bacterial
7 Hazards generated by neglecting ergonomic principles

Source: CEN, 1991.

5.5 Logical tree models

Logical models aim at analysing the causes of accidents in terms of logical relations between events and conditions in the affected system. The models are based on the systems concept, which means that a production system is understood in terms of its components and the relationship between these. They are typically applied in theoretical analyses to estimate the risk of accidents (risk analyses) but may also be employed in diagrams to display the results of an accident investigation.

Fault Tree Analysis employs an analytical tree to display the results of an analysis (Suokas and Rouhiainen, 1993). It starts with the top event (injury or damage). The analysis proceeds backwards in order to identify all events and conditions that have caused the injury or damage. Logical relations (necessary and/or sufficient conditions) are established. Fault-tree analysis is not an accident model *per se* and gives limited support in the identification of causal factors.

The **INRS Model** builds on the principles of fault-tree analysis (Leplat, 1978). The model focuses on variations or deviations from the usual course of work at the work-systems level. There are four classes of variations, those related to the individual, the task, equipment and the environment respectively. We here see a clear relation to ergonomic models of work systems. The findings from an accident investigation are displayed in an analytic tree, showing causal relations, Figure 5.9. It gives a schematic presentation

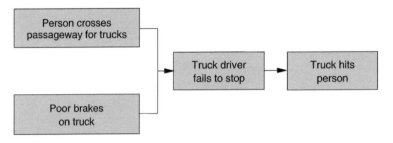

Figure 5.9 Example of a typical combination of deviations likely to lead to a traffic accident.

of the accident sequence and causal factors in order to identify logical errors/missing information and to support the communication of results of the investigation.

5.6 Human information-processing models

Human information-processing models focus on the interaction between the human operator and the environment in a disturbed system. This interaction is analysed from the human operator's point of view. The operator is viewed as an information processor who responds to deviations and hazards in the environment, Figure 5.10. In an analysis of an accident, the aim is to identify human failures in identifying and evaluating the situation and in taking the appropriate measures.

Figure 5.10 is a typical outcome of the research into human perception and cognition that made great progress in the 1960s. The human operator was viewed as an information processor that is exposed to a demanding environment. Accidents result mainly from the human operator's inability to handle the information in complicated situations. We will come back to other theories and models of human behaviour in accident situations in Chapter 8. There we will see how other theories and models have substituted

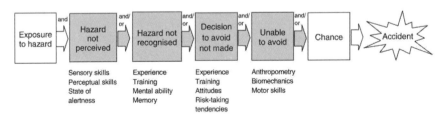

Figure 5.10 Accident sequence model showing failures in the various human information-processing stages leading to an accident.
Source: Adapted from Sanders and McCormick, 1992.

this view of the human operator as a victim of an excessively demanding situation. They have been developed to account for research findings, especially from the traffic-safety field, showing that the human operator has a certain degree of control over the demands from the environment and hence over the risks of accidents.

A problem in the application of the human information-processing models is that internal information processes are not readily available. They have to be interpreted from observations of actual behaviour and from interviews. This usually requires expert knowledge. The application of information ergonomics models in safety practice is thus mainly limited to in-depth investigation with participation of human-factor experts.

A second concern is that the models focus on 'cold' variables related to the cognitive processes of the human being. In a real situation, 'hot' or emotional variables such as the feeling of embarrassment and threat will have significant impact on the individual's ability to handle the situation and to avoid accidents. These types of variables are de-emphasised in the human information processing models. We will come back to these types of models in Chapter 8.

5.7 Moving the perspective to the organisational context

5.7.1 SHE management models

SHE management models are qualitatively different from many of the accident models presented in Sections 5.2 to 5.6. They expand the analysis of accident causes beyond the immediate causes in the work system by exploring contributions from management factors. The basic idea behind these models is that there exists an ideal SHE management system as specified in industry standards or handbooks. In this sense, they are prescriptive. By comparing the actual conditions with the ideal model, the analyst is able to identify gaps that represent so-called root causes of accidents. Typically, the SHE management models are based on a structural perspective on organisations.

The **Management Oversight and Risk Tree (MORT)** was first developed in the 1960s by the US Atomic Energy Commission. It has had large influence on the developments in SHE management since then. The idea at that time was to formulate an ideal SHE management system from a synthesis of the best accident models and quality assurance techniques then available (Johnson, 1980).

Figure 5.11 shows the main elements of MORT (Johnson, 1980; Knox and Eicher, 1992). They are displayed as events in a logical tree of, in total, about fifteen hundred basic events. Each event corresponds to a question to be addressed in accident investigations and SHE audits. The MORT diagram is not based on systematic research, but is rather a codified collection of the practical experience and judgement existing at that time.

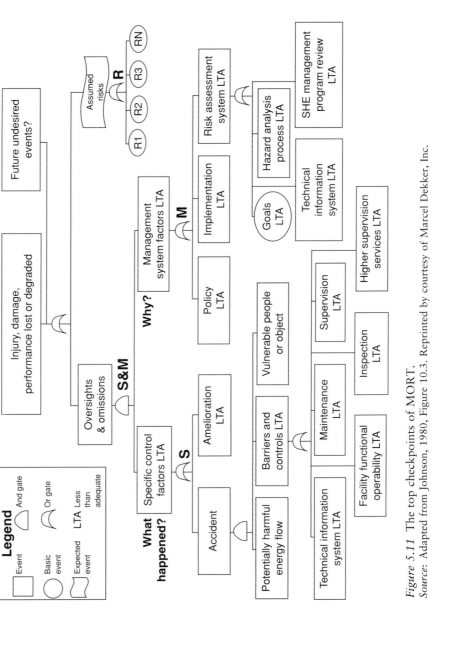

Figure 5.11 The top checkpoints of MORT.
Source: Adapted from Johnson, 1980, Figure 10.3. Reprinted by courtesy of Marcel Dekker, Inc.

The top event of the tree is made up of the accidental loss (experienced or potential) to be analysed. The branches below the top event are built up around four principles:

- Assumed risks
- The energy model
- The feedback control cycle
- A system's life cycle.

First, the analyst has to address whether the accident represents an assumed risk or whether it is the result of oversights and omissions. *Assumed risks* are accident risks that have been identified, evaluated and accepted at a proper management level prior to the MORT analysis of an accident. The analyst will terminate the investigation if all these three conditions are fulfilled.

In all other cases, the analysis will continue along two branches, called Specific oversights and omissions (S-branch) and Management oversights and omissions (M-branch) respectively. The *S-branch* of the MORT diagram focuses on the events and conditions of the accident occurrence (actual or potential). Here, time develops from left to right. Causal influences go from bottom to top.

We recognise Haddon's energy model as a key element in the S-branch (Haddon, 1980). An event is denoted an accident when a target (a person or object) is exposed to an uncontrolled transfer of energy and sustains damage. Accidents are prevented through barriers. There are three basic types of barriers: (1) barriers that surround and confine the energy source (hazard), (2) barriers that separate the hazard and the target physically or in time or space and (3) barriers that protect the target. We find these different types of barriers in the branches below the accidental event. Amelioration relates to the actions taken after the accident to limit the losses (rescue, fire fighting, etc.).

At the next level of the S-branch, we recognise factors related to the different phases of the life cycle of industrial system: the project phase (design and plan), start-up (operational readiness) and operation (supervision, maintenance). The idea here is to link barrier failures to when they first occurred in this life cycle.

When proceeding along the *M-branch*, the analyst is interested to know why these inadequacies have been allowed to occur. The queries are directed at the main elements of the management system:

- The standard, i.e. policy, goals, requirements, etc.
- Implementation, and
- Measurement and follow-up.

These are the same basic feedback control elements that we find in the Quality Assurance principles of, for example, the International Standards

Organisation ISO 9000-series. Influences can be traced back to Juran's work in the 1960s (Johnson, 1980).

This part of the analysis helps the analyst to generalise the specific findings from one specific accident investigation or safety-programme evaluation. Events and conditions of the S-branch thus often have their counterparts in the M-branch. By querying the M-branch, the analyst is able to move his or her focus from the specific circumstances at the accident site to the total management system. The recommendations from this analysis will affect many other accident risks as well.

On the bottom line, MORT has a collection of questions whether specific events and conditions are satisfactory or 'less than adequate' (LTA). In the detailed questions we find elements from such different fields as risk analysis, human-factor analysis, SHE information systems, and organisational analysis. The judgements made by the analyst are partly subjective. It has proved necessary to train and certify analysts to ensure an adequate quality.

The MORT diagram is used in the planning of accident investigations and SHE audits. It is also used when reviewing the findings from an investigation. In this application, MORT assists in checking the completeness of the investigation and in avoiding personal biases.

The literature on evaluations of MORT is sparse. Johnson reports a better coverage of supervisory and management deficiencies in the accident investigations after the introduction of MORT (Johnson, 1980). Experiences have also been gained from evaluations of MORT applications within Finnish industry (Ruuhilehto, 1993). These point at MORT's significance in building bridges between SHE and production management. The application of MORT assists in general planning and control and affects the frequency of production disturbances as well. Some limitations have also been identified in the Finnish studies. MORT is not well suited for the identification of immediate risks due to failures and disturbances. Another problem is that there is no scheme for building priorities into the MORT concept.

SMORT (Safety Management and Organisation Review Technique) was developed from MORT in order to make it usable as a tool in the data-collection phase of accident investigation as well (Kjellén *et al.*, 1987). This model is a basic component of the accident-analysis framework of Chapter 6. The basic elements of MORT have been re-arranged in order to provide for a logical development of an accident investigation process, Figure 5.12. This starts with the mapping of the sequence of events (level 1). Explanations for the findings at this level are sought at the workplace level (level 2), in project work, i.e. design, fabrication and start-up of new systems (level 3) and in SHE and general management (level 4). SMORT contains a number of checklists for use during the investigations at each level. We will come back to applications of SMORT in Part III.

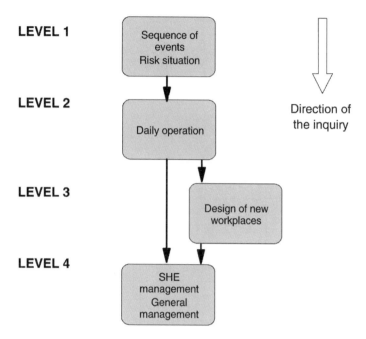

LEVEL 1 — Sequence of events / Risk situation

LEVEL 2 — Daily operation

Direction of the inquiry

LEVEL 3 — Design of new workplaces

LEVEL 4 — SHE management / General management

Figure 5.12 The levels of a SMORT analysis.

In Section 5.2, we mentioned the **ILCI model**. We identified the three SHE management elements of this model: inadequate programme; inadequate programme standard; and inadequate compliance with standard. These elements are built around the three basic elements of administrative systems for feedback control, i.e., goal, implementation and follow-up. A factor that has contributed to the success of the ILCI model in industrial practice is the fact that it prescribes a standard SHE programme with the necessary elements and activities. For each element, there is a set of detailed criteria against which the actual situation must be compared. Value judgements as to what is 'less then adequate' are to a lesser extent left to the discretion of the analyst than in MORT and SMORT analyses. The application of the ILCI model in accident investigation is not dependent on the analyst's experiences and skills to the same extent as in MORT or SMORT. The downside is the uncertain validity of the findings to the industrial organisation.

E&P Forum's SHE management model represents a typical trend of the 1990s. The ISO 9000 and ISO 14000 standard families on quality management and environmental management have influenced it (ISO, 1994; ISO, 1996). The model represents the backbone of E&P Forum's guidelines for SHE management systems. It identifies basic elements of a SHE management system, including:

Figure 5.13 E&P Forum's SHE management model.
Source: E&P Forum, 1994. Copyright 1994 by E&P Forum and reproduced by permission.

- Leadership and commitment, i.e. commitment from top management downwards and company SHE culture,
- Policy and strategic objectives, i.e. management intentions, action principles and aspirations with respect to SHE,
- Organisation, resources and documentation, i.e. organisation of people, responsibilities, resources and documentation of the SHE management system,
- Evaluation and risk management, i.e. identification and assessment of hazards and development of remedies,
- Planning, i.e. activity planning, including planning for changes and emergency response,
- Implementation and monitoring, i.e. implementation and monitoring of activities and results including incident reporting and the follow-up and taking of corrective actions when necessary, and
- Auditing and review, i.e. periodic assessments of the effectiveness of the SHE management system and its suitability.

There is a certain degree of arbitrariness in the development of SHE management models. The number and types of elements vary from model to model. There is no comprehensive theory behind any of the models, Rather, each model represents the collective experiences and perceptions of the expert team responsible for developing the model. Typically, these experts are inclined to believe in a structural perspective.

We also see different approaches as to how detailed the guidelines are, which define an appropriate implementation of each element. This has to do with the general difficulty in balancing between perspective and goal-oriented recommendations. Detailed and concrete advice is often easy to implement without requiring substantial management effort. The applicability and validity of these recommendations to the company in question may, however, be questionable.

5.7.2 The SHE culture

The SHE culture is a perspective for the analysis of organisations seen from the SHE point of view rather than an accident model. Whereas the SHE management models focus on the 'cold' aspects of SHE management programmes, we here focus on the 'hot' variables of shared beliefs, attitudes and norms within the industrial organisation. We will here regard the SHE culture to be an integrative concept. It addresses different organisational aspects such as SHE related values, attitudes, perceptions, competencies and resulting patterns of behaviour at the different organisational levels (see e.g. ASCNI, 1993). We here include provisions for workers' involvement and organisational learning in the SHE-culture perspective. This interpretation of the concept of SHE culture has a clear parallel in the human resource and symbolic perspectives of organisations.

The **TRIPOD model** includes SHE culture elements (Reason, 1991). In an accident investigation, the model is used to classify top management commitment on a scale from 1 (pathological) to 7 (generative-proactive). We will treat this aspect of the TRIPOD model in more detail in Section 6.5.

There is a relationship between the 'organisational culture' and the organisation's ability to learn from accidents to prevent recurrence. **Lucas' framework of organisational cultures** distinguishes between three different types of organisations in this respect (Van der Schaaf *et al.*, 1991). His framework does not represent a new accident model. Rather, it analyses the types of accident or human-error models that are predominant within an organisation. Such shared models will determine the organisation's ability to learn from experience and to prevent the recurrence of accidents. Lucas distinguishes between three different types of organisational safety cultures and associated human-error models:

1 Traditional 'occupational safety management' culture, i.e., a culture where the causes of errors and accidents are attributed to inattention and carelessness on behalf of the workers. Disciplinary measures will dominate the remedial actions.
2 'Risk management' culture, where an engineering view of human error causation is dominant. Errors and accidents are analysed in terms of mismatches between the operator and his environment. Remedial actions typically include design changes and provisions of procedural support.

3 'Systemic safety management' culture, where the causes of errors are analysed in relation to the total work context. Not only are traditional causes such as poor design and procedures considered, but also such aspects as unclear responsibilities, lack of knowledge and a low morale. These in turn are traced back to management liability issues.

It is relevant to apply the approaches by Reason and Lucas to cases where an independent body (e.g. an accident commission) carries out the accident investigation. The commission will ask questions about previous incidents of a similar type and the organisation's ability to learn from them. The commission may judge the failure to learn from previous experience as a root cause of the incident.

6 Framework for accident analysis

6.1 Characteristics of the accident sequence

This chapter presents a framework for the collection and analysis of data on accident risks, Figure 6.1. It borrows aspects from the different accident models presented in Chapter 5. We will apply this framework in a review of classification systems and variables used in the collection and analysis of accident risks. It will also be a common denominator in our reviews and analyses of different SHE management methods and tools in Parts III to V.

Let us analyse the framework by starting at the output side. There are two different types of unwanted results of accidents:

- 'Material' losses including injury to people, damage to the environment and property damage, and

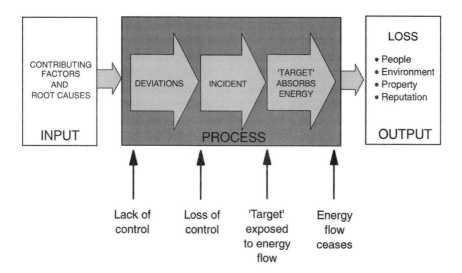

Figure 6.1 Accident analysis framework.

- Immaterial losses such as damage to the reputation of the responsible organisation (company), pain and suffering of the involved persons and a reduced quality of life in general in the area affected by an environmental accident.

We recognise the energy model in the parts of the model starting with the loss of control. Let us focus on personnel injury. The energy flow originates from an energy source or a hazard. An *incident* is here defined as a loss of control of the energies in the system or of body movements in relation to these and the subsequent uncontrolled energy flow/body movements. Development of loss occurs when the victim's body gets in contact with the energy flow and the exposure exceeds body-injury threshold. The *danger zone* is the area around the energy source that may be affected by the energy flow. By replacing a person with other types of 'targets', the model applies to damage of material assets (buildings, machinery vehicles, materials) and the environment as well.

There are different types of loss of control:

1   A structural failure or a failure in the control system of a machine can illustrate a purely technical event. There is no immediate human intervention.
    Example: The crane load is lost due to a brake failure.
    Example: A robot makes uncontrolled movements due to a programming error.
2   A human operator loses control of an external energy source.
    Example: A typical example is driving, where the driver may lose control over the car. The energies in this system are generated by the motor.
3   A human operator loses control of a hand tool that is powered by his muscular energy.
    Example: A carpenter misses the nail with his hammer and hits his thumb.
    Example: A butcher cuts himself with the knife.
4   A person loses control of his or her own body movements.
    Example: A person slips and falls on the floor.
    Example: A person slips and falls against a rotating saw blade.
    Example: A person lifts a heavy box in an awkward position and overstrains a muscle.

Incidents rarely occur by pure chance. In the analysis of how incidents occur, we rely on the OARU model of Section 5.3. An incident is usually preceded by deviations at the workplace that increase its frequency and/or the consequences of it. SHE measures that eliminate existing deviations (e.g. repair of faulty safety equipment) will have an immediate effect on the risk of accidents. They will, however, not have lasting effects if the deviation may occur again.

The input to the accident process consists of contributing factors and root causes. *Contributing factors* are more stable conditions at the workplace. By changing such factors, more lasting effects will be achieved. In practice, it is sometimes difficult to separate contributing factors (input) and deviations in the accident sequence (process).

Root causes are the most basic cause of an accident/incident, i.e. a lack of adequate management control resulting in deviations and contributing factors. Stop rules have to be applied in the investigation into the accident sequence in order to avoid the 'garden-of-Eden' problem where we look for accident causes far away from the accident site in time and space (Rasmussen, 1993).

To summarise, the following characteristics of the accident sequence represent the basis of this analytic framework (Kjellén and Hovden, 1993):

1 The *time dimension*, i.e. the understanding of the accident as a process rather than as a single event or as a chain of causal factors. The process developed through consecutive phases, where there is a transition from lack of control, through loss of control of energies in the system (incident) and further to development of loss.
2 The initial phase of the accident sequence is described in terms of *deviations* from the normal and/or faultless production process.
3 The injury or damage is the result of an *uncontrolled flow of energy*, where the victim (a person, the environment, property) is exposed to this energy flow. In the case of injury to a person, the energy flow exceeds body-injury thresholds.
4 The injury or damage results in *losses*. We will here primarily be concerned with 'material' losses, i.e. injury to persons and damage to the environment and to material assets.

Figure 6.2 illustrates this analytic framework by presenting an example. This framework helps us in defining the start and the end of the accident sequence. The accident sequence *starts* with the occurrence of the first deviating event, which is logically and chronologically connected to the outcome (injury or damage). It *ends* when the energy flow ceases. The focus on deviations also helps in making priorities in data collection. It is important that data about deviations are secured immediately, when there is a risk that the information otherwise gets lost.

The framework also helps us in defining *near accidents*. This is a sequence of events which includes an incident, but where injury was avoided due to pure chance.

The contributing factors are technical, organisational and social conditions and individual circumstances at the workplace in question that affect the accident sequence. They change only slowly in time as compared with

DEVIATIONS A valve lever blocks the gangway Oil spill on the floor Gloves are not used	INCIDENT The operator slips when he ducks to pass below the lever, he loses his balance and falls	DEVELOPMENT OF INJURY The right hand is squeezed between the pipe and the floor No protection from gloves
DEVIATIONS - Work material - Personnel - Instructions - Technical equipment - Human action - Influence from other work	INCIDENT - Loss of control - Active safety system failure - Fixed protection failure - Personnel inside danger zone	DEVELOPMENT OF LOSS - Personal protection failure - Failure in limitation of loss

Figure 6.2 Framework for the analysis of an accident sequence. The example shows an accident in a process plant, where an operator is carrying a heavy load (pipe).

Source: Kjellén, 1992. Copyright 1992 by Tiden Norsk Forlag and reproduced by permission.

the rather rapid sequence of events. There are different types of contributing factors. Design of machines and maintenance routines, for example, will affect the likelihood of technical deviations and incidents. Safety barriers such as guards will, on the other hand, affect the consequences of such incidents.

In an accident investigation, the first step is to map all relevant facts. This step of the investigation will focus on the losses and the accident sequence, i.e. the extent of injury and damage, the incident and preceding deviations. In the next step of the investigation, the investigator interprets the evidence in order to identify contributing factors. Here, subjective judgements are a necessary ingredient. This is especially the case when human factors have played a central role in the accident sequence. Differences in focus as well as in opinion may occur and need to be resolved.

This distinction between facts and interpretations is not always made in the causal sequence models of accidents. Such linear accident causation models as the ILCI model do not make a clear distinction between observable facts and conditions on the one hand and opinions about effects of personal and job factors on the other hand.

Exercise: Establish an accident-investigation team of a group of students. Each student should represent one of the following persons holding different accident perceptions and views on how to prevent accidents:

- *A line supervisor, representing a traditional occupational safety management view;*
- *A safety engineer, representing the barrier approach according to Haddon (Section 5.4);*
- *An expert on ergonomics, representing a human information-processing approach (Section 5.6); and*
- *A maintenance supervisor, representing an industrial engineering view.*

The task of the team is to discuss possible contributing factors to the accident in Figure 6.2. Each member contributes by sharing his accident perception with the others. Discuss what aspects each of the members will focus on. Will any one of the perspectives be more likely to contribute to effective countermeasures than the others?

In Table 6.1, we have applied the analytic framework of Figure 6.1 to the analysis of the development of different types of losses. The analytic framework is well suited to the analysis of accidents resulting in personal injury, material damage or acute environmental pollution. By definition, these types of occurrences involve an incident. We also know that 'low-intensity' incidents may contribute to occupational diseases such as hearing losses, back injuries and lung diseases.

Deviations are common, but not always present in the accident sequences. We know that deviations contribute to the occurrence of all different types of losses. The 'control climate' at the company has a significant influence on the risk of accidents (Grimaldi, 1970). An efficient control of deviations is thus a key strategy for the control of losses in general.

We will later apply the accident-analysis framework in a review of different types of methods used in the collection and analysis of data of accident risks. We will start at the output side of the model by reviewing the different types of classification systems used to document the consequences of accidents and different measures of loss. We will then continue by looking into the classification systems used to document incidents and deviations. Finally, we will review the different classification systems for contributing factors and root causes. Our aims will be twofold: first, to be complete, i.e. by presenting all alternative means of measuring and classification, and second, to give specific advice on the preferred method. The reader will find recommended alternatives in shaded tables and checklists.

Table 6.1 Application of the accident-analysis framework to illustrate the development of different types of losses

Type of event	Initial phase (deviations)	Incident	Development of loss
Occupational accidents	Deviations are common	An incident will always precede the loss	Person is exposed to sudden and uncontrolled energy flow
Occupational diseases	Deviations may result in increased exposure	Low-intensity (and not reported) incidents may contribute	Person is exposed to low intensity energy flow ('planned' and unplanned) over a long period
Material damage	Deviations are common	An incident will always precede the loss	Material is exposed to sudden and uncontrolled energy flow
Poor product quality	Deviations are common	Not relevant	Production outside quality norm
Production regularity	Deviations are most often present	An incident may precede the loss	Stop in production
Acute environmental pollution	Deviations are common	An uncontrolled release of toxic substances will always precede	The environment is exposed to sudden release of toxic substances
Long-term environmental pollution	Deviations may occur	Small uncontrolled releases may precede	The environment is exposed to low intensity release of toxic substances over a long period

6.2 Consequences of accidents

6.2.1 Types of consequences

Information on the immediate consequences to the victim of an accident is usually readily available and is well suited for classification. There are standard schemes for the classification of the nature of the injury and part of body affected. Table 6.2 shows an example of a common injury classification scheme (ILO, 1998; Eurostat, 1998).

Statistics on accident distributions by type of injury and part of body affected find many practical uses.

Table 6.2 Type of injury and part of the body injured

Type of injury	Part of the body
• Contusions, bruises • Concussions and internal injuries • Open wounds, including cuts, lacerations, abrasions, severed tendons, nerves and blood vessels • Amputations • Open fractures • Closed fractures • Dislocations • Distortions, sprains, torn ligaments • Asphyxiation, gassing, drowning • Poisoning, infections • Burns (including chemical burns), scalds, frostbites • Effects of radiation • Electrocutions • Other	• Head, including facial area, eyes, ears and teeth • Neck • Back • Torso and organs, including rib cage, chest area, pelvic and torso • Upper extremities, including shoulder, arm, hand, fingers and wrist • Lower extremities, including hip, leg, ankle, foot and toes • Other part of body • Whole body and multiple sites

Example: The yearly accident-statistics summary from a mechanical workshop showed that 25 per cent of the injuries were burns or cuts in the eyes. It was decided to introduce mandatory eye protection while working in the production halls in order to reduce this type of injury.

Adverse environmental effects are usually not recorded in a similar way, due to the difficulties in identifying the damage. They are local, regional and/or global and acute and/or delayed. Rather, the type of release is described, such as oil spill.

Insurance companies apply elaborate classification schemes for material damage. In industry, it is common to distinguish between material damage due to fires and explosions from other types of material damage.

There are no generally accepted schemes for the classification of social and political losses following an accident.

6.2.2 Consequence measures

The size of the losses is an important parameter in making priorities on safety measures. Different types of scales are here applied. Table 6.3 gives an overview.

Common consequence measures such as LTI and RWI are easy to record but are insensitive to the actual size of the loss. An eye injury resulting in a few days of absence, and an amputation of an arm, are both recorded as a LTI. The number of days of absence better reflects the actual losses but is not readily available at the time when the accident is recorded. This is especially a problem in keeping accident records, when the duration of the sick leave extends over several recording periods (months, quarters, etc.).

Table 6.3 Overview of measures of the severity of losses due to injury or damage

'Target'	Type of scale[1]	Type of measure
Person	Nominal	LTI = Lost-Time Injury[2] (yes/no)
	Nominal	Recordable injury[3] (yes/no)
	Nominal	RWI = Restricted Work Injury[4] (yes/no)
	Nominal	Fatality (yes/no)
	Ordinal	Consequence categories (e.g. first aid, temporary disability with a few days of absence, temporary disability over an extended period of time, permanent disability, fatality)
	Ordinal	ILO's consequence categories (temporary incapacity to work, less than one day lost, 1–3 days lost, 4–7 days lost, 8–14 days lost, 15–21 days lost, 22 days–one month lost, one to three months lost, three to six months lost, permanent incapacity to work or more than 183 days lost, fatal injury)
	Ordinal	Classification according to extent of disability (no permanent disability, < 5%, 5–10%, 10–20%, 20–50%, > 50%, death)
	Interval	AIS = Abbreviated Injury Scale (AAAM, 1985)
	Ratio	Number of whole days lost compensated by the insurance company/employer[5]
	Ratio	Number of fatalities
Environment	Ratio	Amount of release in m^3, tons, etc. by substance
Material	Ratio	Losses in monetary unit (e.g. Euro or US$)
Production	Ratio	Duration of production stop (e.g. hours)
	Ratio	Lost production (e.g. number of units)
	Ratio	Losses in monetary unit (e.g. Euro or US$)

Sources: ILO, 1998; Eurostat, 1998; OSHA, 1971/97.

Notes:
1 For types of scales, see Chapter 11.
2 A lost-time injury is defined as an accident resulting in injury and where the injured person does not return to the next shift.
3 Injury that involves one or more of the following: Fatality, lost workday(s) loss of consciousness, restriction of work or motion, transfer to another job, medical treatment other than first aid.
4 Injury at work that does not lead to absence after the day of occurrence because of alternative job assignment.
5 Fatality and permanent disability (100%) equal 7500 workdays.

Consequence measures based on medical evaluations, such as the AIS and the extent of disability, are more valid loss measures than the LTI. AIS measures the probability that an injury may result in a fatality. Category AIS 1 is used for small cut injuries, finger fracture, small burn, etc. AIS 4 is an injury which is life-threatening but where survival is likely and AIS 6 is a fatal injury.

Consequence categories based on assessments of the severity of injuries are applicable immediately after the accident. The problems with delayed recording are thus avoided. They are also well suited for statistical summaries. A current trend is to apply common measures for the severity of

Table 6.4 Common consequence measures for different types of losses

Grade	Personnel	Environment	Material production	Delivery	Image
1	First-aid injury	Insignificant damage	< 1k Euro	Insignificant internal delay	Insignificant impact
2	Lost-time injury	Moderate damage, restitution time < one month	1–10k Euro	Internal delay that affects production plans for less than one week	Slight impact
3	Permanent disability	Severe damage, restitution time < one year	10–100k Euro	Internal delay that affects production plans for more than one week	Limited impact
4	Fatality, one person	Locally irreversible damage, restitution time 1–10 years	100k Euro– 1000k Euro	Delay that affects the client's production plans	Considerable impact
5	Fatality, two or more persons	Danger of exterminating fauna and flora, restitution time > 10 years	> 1000k Euro	Extensive delay, risk of losing major client	Major impact

different types of consequences. Table 6.4 shows an example with five severity classes. The matrix in the table has been influenced by the so-called Injury Potential Matrix (Booth, 1991).

6.2.3 Economic consequences of accidents

The costs of accidents are shared between the individual, the company responsible for the accident, the insurer and the public sector. Company's costs are normally not visible in their books of account. A traditional approach in accident research has been to develop models for use in the recording of accident costs to the company. Average results are then used in estimations of costs of future accidents. The aim is to make the losses associated with accidents visible to the management of the company.

Early research on accident costs applied the *market-pricing model*. Here, the analyst registers the actual losses due to accidents for different production factors such as lost working hours, materials and production. These are assessed in monetary units by applying the market prices for each factor. The cost of lost working hours, for example, is set as equal to the hourly wage. Heinrich pioneered this work. He distinguished between direct and indirect costs, where direct costs are those paid by the insurer to the victim (Heinrich, 1959). Indirect (or hidden) costs include costs directly carried by

Table 6.5 Company cost elements of an accident

Market-pricing model	Accounting model
• Lost work hours, victim • Work hours spent on changing work routines • Work hours spent on the investigation • Work hours spent on repair of damaged equipment • Lost work hours due to interrupted production or reduced productivity	• Increased costs for personnel (replacement, overtime, etc.)
• Costs of replacing damaged material	• Costs of replacing damaged material
• Costs for transportation (of victim etc.)	• Costs for transportation (of victim etc.)
• Capital costs (for machinery etc.) during production stop • Insurance expenditures • Loss of income • Costs of safety measures • Company's costs for medical treatment	• Insurance expenditures • Loss of income • Costs of safety measures • Company's costs for medical treatment

Source: Sklet and Mostue, 1993.

the company such as lost working hours, reduced productivity, costs of property damage and overhead costs. There are variations of the direct/ indirect cost split in the later research literature such as insured/uninsured costs and costs of accidents that can/cannot be controlled by management (Grimaldi and Simonds, 1975; Laufer, 1987).

Heinrich's empirical work showed a one to four ratio between direct and indirect or hidden costs (Heinrich, 1959). The implications of these figures are that significant economic benefits of safety work are overlooked. However, later studies of accident costs in Britain, Finland, Israel, Norway and USA, applying variations of Heinrich and later accident cost models, have come to other results. The ratio between direct/insured costs and indirect/ uninsured costs vary from less than one to over 10, see Rognstad (1993) for an overview. Differences in the applied methods and between industries and countries may explain the varying results.

The market-pricing model has been criticised, because it is based on unrealistic assumptions, i.e. perfect market conditions and full capacity utilisation (Rognstad, 1993). An alternative model, *the accounting model*, studies the effects of accidents on the company's contribution margin (Matson, 1988). These are made up of changes in revenues and variable costs (including wages) because of the accident. Table 6.5 gives an overview of the different cost elements of the two models.

The different cost elements of the market-pricing model have been further systematised in the *Accident-Consequence-Tree method*. It displays the different consequence elements related to the individual, the company and the national economy as branches of a standard tree (Aaltonen *et al.*, 1996).

In Norway, a series of ex post-accident cost studies have been performed within the chemical, metallurgical and mechanical industries (Sklet and Mostue, 1993). When the market-pricing model was applied, average costs to the company of accidents were about 1000 euro. They were lowest in the chemical industry (800 euro) and highest in the metallurgic industry (1100 euro). Costs increased with the severity of the accident. Costs to the company for permanent disabilities and fatalities varied between 1500 and 10000 euro. Uninsured salary to the injured worker accounted for between 74 per cent and 90 per cent of the total costs for the temporary disabilities. In-plant accident costs were significantly less than 1 per cent of the salary costs.

When the accounting model was applied, the costs were even lower. For temporary disabilities of up to 10 days of absence, they varied between 10 and 350 euro. For accidents resulting in more than 10 days of absence, the company actually experienced economic benefits of up to 3400 euro. This is because the company could produce at normal level with a reduced work force, while the insurer paid the wage costs for the victim.

When these studies were carried out, companies in Norway paid between 20 and 40 per cent of the total costs of accidents. After changes in the legislation in Norway, companies pay between 60 and 90 per cent of the costs. Even when these changes in the legislation are accounted for, the results of the studies are discouraging as to the economic incentives to the company of preventing occupational accidents. Usually, the costs for SHE work exceed the in-plant costs of occupational accidents (Rosness, 1995).

In spite of these results, many large companies, also in Scandinavia, have developed SHE policies, on the argument that these aspects are of equal importance to economy. These statements are not only based on ethics. Evaluation research shows that, under certain circumstances, SHE work is a value-adding activity that promotes management's control of production (Grimaldi, 1970; Kjellén *et al.*, 1997). It contributes to a reduction in the probability of all types of losses resulting from unwanted events and deviations.

6.2.4 *Actual versus potential losses*

Both accidents and near-accidents may have a potential of resulting in more severe losses than the actual outcome. There are two advantages in evaluating the potential losses of the occurrences:

1 Improved possibilities of learning from experience before the occurrence of severe accidents. In practice, this means that certain events are given a more rigorous treatment than would have been the case if only the actual outcome was considered.
2 A higher number of events are evaluated. This gives an increased statistical basis for the identification of trends and needs of remedial actions.

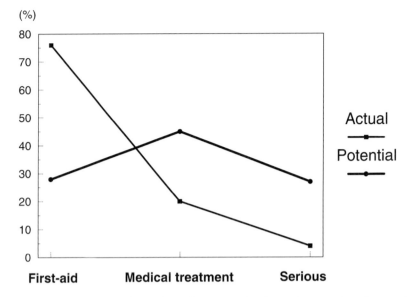

Figure 6.3 Distribution of accidents from an oil refinery by actual and potential losses.
Source: Adapted from Booth, 1991.

In assessing the potential losses of accidents and near-accidents, the worst consequence that realistically may happen is identified. This is to a large extent determined by the energies involved (Section 5.4). In doing this assessment, existing safety measures and other conditions are considered, but pure luck is disregarded. Consequence categories such as those in Table 6.4 are used in assessing the potential losses. Figure 6.3 shows the result of an evaluation of potential losses as compared to actual losses for an oil refinery. The assessment results in a redistribution of the accidents in the direction of a higher severity.

This technique of assessing the potential of accidents can be further developed by also assessing the expected frequency of a reoccurrence of the events. By doing so, a risk score is achieved (low, medium or high), see Table 22.4. A high-risk score means that measures must be taken to reduce the risk of the event to a lower level by reducing frequency and/or consequence. A medium-risk score means that measures ought to be taken to reduce the risk further, based on the ALARP (as low as reasonably practicable) principle. A low-risk score means that the conditions are acceptable and no measures are necessary.

Experiments have shown that judgements of the potential of accidents made by safety experts are reasonably reliable. At the workplaces, the supervisors may be assigned the task to make such judgements. Experience shows that a number of subjective factors influence the supervisor's assessments. There are biases in the direction of lower-risk scores, when there are 'negative' consequences to the supervisor of making a high-risk score. This has to do with the fact that the supervisors are responsible for following up incidents with action. A high-risk score will result in more attention from upper management and the authorities, and requirements to implement rigorous safety measures. It is thus recommended that an independent person, e.g. a safety expert, makes the assessments.

> *Exercise: Select about five accident reports. Assess actual and potential losses for each accident. Compare the results between the different groups of students and discuss the differences in the assessments.*

6.3 Incident (uncontrolled energy flow)

Sources of energy in the working environment represent potential hazards (Section 5.4). The energy sources are either controlled directly by the operators or by technical systems. An incident involves a sudden and uncontrolled release of energy. The human operator may be directly involved in this event by losing control of the energies. Failures of technical control systems may also result in uncontrolled energy release. An uncontrolled movement of a robot arm due to a programming error is a typical example. Another possibility is that man loses control over his body motions in relation to the energy flow (e.g. falls against a rotating saw blade).

The injury or damage develops when the energy flow reaches the target (the human body, the environment, material assets such as buildings, machines and material). The severity of the injury or damage is dependent on the type and amount of energy and the way it reaches the target. We have seen that fatalities were associated with a few types of events, where large amounts of energies are involved (Section 5.4). These included movements of vehicles (kinetic energy), fall to a lower level or being hit by falling objects (potential energy).

The investigator documents the loss of control and subsequent uncontrolled energy transfer in the free-text description of the sequence of events. In addition, it is common for statistical purposes to classify the event by 'accident type'. Table 6.6 shows two different schemes for classification of this energy transfer. The first is a traditional accident type classification applied by OSHA in the USA (OSHA, 1971/97). The second is taken from a proposed scheme for the classification of occupational injuries (ILO, 1998).

It is also common to register the object or substance involved in the energy transfer, the so called 'injury agency'. Table 6.7 shows the classification of

Table 6.6 Two examples of accident type classifications

OSHA	ILO
• Fall from elevation • Fall on same level • Struck against • Caught in, under or between • Rubbed or abraded • Bodily reaction • Overexertion • Contact with electrical current • Contact with temperature extremes • Contact with radiation • Contact with caustics, toxic and noxious substances • Motor vehicle accident	• Contact with electrical voltage • Contact with temperature extremes • Contact with hazardous substances • Drowned, buried • Fall, crash into (victim in motion) • Struck by (object in motion) • Collided with (victim and object in motion) • Contact with sharp, pointed, rough or coarse element • Trapped, crushed • Acute overload of body • Bites, kicks

Table 6.7 Injury agency classification

OSHA	ILO
• Machine • Conveyor, elevator, hoist • Vehicle • Electrical apparatus • Hand tool • Chemical • Working surface, bench, table, etc. • Floor, walking surface • Bricks, rocks stones • Box, barrel, container • Door, window • Ladder • Lumber, woodworking material • Metal • Stairways, steps	• Building, constructions and areas • Prime movers • Distribution systems • Hand tools • Machines and equipment • Conveying, transport and packaging equipment • Parts of machines, vehicles, etc. • Agents, substances, materials and objects, including radiation • Living organisms and human beings • Physical and natural phenomenon, including natural disasters • Household equipment, aids for schools and instruction and personal articles

injury agency used by OSHA. Such a list of types of objects and substances does not represent a taxonomy, since the list is not exhaustive and the classes are not necessarily mutually exclusive (see Chapter 11).

Registrations and coding of accident type and injury agency are based on physical evidence from the accident site and on interviews with witnesses. Such registration is reliable, provided that it is based on facts. There may, however, be a certain arbitrariness involved in the coding of accident type and injury agency in case of complex accident scenarios.

6.4 Deviations

Nonconformities are defined as the nonfulfilment of specified requirements in the ISO 9000 standards for quality management systems (see ISO, 1994). To avoid confusion with this definition, we will here use the term '*deviation*' instead. It covers those aspects of the accident sequence that represent a mismatch between our norms for a faultless production process and what actually happened. Nonconformities are one type of deviations, where the norm has been defined and documented.

Deviations are social constructs, whose identification and evaluation is dependent on the types of norms that are in use at any time. The focus on deviations helps us in identifying and evaluating the transient and specific circumstances at the accident site. In the next step, we look into how these circumstances can be explained by failures in the prevailing conditions at the workplace and in the company's management systems.

In Section 5.3 we reviewed different norms of this kind. These included the type of norm according to ISO 8402, i.e. specified requirements in regulations, standards, rules, etc. (ISO, 1994). There are also other types of norms such as what is planned/intended, what is normal/usual and what is acceptable. This extension of the definition has been necessary in order to accommodate types of production where the detailed planning and method of work is decided by the individual worker or work team.

The identification of deviations may be problematic due to differences in opinion, for example, among workers, supervisors, managers and systems designers, about what is normal. Another problem is the lack of norms in situations which have not been encountered before. These differences of opinion and lack of norms may in themselves contribute to an increased risk of accidents. One intention of focusing on deviations is to stimulate discussions inside companies, where such differences in opinion are highlighted and preferably resolved. The aim is to arrive at mutually shared norms of what constitutes a faultless production process.

Studies of accident reports show that the investigators often fail to document deviations (Kjellén, 1982). In order to support the identification and classification of deviations, different checklists have been developed. Table 6.8 shows examples of different aspects applied in the classification of deviations.

6.4.1 Heinrich's classical man–environment taxonomy

We recognise the classical man–environment taxonomy according to Heinrich in Table 6.8. It consists of two classes, i.e. unsafe acts by person and unsafe physical conditions.

Bird and Germain (1985) have further developed this taxonomy in the so-called ILCI model. They start from Heinrich by distinguishing between substandard acts and substandard conditions. Based on practical experience,

Table 6.8 Overview of different schemes for the classification of deviations

Basis for classification	Classes of deviations	Source
Man–environment	• Act of person • Mechanical/physical condition	Heinrich, 1959
Man–environment	• Substandard practice • Substandard condition	Bird and Germain, 1985
Ergonomics systems view	• Personnel • Task • Equipment • Environment	Leplat, 1978
Industrial engineering systems view	• Material • Personnel • Information • Technical • Human act • Intersecting/parallel activities • Environment • Guards	Kjellén and Larsson, 1981; Kjellén, 1984

Type of loss of control	• Deviation outside the injured person
	  ◆ Electrical
	  ◆ Explosion
	  ◆ Fire, flare
	  ◆ Overflow, overturn, vaporisation, leak
	  ◆ Break, fracture, deformation of material
	  ◆ Slip, fall, collapse of elements
	• Injured person has lost control
	  ◆ Loss of control of machinery
	  ◆ Loss of control of equipment
	  ◆ Loss of control of tools
	  ◆ Fall of person on same/to a lower level
	  ◆ Fighting, violence
	• Injured person is actively involved in the accident
	  ◆ Tread
	  ◆ Lean against
	• Injured person loses of control of own body movements
		ILO, 1998

they have further broken down each class of deviations into subclasses. They hence distinguish between such substandard acts as operating equipment without authority, failure to warn, failure to secure, operating at improper speed, using equipment improperly, etc. Amongst the substandard conditions we find inadequate guarding, defective tools, congestion, inadequate warning system, etc. It is here no longer a question of a taxonomy. The different classes of deviations are not mutually exclusive and one act may be classified in two or more different ways. In spite of these shortcomings, the ILCI classification scheme is in wide use in industry.

6.4.2 Ergonomics and industrial-engineering systems views

Systems ergonomics is concerned with the man–equipment–environment system and the tasks carried out by man as a systems component. A deviation may occur in any of the systems components (man, equipment and environment) or in the execution of the task.

Systems ergonomics has developed from an industrial-engineering systems view. Industrial engineering is concerned with the design and operation of industrial systems to promote productivity, quality and safety. Table 6.9 shows an example of the classification of deviations according to this systems view (Kjellén and Larsson, 1981). Here, the human actions, the necessary information and instructions, the work material and the technical equipment in use define the tasks. The environment is split into intersecting/parallel activities (i.e. influences from other work groups) and physical environment (noise, illumination, climate, etc.). Deviations in any of these respects will affect the risk of accidents. There is a system for control of production linked to each type of deviation, see Table 5.1. Below is a modified version of the original checklist.

The ILO scheme for classification of deviations takes into account whether the injured person controls the energies that cause harm or not. The latter is the case when the energies are controlled e.g. by a technical control system. Control-system failures are here a concern. There is no clear distinction between deviations and incidents in this scheme.

We will revert to the different types of human-error taxonomies in Chapter 8.

Table 6.9 Checklist of deviations

Work situation
- *human error*, e.g. wrong action, wrong sequence, omission
- *technical failure*, e.g. machine breakdown, missing equipment or tools
- *disturbance in material flow*, e.g. bad raw materials, delays
- *personnel deviations*, e.g. absence, temporary personnel, indisposed
- *inadequate information*, e.g. instructions, work permits

Environment
- *intersecting or parallel activities*, e.g. other work team
- *bad housekeeping*
- *disturbances from the environment*, e.g. excessive noise, high temperature

Safety systems
- *failure of active or passive safety systems*
- *inadequate guarding*
- *inadequate personal protective equipment or clothes*

6.5 Contributing factors and root causes

In recent developments of SHE information systems, the identification of causal factors has been in focus. Accidents and deviations are considered as symptoms of the underlying SHE problems in the organisation and the technical systems. The idea is that the identification and amelioration of these basic causes will have lasting effects on the SHE level. Different terms are used in the research literature to label these underlying problems such as determining factors, latent failures and root causes (Kjellén and Larsson, 1981; Reason, 1991; Cornelison, 1989; Bird and Germain, 1985). Different conceptions of how and to what extent psychological, social, organisational and technical factors affect the risk of accidents lie behind these different terms.

Reason sees '*latent failures*' as dormant dysfunction in the design and management of the industrial system. In this sense, they are similar to the medical concept of illness inducing so-called resident pathogens in the human body. Combined with local triggering factors in the work system, the latent failures will overcome the system's defences and produce accidents. Accidents thus do not arise from single causes but from a combination of latent failures and local triggering factors. We recognise the link between an accident investigation and a medical diagnosis in this analogy. The aim of an accident investigation is to identify the underlying 'illness' in order to be able to prescribe an adequate 'cure'. In practice, however, we do not have an overview of the different combinations of latent failures or 'illnesses' that result in accidents. The effects of remedial actions on such latent failures are often uncertain.

The MORT system uses the term '*root cause*' in the sense of the most basic cause of an accident or incident (Cornelison, 1989). It can be traced back to a lack of adequate management control that results in substandard practices and conditions and subsequently in an accident.

In the research literature we find many different checklists of causal factors. Such checklists are used in accident investigations for the purpose of ensuring that all relevant causal factors are identified. They are also used in the coding of causal factors as input to accident statistics. These statistics are in turn used in identifications of common causal factors, which are focused on action plans to improve safety.

A review of the accident research literature shows that many of the checklists are based on a hierarchical conception of the relations between causal factors and the risk of accidents. This hierarchy of causal factors usually follows the hierarchical levels of a traditional industrial organisation, Figure 6.4.

Table 6.10 shows the causal hierarchy of different models of accidents. Here we recognise the two SHE management models of Section 5.7, MORT and SMORT. Also the ILCI model and TRIPOD include upper-management elements. TRIPOD is unique in the sense that it analyses the relations between human errors at different hierarchical levels.

Table 6.10 Examples of accident models applying a causal hierarchy

Level	ILCI model (Bird & Germain, 1985)	SMORT (Kjellén et al., 1987)	MORT-based root-cause analysis (Cornelison, 1989)	TRIPOD (Reason, 1991)
General and SHE management	Lack of control • inadequate programme • inadequate standards • inadequate compliance	Higher management Health and safety work Control of projects • design • construction • start-up	Policy and implementation Risk assessment • information systems, • hazard analysis, • etc. Bridge elements • directives • budget • etc.	Source failure types • Pathological • Incipient–reactive • Worried–reactive • Repair–routine • Conservative–calculating • Incipient–proactive • Generative–proactive
Functional department	Immediate causes • personal factors • job factors	Daily operation • organisation • planning • technical systems	Specific factors • maintenance • supervision • etc.	General failure types • Hardware • Design • Maintenance • Procedures • Error enforcing conditions • Housekeeping • Incompatible goals • Organisation • Communication • Training • Defences (barriers)
Work system			Task performance	

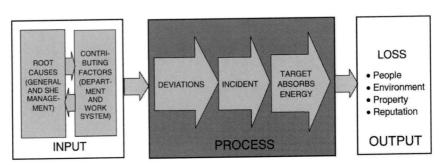

Figure 6.4 Hierarchical relation between contributing factors and root causes (input).

In Table 6.10, we have deliberately limited ourselves to factors inside the company. This is a natural boundary for limiting the scope of an accident investigation. We are concerned with those types of causes that people in the organisation, and management in particular, can do something about. We are thus primarily concerned with the *preventive value* of the causal information. If the *explanatory* or *predictive values* were in focus, we might well have expanded the analysis to conditions outside the boundary of the company, such as the domestic situation of those involved or the conditions in society in general. Under those circumstances, we would have had to define another boundary to avoid extending the analysis back to Adam and Eve.

In the subsequent sections, we shall review the contents and underlying theory of some of the checklists. The intention is to arrive at an understanding of the rationale behind the different checklists and their potentials and limitations. The different approaches are complementary, and there is no single 'true' accident causal model.

6.5.1 Contributing factors at the functional department and work-system levels

An early example of a causal factors checklist is Swain's checklist on human-performance-shaping factors (Swain, 1974). It is human-factor oriented and lists factors that affect the quality of human performance and thus the likelihood of human errors. Table 6.11 shows an extract from this checklist.

The checklist has been used as a design tool as well as an accident-investigation tool in cases where human errors have played a central role. It draws attention away from blaming the person that made the error. Instead, it focuses on the identification of dysfunctions in the design of the man–machine system from an ergonomics point of view. The intention is to identify accident-prone workplaces rather than accident-prone persons.

Table 6.11 Extracts from Swain's checklist on human-performance-shaping factors

Human-performance-shaping factors

1   *Situational characteristics* such as physical environment (temperature, air quality, noise), manning, work hours, supervision, rewards, organisational structure.
2   *Task and equipment characteristics* such as perceptual, anticipatory, decision-making and motor requirements, feedback and knowledge of results, man–machine interface.
3   *Job instructions* such as procedures, communication, work methods.
4   *Psychological stresses* such as task speed and load, threats in case of failure, monotonous work, conflicts of motives, negative reinforcement.
5   *Physiological stresses* such as fatigue, discomfort, hunger and thirst, temperature extremes, vibration, lack of physical exercise.
6   *Individual factors* such as knowledge and experience, skills, intelligence, attitudes, physical condition, influence of family, group identification.

Source: Swain, 1974.

Behind many of the items in Swain's checklist are objective data on pre-requisites for optimal human performance. Human-factor experts are its intended users.

Swain's checklist on performance-shaping factors has had great influence. The ILCI checklist on basic causes, for example, includes many of its items in the part covering personal factors. It is in wide use in industry in accident and incident investigation.

The ILCI checklist on causal factors at the workplace level is based on a man–environment dichotomy, because it distinguishes between individual and workplace-related causal factors. The latter category also includes management factors such as leadership and supervision. It has been developed from practical experience and does not represent a taxonomy. Some of the items or classes in the checklist are overlapping, for example, inadequate equipment and inadequate engineering. This causes problems of reliability in the classification of causal factors. Whereas Swain's checklist is intended for use by human-factor experts, the ILCI checklist is used by the supervisors in their first report on accidents and incidents. This also causes problems of reliability in applications of the checklist, because especially the individual factors are subject to interpretation and value judgements.

The OARU model offers another example of a causal-factor checklist at the workplace and department levels (Kjellén and Larsson, 1981). The checklist is based on an industrial-engineering systems view. It distinguishes between three main classes of factors describing such systems, i.e. technical, organisational and behavioural. Table 6.12 shows a revised version of the original checklist. It is intended for use in accident investigations by a problem-solving group rather than by the individual supervisor. The aim is to overcome problems with individual interpretations and judgements.

Table 6.12 Checklist of contributing factors at the workplace and department levels

Physical/technical	Organisational/economic	Behavioural
1　Workplace layout 　a　Access to equipment 　b　Walkways, transportation routes 　c　Safe distances between moving equipment 2　Design of equipment 　a　Physical hazards 　b　Reliability 　c　Man–machine interface 3　Physical working environment (lighting, inner climate, noise) 4　Protective equipment 5　Work materials, chemicals 6　Safety equipment and systems	1　Work organisation, manning 2　Activity planning 3　Methods of work, work pace 4　Maintenance routines 5　Education, training of personnel 6　Systems of remuneration, promotion, sanctioning 7　Controls of other type, e.g. economic, 'third party' 8　Systems of shift, work-time 9　Instructions, rules 10　Routines in safety work 11　Organisation of first aid	1　Supervision, instructions 2　Informal information flow 3　Workplace norms 4　Individual norms and attitudes 5　Individual knowledge and experience 6　Special circumstances

Example: We will illustrate this approach with an accident during construction of a pressure tunnel to a hydroelectric power plant. While transporting a 12-ton pressure pipe in the tunnel, the operators lost control of the pipe. It skidded several hundred metres down the tunnel and caused substantial equipment damage. Figure 6.5 illustrates the different deviations and contributing factors.

This example illustrates how decisions in separate units of the organisation contribute to an accident. Each decision-maker is not likely to see the consequences of his decisions in the total operational context. The example also illustrates how a combination of different decisions has resulted in a specific accident. Even if this particular combination is not likely to occur again, each erroneous decision may 'cause' other types of accidents. This type of analysis thus leads to questions about improvement needs in management's decision-making routines.

In the TRIPOD model, the underlying factors at the workplace or department level behind accidents are called latent failures. Based on research into accident causation, 11 different so-called *General Failure Types* have been identified (Wagenaar in Feyer and Williamson, 1998). They comprise

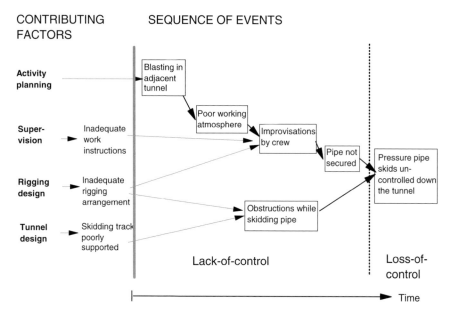

Figure 6.5 Deviations and contributing factors of an accident during transportation of a 12-ton pressure pipe in a tunnel.

hardware, design, maintenance, procedures, error-enforcing conditions, housekeeping, incompatible goals, organisation, communication, training and defences. The presence or absence of failures is identified in accident investigations. A Failure State Profile is developed on the basis of accident statistics.

Tinmannsvik distinguishes between specific and general safety factors (Tinmannsvik, 1991). Management's manipulation of specific safety factors has as its main objective the promotion of safety. These factors include attitudes to safety, guarding, emergency preparedness, safety work and experience exchange. Similarly, the objective of manipulating the general safety factors is to promote production goals. Examples of such factors are training programmes, design of machinery and equipment, maintenance routines, transportation and storage facilities, housekeeping and supervision. Empirical studies of fourteen companies showed a positive correlation between the accident frequencies and the general safety factors only.

In conclusion, there is a certain degree of arbitrariness in the selection of elements to be included in the causal factors checklist. Industrial systems are human constructs and there is no unifying theory on how the different aspects of such systems interact to produce harm. Any checklist must include elements from different accident models.

Table 6.13 The different elements of a root-cause analysis

Main categories	Subcategories
• Policy	
• Policy implementation	• Line/staff responsibility • Accountability • Vigour and example • Methods and criteria analysis
• Risk assessment	• Safety-information systems • Hazard-analysis process • Safety-programme auditing
• Bridge elements	• Management services • Directives • Budget • Information flow

Source: Briscoe, 1991.

6.5.2 Root causes at the general and SHE-management-systems levels

6.5.2.1 Causes derived from quality-assurance principles

The theoretical basis of the different causal models becomes more obvious at the upper management level. MORT was the first comprehensive model to include organisational and individual factors at the top management level. At this level, it draws from quality assurance management principles. The SMORT and ILCI models have been influenced by this pioneering work and represent variations on the same theme. The concept of root causes originates from the MORT model. The checklist above shows the different items of a root-cause analysis.

We here recognise the three main elements of the quality-assurance management model in the three first main categories of root causes. In the third category, risk assessment, we find the types of themes that are brought up in Parts III to V of this book. 'Bridge elements' (category number four) show how upper-management and SHE-management principles are implemented at the department and workplace levels.

Example: In the accident with the pressure pipe in Figure 6.5, we identified four different contributing factors. Each factor was linked to a decision in a separate part of the organisation. We can proceed with the investigation by analysing deficiencies in the decision-making. A poor tunnel design was identified as one of the contributing factors. Had adequate criteria been applied in the design of the support to the skidding rails? Was implementation adequate? Had the results been checked and followed up adequately?

The SMORT and ILCI models represent further developments of the original MORT concept mainly for the purposes of improving user-friendliness. In the application of both models, the analysis proceeds from the specific accident occurrence and deficiencies at the workplace and department levels to the general management level in a step-by-step process.

SMORT has in common with MORT that it brings up questions concerning deficiencies in the design of industrial systems. The intention here is to learn from accidents in order to improve the design processes and thereby to acquire safer industrial systems in the future. At the top level, we find similar elements in SMORT as those of the quality-assurance management model.

In the ILCI model, three common 'reasons' for the lack of control at the workplace level are identified: (1) inadequate SHE programme, (2) inadequate programme standards and (3) inadequate compliance with the standards (Bird and Germain, 1985). These elements also coincide with the main elements of the quality-assurance management model.

6.5.2.2 Causes derived from elements of the safety culture

The TRIPOD model has another focus (Reason, 1990). Source-failure types relate to top-management commitment and competence and to accident perceptions and the safety culture at the company. A seven-point scale for the evaluation of top-management commitment is suggested:

1   *Pathological*, i.e. no top-level commitment, safety practices at industry minimum.
2   *Incipient–reactive*, i.e. keeping one step ahead of the regulators.
3   *Worried–reactive*, i.e. beginning to be concerned with incidents and accidents.
4   *Repair–routine*, i.e. safety problems are recognised but only dealt with locally.
5   *Conservative–calculating*, i.e. involved in various safety management techniques, but limited to technical and human error safety issues.
6   *Incipient–proactive*, i.e. involved in active searches for better solutions, acknowledging the importance of organisation and management factors.
7   *Generative–proactive*, i.e. top-level commitment to improve safety culture, absence of complacency.

At this level, we may also refer to Lucas' framework of organisational cultures, see Section 5.8 (Van der Schaaf *et al.*, 1991). According to Lucas, the prevailing accident models within an organisation will determine its ability to learn from accidents. Management's failure to learn from accidents and as a result allowing the same type of accidents to occur again is here considered as a root cause of accidents.

6.5.3 Problems in identifying causal factors

In accident investigations, we identify causal factors as part of a diagnostic process, Figure 6.6. This starts with the *symptoms*, i.e. the specific occurrence

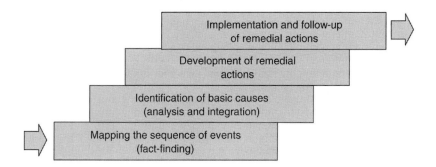

Figure 6.6 Accident-investigation stairs.
Source: Adapted from Wig, 1996.

(the loss of control of energies in the system and the subsequent develop-ment of losses) and the associated deviations at the work-systems level. In the step that follows, contributing factors and root causes, i.e. deficiencies at the department and higher-management levels that may explain these symptoms, are identified. In principle, there are three different methods to establish such links between accidents and causal factors:

1   *Analytic methods* are used when there are logical relations between the factors at the different levels, i.e. a causal factor is a necessary condition for the occurrence of a deviation in the accident sequence. A typical example is breakdown of equipment. It is often possible to trace such failures back to inadequate design, which in its turn can be traced back to inadequate design standards, inadequate design tools such as calcula-tion programs and/or inadequacies in the quality control of design draw-ings. *Example: This technique was applied in the accident investigation following the loss of the gravity base structure (GBS) for the Sleipner offshore installation in 1991. It was possible to trace a structural failure in the GBS back to deficiencies in the design process.*
2   *Statistical methods* use accident statistics in the identification of causal factors. They may show that the probability of an accident will increase, if certain conditions are prevailing. These methods will be further dis-cussed in Part III.
3   *Subjective or expert judgements*, i.e. judgements that it is reasonable to assume causal relations. Persons with direct knowledge about the circumstances around the accident make these judgements. It is here a question of establishing causal relations by interpreting facts about the accidents in the light of experience with the particular workplace or management system in question.

We are here concerned with the identification of causal factors in indi-vidual accidents and near accidents for use in accident prevention. The analytic method is only valid under certain conditions, where physical causal relations are studied. This is not the case for psychological or organisational factors. For individual accidents, subjective judgements must play an important part in the evaluation of causal factors. This means of arriving at conclusions is especially relevant when we deal with interpretations of people's behaviour in terms of reasons rather than causes. It is also relevant when we analyse how decisions affect the accidental outcome and have to review the distribution of responsibility, authority and accountability within the organisation. In any collection of data on accident risks, a clear distinction must thus be made between what are facts and what are *interpretations* based on expert judgement. In making interpretations, it is important to corroborate these by checking different sources of information as far as possible.

Checklists are applied in accident investigations in order to ensure that all relevant causal factors are considered. In practice, the checklist-supported identification of causal factors is problematic. Items at the top of the checklist are more often addressed than items further down (Hale *et al.*, 1997). Experience also shows that the identification of causal factors is affected more by the items on the checklists and the underlying accident model than by factual circumstances. A detailed presentation of certain causal aspects will result in an overestimation of the importance of such aspects.

Supervisors have a tendency to choose causal factor alternatives that are not possible to verify and that involve limited management responsibilities and obligations to implement remedial actions (Kjellén, 1993). Typically, causal-factor alternatives related to human factors such as 'improper motivation' are selected. The link between the identified causes and the selection of remedial actions is often weak.

Example 1: A comparison of the results of investigations of the same accidents by two different companies, applying different causal models, showed large discrepancies between the results. One of the companies, the operator of an offshore oil installation, applied the ILCI model on accident causation. The causal analysis especially pointed at carelessness, lack of motivation and lack of compliance with rules as causes of dropped objects on offshore drilling. The drilling company applied a technically oriented 'model', including such categories as wear and tear, inadequate design, wrong use of equipment and failure to secure. In this company's statistics, the most common cause of dropped objects was wrong use of equipment.

Example 2: In the same study, ten safety delegates were independently asked to code a written description of an accident by means of the ILCI model. No two such causal classifications were similar.

Example 3: An evaluation of the accident-reporting system of an offshore installation showed that remedial actions were identified before and independently of the identification of causal factors based on the ILCI model. This classification of causal factors was done in order to satisfy formal requirements rather than as a tool in order to come up with better safety measures.

The subjective biases of the decision-makers are a concern. In **attribution-theory** research, people's judgements in determining accident causes and selecting remedies are studied (DeJoy, 1994). Due to the fact that accidents often are causally ambiguous and emotionally charged, subjective biases play an important role in the attribution of causes. Self-protective biases on behalf of the supervisor makes him/her likely to deny his/her own responsibility for an accident. Instead, the accident is often attributed to causes beyond the supervisor's control by blaming the workers involved. They, on the other hand, are likely to favour situational causes in the working environment.

In general, people tend to explain accidents that other people have been involved in by referring to their stable personal characteristics. This corresponds to the third quadrant in Figure 6.7. However, people explain accidents that they themselves have been involved in by referring to transient external causes, i.e. the second quadrant. This so-called 'fundamental attribution error' has great implications for the routines for accident investigations.

Research has also shown that decision-makers apply shorthand methods or rules derived from earlier experiences in the analysis of causes of accidents. For instance, many decision-makers have developed rules implying that unsafe behaviour on the part of the workers is the important feature of accident causation. A decision-maker may thus stop looking for more evidence after detecting a human error, even if more important causal factors are at hand.

We also know that incentives and motivational factors play an important role in causal attributions and the selection of remedies. An illustration of this is when decision-makers attribute accidents to unstable causes rather than stable causes. In the ILCI causation model, we find examples of both types of causes. Mental stress is an example of an unstable factor as opposed to inadequate equipment, which is a stable characteristic. Unstable causes produce less certain predictions about the efficiency of remedial actions. By selecting such causes, the decision-maker may escape from the obligation to decide on binding and resource-demanding solutions to the safety problem (Kjellén, 1993).

	PERSON	ENVIRONMENT
TRANSIENT	1. Inattention	2. High wind speed
STABLE	3. Insufficient crane-driving skills	4. Poor sight from crane cabin

Figure 6.7 Two dimensions in the attribution of causes. They are illustrated by an example showing different causes of a crane accident.

It is a well established principle in crisis management to limit the negative public-relations effects of a severe accident by focusing on the situational factors when dealing with the media. These tactics will help to fend off outside forces that demand change. They will, however, when applied within a company, prevent adequate learning.

Example: In a rigging accident at a construction site, two riggers were seriously injured when a pull chain ruptured. The investigation showed that the riggers had violated the safety rules when they stayed in the danger zone while the winches were operated. The investigation also identified serious deficiencies in the planning of the work and in the technical control of lifting equipment. Site management in their communication with the press explained that the accident was caused by human error on the part of the operators. Managerial errors were de-emphasised.

Expert judgements are in many cases the only practical way to establish basic causal factors. This condition has some important implications. The person with direct responsibility for safety at the site of the accident, i.e. the supervisor, should not make the judgements alone. A consensus process is recommended, where persons with direct knowledge of the circumstances and persons with different responsibilities and perspectives take part in the investigation into causal factors.

The identified causal factors should not be treated as objective facts but as subjective judgements based on available evidence. The value of accident statistics, based on identified causal factors, is questionable. This type of statistics may tell us more about the accident perceptions of the persons involved in the accident investigation than about the actual circumstances. Such statistics should not be used in a context where the responsible persons lack knowledge about the circumstances under which the data was collected and analysed.

Statistics on remedial actions give valuable information and may replace statistics on basic causes. The same checklist items may be employed in this analysis as those employed in the identification of causes. Such statistics give information about management's comprehension of the need for changes at the workplaces in order to improve safety. It also gives information about the level of learning from experience at which management operates.

To sum up, rather than 'objectivity', what matters is the organisation's confidence in the established relations between causal factors and safety measures on the one hand, and accident risks on the other. Examples of factors that will influence the organisation's confidence are:

- The status of the experts in the organisation;
- The consensus processes that lead to the results and the extent to which key persons have participated; and
- The extent to which the expert judgements are transparent and easy to follow.

7 Accident counter-measures

7.1 Barriers against losses

Accidents are a physical phenomenon. Losses occur due to a transfer of energy to a victim or other 'target'. The energy may be mechanical, electrical, involve hazardous substances, etc. A loss of control of energy has preceded this uncontrolled energy transfer. It follows that the immediate accident counter-measures also have to be physical. Haddon's ten accident prevention strategies of Section 5.4 are primarily physical constructs and will here serve as the starting point in our review of different types of measures. We will use the term 'barriers' rather than 'strategies' to denote the physical counter-measures that intervene in the accident process to eliminate or reduce the harmful outcome.

Figure 7.1 illustrates that, in principle at least, there are many opportunities to interrupt or change an accident process before it evolves into a loss. First, there are possibilities of totally changing the preconditions for an accident to occur by eliminating the energy source, modifying the energy characteristics or limiting the amount of energy. Second, barriers may interrupt, dilute or redirect the energy flow during the latter part of the accident process. This could include prevention of the uncontrolled release of energy, dilution of the energy transfer, separation of the victim from the energy flow or improving the victim's ability to endure the energy flow. Finally, the losses may be reduced after the occurrence of the accident through stabilisation, repair and rehabilitation.

7.1.1 Prevention of occupational accidents

In the case of occupational accidents, few of the opportunities for accident prevention are usually utilised. Take the classical example of the carpenter hitting his thumb with a hammer. A moment of lack of concentration may result in a sore thumb. The only barrier against such an accident is the carpenter's ability to control the energy as represented by the kinetic energy of the hammer. There are many other such examples of hazards in our daily lives that are controlled by one barrier only. When the barrier is controlled

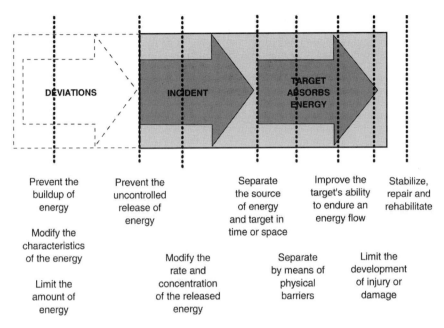

Figure 7.1 Illustration of when the different barriers according to Haddon intervene in the accident sequence. The first three barriers (eliminate or limit energy, modification of its characteristics) are not necessarily a logical and chronological part of it.

directly by the operator performing the work, we must expect accidents to happen due to the inherent variability in human performance. Usually, the consequences are minor and have been tacitly accepted.

The European directive on the protection from chemical hazards illustrates a more systematic approach in the application of the barrier philosophy to prevent occupational accidents and diseases (European Council, 1998). To prevent workers from being exposed to hazardous chemical agents, the employer must take actions in a priority from left to right according to Figure 7.2.

In Section 7.5, we will see how accidents with machinery are prevented through the application of similar principles.

7.1.2 Prevention of major accidents due to fires and explosions

At the other side of the spectrum we find the risks of large-scale or *major accidents*, such as a fire or explosion in a chemical plant or a nuclear power-plant melt-down. We are here speaking of accidents that may result in multiple fatalities inside and outside the plant and extensive material or environmental damage. A major accident may threaten the survival of the

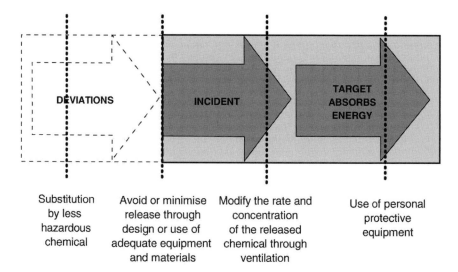

Figure 7.2 Measures to prevent exposure to hazardous chemicals.

company responsible for the operation. Fortunately, the associated hazards are well defined and are possible to identify and evaluate at an early stage. Management of the company will marshal all feasible barriers in order to avoid losses.

Figure 7.3 illustrates how fires and explosions in a refinery or an off-shore installation are prevented through the implementation of a number of different barriers, based on the so-called philosophy of '**defences in depth**'. The rationale behind this philosophy is to establish independent barriers between the hazard (i.e. hydrocarbons) and the victim. A major accident can only occur under those rare circumstances when there is an improbable combination of barrier failures.

Let us analyse how these barriers coincide with Haddon's strategies. Oil and gas processing is inherently hazardous due to the presence of flammable hydrocarbons in the process. It is thus not possible to prevent the build-up of energy or to modify its characteristics. An important measure is to limit the amount of oil and gas in the process, for example by dividing the process into separate units, and to isolate these from each other in case of a process upset.

The uncontrolled release of oil and gas is prevented primarily by different means:

- A process control system that keeps process parameters such as pressure and temperature within acceptable limits and
- High quality containment (quality of material/equipment, thickness of piping/vessels).

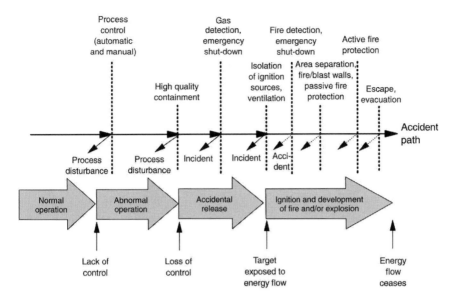

Figure 7.3 The prevention of fires and explosions in a plant for the processing of hydrocarbons (refinery or offshore installation) based on the so-called defences in depth philosophy. The different barriers are reflected in the European legislation on health and safety related to chemical agents at work.

In the case of an escape of oil and gas from the process containment, gas detection and emergency shut-down prevents harmful consequences. The aim is to modify the rate and concentration of the release of oil and gas. The different parts of the process will be separated from each other through the closure of isolation valves. Hence, only oil and gas from the part of the process where the loss of containment has occurred will be able to escape. Vessels will be depressurised and the gas will be transferred to the flare tower for flaring.

Normal ventilation and possibly additional emergency ventilation will dilute the gas, and potential ignition sources will be shut down to avoid ignition of the hydrocarbons and subsequent fire and explosion. Should a fire or explosion occur, physical barriers such as fire and blast walls will avoid injury or damage outside the immediately affected area.

Finally, there are a number of barriers aiming at limiting the development of injury and damage:

- Passive fire protection of the building and process equipment, and explosion ventilation will improve the ability of the building and the equipment to endure the explosion and heat from fires.

- Active fire-protection systems such as deluge systems and the fire brigade will have the same effects by combating the fire.
- Alarms and messages over the personal address system will notify the personnel, who will escape from the affected process areas to safety through marked escape ways. If it is not possible to get the fire or explosion under immediate control, evacuation will follow.

7.2 Active and passive barriers

We distinguish between passive and active barriers, although the borderline is not always clear. *Passive barriers* are embedded in the design of the workplace and are independent of the operational control systems. They will, however, require regular inspection and maintenance to function appropriately in the long term. A typical example is the safety cage surrounding the driver (and passengers) of a car. Other examples of passive barriers are fixed-mounted machinery guards and fire and explosion walls in a process plant.

Active barriers are dependent on actions by the operators or on a technical control system to function as intended. The carpenter's control of the hammer movements in the example above is, in principle, such a barrier. It is very vulnerable to changes in behaviour of the carpenter. Several of the barriers against fires and explosions in the example above are of the active type. Process control is primarily maintained through the process control system, but the operator may intervene and take over the control. If gas is detected, shut-down is done automatically by the emergency shut-down system. Man is a very sensitive gas detector, and manual shut-down from the control room after a field operator has detected a gas leak but before it has been picked up by the gas detectors is not an uncommon event. Active fire-protection systems are both automatic (e.g. sprinkler systems) and manual (the fire brigade). Escape and evacuation is an active barrier.

7.3 Different time frames in the implementation and maintenance of barriers

We have seen how accident prevention is accomplished through the establishment and maintenance of active and passive barriers. Inside the company, this is accomplished at different levels of the organisation and within different time frames. There is an ongoing and immediate control of barriers by the operators in their day-to-day operation at the sharp end. In a longer perspective, decisions are made to implement and maintain barriers through design (technical measures) and maintenance and through organisational and individual/social measures in production at the work-system and department levels. Also strategic decisions at the general and SHE management-systems levels of the company will affect the availability of barriers, but this theme will not be discussed further here.

Active barriers are to a large extent dependent on activities in the day-to-day operation. Many of them are established and maintained through the different activities to allocate resources and to control production, including the allocation and supervision of personnel, control of work material and technical systems, co-ordination of activities and house-keeping. The individual operators play a significant role in the immediate control of accident risks by detecting and avoiding hazards. In case of major accident risks, the technical control systems will usually take over from the operators, if they fail to keep the production process within accepted limits.

At the next level, the barriers are implemented and maintained through long-term measures related to design and organisation of the workplaces and measures directed at the work group and the individual operator, Table 7.1.

7.4 The role of experience transfer

An industrial organisation must have sufficient knowledge to be able to establish and maintain adequate barriers. This knowledge is partly acquired from the accumulated external experiences that have been documented in regulations, standards, user instructions of machinery, handbooks, etc. The internal experience transfer also plays a significant role. It is, for example, necessary to find an adequate balance between active and passive barriers. The barriers must be feasible and must not obstruct production. Yet the maintenance of adequate barriers must not rely solely on the operations organisation in a way that makes this control task insurmountable. It lies on the organisation itself to find the adequate balance between the different types of passive and active barriers, and a certain amount of trial and error is necessary in this process.

More explicitly, experience transfer from the day-to-day operation to the management levels of a company will support in:

- The identification and elimination or reduction of hazards;
- The identification of the required barriers and information on whether they are manageable from an operations point of view; and
- Feedback on barrier efficiency.

Experience feedback on barrier efficiency is a challenge in well defended systems, where there are many independent barriers operating simultaneously. This applies especially to the passive barriers. The operators will have limited contact with many of them in their day-to-day activities, and their efficiency will only be tested in rare circumstances, when there is an accident. We will discuss this problem further in Section 8.3.

Table 7.1 Examples of long-term measures for the implementation and maintenance of barriers against accidents

Category	Examples
Technical measures	• Elimination or reduction of hazards in the design of equipment. • Design of reliable machinery and equipment in order to minimise the personnel's exposure to hazards in connection with correction of disturbances and repair. • Introduction of guards and other types of barriers to prevent people from coming in contact with the danger zone. • Introduction of personal protective equipment. • Design of a suitable workplace layout that allows for a separation of people from transportation routes and other types of danger zones, adequate gangways and escape ways, easy and safe access for operation and maintenance, good viewing conditions, etc. • Design of an adequate physical working environment with respect to illumination, air quality, temperature and noise. • Selection of non-hazardous working materials. • Design of adequate transportation facilities to avoid manual material handling.
Organisational measures	• Planning for adequate manning with competent personnel. • Development of adequate work instructions and safety rules. • Development of adequate technical information and documentation of machinery and equipment. • Development of adequate routines for inspection and maintenance of equipment and buildings including passive barriers. • Introduction of a permit-to-work system to avoid maintenance work on equipment that is energised. • Planning for emergencies.
Social/individual measures	• Education and training of the personnel. • Implementation of motivational campaigns such as the distribution of written material, posters, video, safety meetings, etc. • Feedback and behavioural modification through remuneration of safe behaviour and punishment of unsafe behaviour.

Table 7.2 Strategy for selecting safety measures according to EN 292. The duties of the designer are shown as well as the priorities for selection of safety measures in design (from left to right).

	Safety measures taken by the designer		
Risk reduction by design	*Safe guarding*	*Information for user*	*Additional precautions*
	Provision of personal protective equipment	• Training • Safe working procedures • Supervision • Permit-to-work system	
	Safety measures taken by the user		

Source: CEN, 1991.

7.5 Designing for safety of machinery

An 'ideal' barrier against accidents is of the passive type, since it will not be dependent on the more or less predictable day-to-day operation to function appropriately. Such barriers have to be implemented during the design phase of industrial systems and equipment. This is the safety philosophy behind the European legislation on machinery safety. We use it here to illustrate the significance of design in the prevention of accidents.

Many machines represent significant potential hazards and have been subject to machinery safety regulations to reduce and eliminate the risks. One of the four freedoms within the European Union is the free float of products. It has thus become necessary to harmonise the machinery safety legislation in Europe. The Council Directives on Machinery is the cornerstone in this harmonisation (European Council, 1989/98). So-called harmonised European standards detail the requirements. This legislation fits well into the barrier framework presented here.

In the harmonised standard EN 292 concerning basic concepts and general principles for design, a strategy for selection of safety measures is described, Table 7.2 (CEN, 1991). This strategy reflects the general conception among international safety experts that hazards should be prevented at the design stage. Residual hazards may be accepted if design solutions are unfeasible. It is, however, the responsibility of the producer to inform the user about these hazards.

The standard defines the duties of the designer in order to identify and select the necessary safety measures. A stepwise procedure shall be followed, including:

1   Determination of the intended uses of the machinery, life span, space requirements, etc.;
2   Identification of the hazards and assessment of the risks;
3   Removal or limitation of the hazard;
4   Design guards and/or safety devices against any remaining hazards;
5   Information and warning of the user; and
6   Consideration of any necessary additional precaution.

The first three steps involve the execution of a risk assessment. An example of how this may be carried out will be presented in Chapter 24. We will here focus on the principles behind the legislation in the selection of safety measures or barriers. There is a clear relationship between the different types of safety measures according to EN 292 and Haddon's strategies for accident prevention (see Table 7.3).

The first priority is to select measures to reduce or eliminate the hazard. EN 292 presents a list of the hazards which machinery is likely to generate (see Table 5.3) and proposes preventive measures. If it is not feasible to eliminate the hazards at the source, guarding should be selected as the second choice. The users should be informed about any residual hazards to be able to implement the necessary additional barriers. These barriers fall outside the scope of the standard.

The European Council Directive of Machinery applies to machinery manufacturers or their representatives (vendors). Their obligations are summarised in Table 7.4.

Table 7.3 Examples of safety measures according to EN 292 and their relation to Haddon's accident-prevention strategies

Examples from EN 292	*Related accident-prevention strategy according to Haddon*
Making machines inherently safe: • Minimum distances between mechanic components to avoid crushing • Limitation of forces or velocity of movable elements • Limitation of noise and vibration	To prevent the creation of the hazard in the first place (Haddon No. 1) To reduce the amount of hazard brought into being (Haddon No. 3)
Fully pneumatic or hydraulic control systems and actuators on machines intended for use in explosive atmosphere	To prevent the release of the hazard that already exists (Haddon No. 4)
Selection of safeguards to allow for safe access to the danger zone during normal operation, e.g. interlocking guard, two-hand control device	To separate, in time or space, the hazard and that which is to be protected (Haddon Nos 6, 7)

Table 7.4 Summary of requirements as to the involvement of an authorised institution in the documentation of the safety of machinery

Not hazardous machines	Hazardous machines	
	For machines that are designed and fabricated in accordance with harmonised standard	For machines that are not at all or only partly designed and fabricated in accordance with harmonised standard
The manufacturer or vendor works out technical documentation of compliance with the Machinery Directive	1 Technical documentation is delivered to an authorised institution, or 2 Technical documentation is delivered to an authorised institution, which verifies that the standard has been correctly implemented and draws up a certificate, or 3 Type-examination is carried out by authorised institution	Type-examination is carried out by authorised institution

Source: European Council, 1989/98.

The manufacturer (or his representative) draws up a declaration of conformity and affixes an EC mark to the machinery. Before this can be done, the manufacturer has to document that the machinery meets the requirements in the Machinery Directives. This technical documentation includes:

- Assembly drawing, drawings of the control circuit and detailed drawing, calculations, etc. where required to document compliance,
- A list of safety requirements, standards and specifications applied in the design,
- A description of the method applied in eliminating the hazards of the machinery,
- If applicable, a technical report or certificate obtained from a competent body or laboratory, and
- User instructions.

To document the method applied in eliminating hazards, a risk assessment has to be carried out. We will revert to this topic in Chapter 24.

The Directives define the types of machines that are regarded as hazardous. This list includes, for example, sawing machines, certain wood-working machines, presses, and machinery for underground working. For non-hazardous machinery, i.e. machinery not included in this list, internal documentation by the manufacturer is sufficient. For hazardous machinery, the manufacturer may choose between designing in accordance with harmonised standards or not. The advantage of applying the harmonised standards lies in the possibility of a simpler procedure for proving compliance with the Machinery Directives. It may be an advantage to the manufacturer to receive a certificate of compliance also in the case where harmonised standards have been applied. This has to do with the manufacturer's liability in case of an accident with his product.

The harmonised standards are of four types:

- *Type A-standards* define basic concepts and terminology, rules for the writing of subordinated standards, requirements as to risk analysis, etc.
- *Type B1-standards* are about certain specific safety aspects including safety of control systems, noise, vibration, safety distances, etc.
- *Type B2-standards* relate to safety equipment (machine guarding, safety mats, etc.).
- *Type C-standards* apply to special types of machines, e.g. cranes, conveyor belts. These standards are all-inclusive and ensure that a piece of machinery that meets the standard also meets type A and B standards and thus the Machinery Directives. For example, the strategy for implementation of safety measures has to be implemented in type C-standards.

7.6 Safety measures in operation

We have in Section 7.4 identified different types of safety barriers that need to be considered during the design of new industrial systems and equipment. Such barriers have to be integrated with the total design. There are also many opportunities for implementing barriers during operation. Below are some examples:

- A simple example is when operators avoid hazards altogether or control them. On-the-job training, job instructions, safety procedures aim to promote this type of 'barriers'.
- The operators use personal protective equipment, such as safety helmet, safety boots, eye protection, ear protectors, respiratory equipment, and safety belts. Instructions, procedures, training and easy access to the equipment aim to promote this 'barrier'.
- Provisions for good housekeeping to avoid stumbling and slipping accidents. Good-housekeeping routines involve such measures as well-defined responsibilities for collection of litter, physical facilities such as garbage cans and regular inspections.
- Provisions for safe scaffolding and guardrails to avoid accidental falls. Companies develop and implement rigorous administrative procedures for the control of scaffolding that include, for example, checking of the scaffolding before it is put into use and regular inspections.

7.6.1 *The permit-to-work system*

The permit-to-work system is an important operational safety measure in industries with severe hazards. It aims at ensuring that work is planned and carried out under adequate safety precautions in cases where the exposed persons do not control the relevant energies themselves.

Two examples illustrate the importance of this procedure:

Example 1: An installation worker should remove a flange during the commissioning of a process plant. He started by mistake to work on a system that was under pressure due to leak testing. When he had removed some of the bolts, the others were ripped off and the flange came flying and hit his leg. The leg was lost.

Example 2: An instrument operator should mount a vibration gauge on a pump. While the man was doing the work, the pump was started from the control room. The rotating shaft caught his arm and he was severely hurt.

An important safety principle behind the permit-to-work system is to ensure that the relevant parts of the system are de-energised during work in the danger zone. The challenge lies in designing and implementing a procedure that is simple and robust and that covers all relevant situations.

The padlock system to avoid electrocution is a simple example. The electrician has his own padlock, which he uses to secure the main switch in a safe position and thereby de-energises the electrical system before he starts to work on it.

In a process plant, work permits are required for such types of work as:

- Work on systems that contain hazardous substances (e.g. hydrocarbons) or pressurised gas;
- Work that requires shut-down of safety systems (e.g. the fire alarm);
- Hot work in hazardous areas, where there is a risk of an explosive atmosphere building up;
- Entering of tanks; and
- Connection and disconnection of electrical equipment.

A permit-to-work system is based on the following principles:

- The energy sources (hazardous substances such as hydrocarbons, electricity, hydraulic pressure, etc.) are controlled by operations. The responsibility for each energy source is well defined and lies with a supervisor from this organisation such as the shift supervisor.
- The responsible person shall at any time have the total overview of energised systems and active work permits.
- The work permits are effective during a well-defined time-period such as during daytime (08.00 to 16.00). There are fixed routines for application and approval of work permits. A special form is used to control

Table 7.5 A checklist of measures for use in evaluating applications of hot-work permits

Typical measure	Type of strategy according to Haddon
• Drainage/emptying and flushing (steam, nitrogen) of process systems containing flammable substances	Prevent build-up of energy
• Depressurise process systems	
• Regular gas measurements	
• Padlock (electrical de-energising)	
• Secure power shafts against accidental start	Prevent the uncontrolled release of energy
• Plugging of hazardous drains	
• Fencing off of affected area	Separate source of energy and target in time or space
• Close and secure valves	Separate by means of physical barriers
• Introduce blind flanges	
• Use of personal protection	Improve the target's ability to endure
• Fireguard	Limit the development of injury/damage
• Fire-fighting equipment	

and check out all activities related to the planning, execution and finishing of work requiring a work permit.

- All work on process and utility systems with few and well-defined exceptions require work permits. The supervisor for the work to be done (e.g. maintenance supervisor) should define the job by type of job and affected systems in the application.
- Operations' responsible person evaluates the application for work permit and identifies the required safety precautions from a checklist of measures before he approves it. Additional measures may also be stipulated. Table 7.5 shows a typical checklist of measures. Usually it is a question of de-energising the affected system and establishing the necessary barriers between the system and the energy source.
- Operations prepare for the work by shutting down the affected system and by introducing the barriers. Checks are also made to ensure that the maintenance operator starts to work on the right system and that he has understood the safety requirements.
- After work has been completed, operations check the workplace and remove the barriers.

The introduction of a permit-to-work procedure should be considered when the potential consequences of an accident are severe and when communication is required between different organisations. Examples of situations with severe potential consequences are installation and maintenance of high-voltage systems, systems that contain large quantities of toxic or flammable gas and mechanical systems where an accidental start may cause substantial harm.

8 The human element in accident control

8.1 Human information processing

The quest for accident causes in the 1920s brought human fallibility into focus. Research 'showed' that 88 per cent of the accidents were primarily caused by dangerous acts on the part of the individual worker (Heinrich, 1959). Heinrich applied an accident model (the 'Domino theory'), where the cause was related to the event that went wrong immediately before the occurrence of injury. These types of results followed from the central position that the human operator held (and still holds) in the control of industrial production processes.

Let us analyse this proposition from the point of view of safety barriers in Chapter 7. Occupational accident statistics are dominated by events where a single barrier failure is sufficient for a hazard to cause harm. Take, for example, the case where an operator handles a tool such as a chain saw. The running chain represents a hazard and may cause harm if the operator does not handle the saw properly. Moving about is also hazardous due to the risk of falling (on the same level or to a lower level) or hitting objects. The operator has to keep track of the hazards in the environment all the time. To avoid accidents, the operator has to control the hazards (avoid uncontrolled release, cf. Figure 7.1) or stay out of the danger zone (separate the source of energy and the victim in time or space). Most of the time the operator will be able to handle the hazards. In rare circumstances, due to a moment of lack of concentration or for some other reason, the operator fails to keep track of and control or avoid the hazard, and an accident follows. Since man is not 100 per cent reliable, we must expect accidents to happen in a working environment where there are hazards present.

In Section 5.6, we discussed a human information-processing model. Surry has further developed this model to account for the development of an accident process through successive phases (Surry, 1974). **Surry's decision model** of the accident process depicts the required perceptive and cognitive skills to avoid accidents. She distinguishes between early warning during danger build-up and immediate threat during danger release. In her model, an accident sequence is displayed as a series of questions concerning the operator's perception of danger, processing of the information and response, Figure 8.1. The danger becomes imminent if the operator fails to perceive or react to its build-up and release. In applications of this model in accident investigations, the aim is to identify where the human information processing has failed. This information is not readily available but must be derived from interviews and observations of the operator's actual behaviour (actions).

Surry's model focuses on human actions rather than errors and the positive contributions of the operators to safety by adequate handling of hazardous situations. Accidents occur when the demands from the environment to handle hazardous situations exceed the information-processing capacity of the human operator. The operator receives information, processes it, makes decisions and acts. The information thus has to pass through several 'filters', where there are possibilities of information loss or distortion. There are perceptual filters such as reduced eyesight or hearing or inadequate

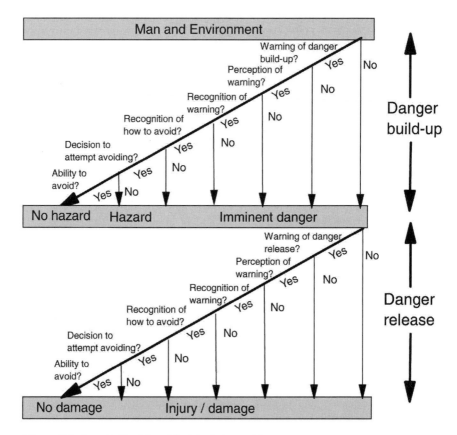

Figure 8.1 Surry's Decision Model of the accident process.
Source: Surry, 1974.

attention. Cognitive filters concern the operator's limited ability to handle large amounts of information simultaneously and to make complicated decisions. Finally, there are limitations in the motor skills of the operator that show, for example, in the reaction time. It follows that improved safety is accomplished through adaptation of the workplaces to the operator's limited capacity.

Surry also brings in another aspect. Under certain conditions, the operator is able to choose between becoming exposed to a hazard or avoiding it. The importance of this aspect was recognised first in the area of traffic safety around 1970. It turned out that many measures aimed at facilitating the driving task did not result in fewer accidents but in the driver adapting to the new situation by e.g. increasing the speed. Wilde has modelled this situation in the so-called **risk homeostasis theory**, Figure 8.2. It describes

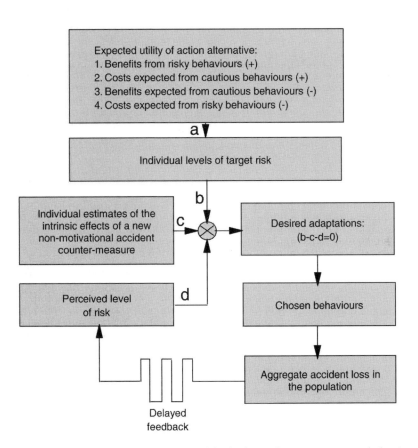

Figure 8.2 Wilde's risk homeostasis model of relation between operator behaviour and accident risk.
Source: Stanton and Glendon, 1996, Figure 1. Copyright 1996 by Elsevier Science and reproduced by permission.

how the operator's caution develops through a process where the operator compares the accepted or target risk with the perceived risk level in a feedback cycle.

This theory is based on the assumption that the operator is able to perceive a risk level in the environment that is related to the 'actual' risk level through a delayed feedback. The operator adapts his or her behaviour in such a way that there is a balance between the hazards that the operator is exposed to and what he or she can accept (target level of risk). It follows from this theory that measures to reduce hazards at the workplace will be followed by changed behaviour on the part of the operators such that their perceived level of risk remains the same. The only way to improve safety is to change the operators' target risk level. The risk-homeostasis theory was primarily developed for situations where people 'voluntarily' expose themselves to accident risks such as in traffic. We must expect, however, that employees at a workplace also to some extent control the risk that they are exposed to based on their own experiences and preferences.

Developments in the 1980s have focused on another aspect, that of learning and the development of skills. The **model of behaviour in the face of danger** takes into account the fact that human information processing occurs at different levels of functioning, dependent on the extent of learning from training and experience that it represents (Hale and Glendon, 1987). It applies **Rasmussen's** famous **skill–rule–knowledge framework**. This framework distinguishes between three different levels of human cognitive control of the environment (Rasmussen *et al.*, 1987):

1 *Skill-based behaviour* At this level, behaviour is automated. Incoming information leads directly to an automatic response without any conscious thought. Skill-based behaviour is established through training and experience.
2 *Rule-based behaviour* At this level, the operator recognises a known situation and applies a pre-stored rule or action-pattern to handle it. The rules have been developed through experiences and may be individual or collective. We find, for example, traffic safety rules to avoid collisions when two cars are on an intersecting course.
3 *Knowledge-based behaviour* When the situation is new or the operator is uncertain of what rule to apply, constructive thinking must take place. He or she must interpret the situation and choose between different action alternatives. Here, errors come about due to the human operator's limited information-processing capacity and due to incomplete knowledge. Behaviour in new situations is knowledge-based. Behaviour evolves from this level to the lower levels through learning and experiences.

Hale and Glendon's model uses Rasmussen's framework, but combines it with stages adapted from Surry's model, Figure 8.3. The model displays what the operator has to do to prevent danger developing into an accident.

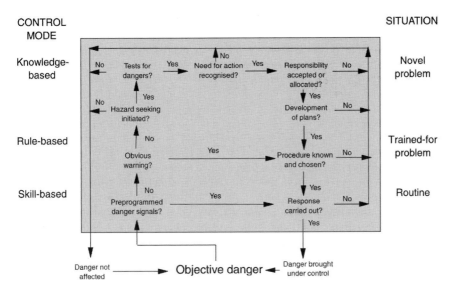

Figure 8.3 Model of behaviour in the face of danger.
Source: Hale and Glendon, 1987, Figure 2.4. Copyright 1987 by Elsevier Science and reproduced by permission.

The starting point of the model is the existence of an objective danger or hazard at the workplace. Most of the time, the operator applies adequate responses learnt from before, and there is not conscious consideration of the hazard. If the danger becomes imminent or changes its characteristics, the operator has at some point to direct his attention more directly and consciously to it. If the situation and the rules to handle it are well known, the operator applies the relevant rules to bring it under control. If, on the other hard, the operator faces a situation where there are no obvious and easily interpreted warnings, the operator must handle the danger through active hazard seeking and development of new plans and action patterns. This is much more demanding to the individual. The model also shows the situation where the operator fails to react to the danger. The situation may then develop into an accident or stabilise itself without any external intervention.

In a real situation with a complex accident sequence, the danger escalates in a series of steps, where there are many opportunities to bring the situation back under control. In an analysis of an accident, the loop is gone through several times in order to identify the different opportunities, either of bringing the situation back under control or of limiting the losses, that were not utilised.

A problem in applying the different models of human information processing is that this information is not readily available. It has to be interpreted from data on observations of actual behaviour and from interviews. This usually requires expert knowledge. The models are rather limited in scope and any one of them will not give us a comprehensive understanding of how the behaviour of operators affects the risk of accidents.

8.2 Human errors

8.2.1 Definition

Let us now move to the more traditional perspective, that of the role of human errors in accident causation. There are many different definitions of human errors. We will here apply a systems perspective in accordance with the definition of deviations in Chapter 5. *Human errors* are here defined as a subset of human actions that transgress some norm or limit of what is planned/intended, normal or acceptable (Miller and Swain in Salvendy, 1987; Kjellén in Rasmussen *et al.*, 1987).

There are some important aspects of human errors. The norms or limits are defined in relation to the context in which the errors occur and their potential consequences. Some systems represent a benign environment by being tolerant towards large variations in human conduct. In such systems, few actions will be defined as human errors. Other systems are more demanding, and a small deviation from the ideal path represents a human error that may result in an accident or other unwanted consequence. It also follows that the consequences of errors are not necessarily correlated with the magnitude of the transgression but rather with the energies involved.

There is a certain degree of arbitrariness in the definition of human errors. In accident investigations, for example, an action may be defined as a human error only in retrospect when the negative outcome is known. The same action may have been carried out successfully many times before, tacitly accepted by operators and management.

Traditionally, the application of the term 'human error' has been restricted to the operators and maintenance personnel at the sharp end, i.e. in immediate contact with the production system. There has also been a tendency to intermix the immediate cause of an accident (i.e. often a human error) with the responsibility for the occurrence, i.e. the person to blame. In a move away from such simplistic conclusions, there is now a trend to extend the analysis to include errors in the decision-making at all management levels of the company. Such analyses may reveal that human errors by operators, in their turn, are caused by poor equipment design, management's tacit acceptance of rule violations, inadequate work instructions and training, etc. It is usually possible to explain an action labelled as a human error fully by the circumstances in which it has occurred.

In 'over-regulated' systems, where the main purpose of the instructions is to allocate blame in case of an accident, violations of written instructions may be a rule rather than an exception. Supervisors tacitly accept these as long as the work gets done. Such conditions are counter-productive to safety and call for a total review of the instruction system.

8.2.2 Human-error taxonomies

Table 6.8 showed an overview of schemes or taxonomies for the classification of deviations. We will here develop this overview further by presenting taxonomies of human errors, Table 8.1.

Swain's taxonomy of human errors is rooted in the field of human reliability. It applies a mechanistic view of the human operator in the sense that the operator is regarded as a systems component in line with machinery. *Human reliability* is thus defined as the probability that a person first, performs an activity required by the system correctly and within a required time period and second, does not perform an extraneous activity that may degrade the system (Rosness in Suokas and Rouhiainen, 1993). Swain's classes of human errors constitute observable behaviour in relation to a planned work sequence. There are data banks on the frequency of typical human errors in different systems contexts. Data on human-error frequencies are applied as input to reliability models to calculate the overall probability of systems failures and accidents.

Table 8.1 Overview of human-error taxonomies

Basis for classification	Classes of human errors	Source
Observable behaviour	• Omission • Commission • Extraneous act • Sequencing error • Timing error (too early/late)	Swain, 1974
Phase in human information-processing	• Detection error • Decision error • Execution error	Surry, 1974

Cognitive stage	• Skill-based error
	  ◆ Slip (execution failures)
	  ◆ Lapse (memory failures)
	• Rule-based mistakes
	  (wrong use of rule)
	• Knowledge-based mistakes
	• Intended action (violation)
		Rasmussen *et al.*, 1987, Reason, 1991

Organisational hierarchy	• Unsafe-act tokens • Function-failure types • Source-failure types	Reason, 1991

Surry's decision model and Rasmussen's framework for cognitive stages have served as bases for the development of human error taxonomies. Reason distinguishes between errors at different performance levels:

- Slips (execution errors) and lapses (memory errors) occur at the skill-based performance levels. They are due to the intrinsic variability of human actions with respect to place, force and time co-ordination.
- Rule-based mistakes have to do with an incorrect recall of a procedure or a misclassification of the situation and the application of a wrong or bad rule. There are also deliberate violations of rules.
- Knowledge-based mistakes are more complex and are often due to an incomplete or inaccurate understanding of the situation.

A problem in applying the human-error taxonomies related to human information-processing and cognitive stages is that the classification must be based on an interpretation of what is going on inside the human brain. This may require detailed interview data, for example from so-called verbal protocols, and interpretation by experts. It is often not easy to attribute a specific action to one of the three performance levels, since an operator usually does not act on one level of performance but mixes the levels in an integrated action sequence (Döös, 1997). Automated action sequences and actions controlled by rule-based behaviour occur intermixed and in combination with knowledge-based behaviour.

Reason's error taxonomy that relates to the organisational hierarchy has been discussed in Section 5.2.

8.2.3 Error recovery

To avoid accidents altogether, a complete elimination of human errors may be seen as the ultimate goal. This goal is not very practical, however, and will have severe side effects. Due to the intrinsic variability in human performance, errors will occur. Errors also provide the operators with task feedback and on-the-job learning about the systems that they operate. This experience is extremely valuable in situations when the operators have to handle unanticipated situations to avoid shut-down or accidents. This has often to be done under tight time constraints and psychological stress.

Error recovery is an important aspect of accident prevention. Research shows that operators manage to recover a significant amount of their errors (Reason, 1990). The recovery actions have different aims, dependent on the severity of the situation (Kontogiannis, 1999). Ideally, there is a full recovery in the sense that the system is brought back to its original state before the error occurred. If this is not possible, the aim may be to bring the system into an intermediate stable state in order to buy time. This is, for example, the case when the operator brings the system to an emergency

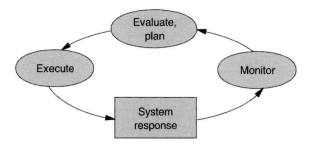

Figure 8.4 Error recovery may occur at different stages in human information processing.
Source: Adapted from Kontogiannis, 1999.

stop in order to avoid the development of a harmful situation. A third possibility is to activate redundant equipment in order to bring the system to the desired goal.

Let us analyse error recovery in relation to a traditional model of human information processing, Figure 8.4.

Monitoring of systems feedback provides an opportunity for error detection and recovery when there is a mismatch between detected and observed outcome of an action. Human errors are not easily detected in complex systems. There may be a significant time lag between action and observed effects that makes detection difficult. The effects of human actions may be masked by actions taken by the technical control system. Biased error attribution on the part of the operator may also impede error recovery. Co-workers and supervisors are important resources for error recovery in this context. There are, however, some important preconditions for colleagues and supervisors to be able to contribute. They must co-operate closely with the erring operator to be able to observe performance and distinguish between erroneous and correct acts. There must also be a climate of trust and willingness to correct each other's behaviour.

The operator may be able to recover errors at the execution stage. This is when the operator catches himself/herself in the act or immediately after it has been executed and corrects it before the system has been adversely affected. Different strategies are applied in accomplishing this error-recovery mechanism. The operator may notice that there is a mismatch between his/her observation of his/her own act and his/her expectations of how it should have been performed. A special case is when the operator is able to avoid an error by observing that the situation is similar to an earlier situation in which he/she committed an error. We also here have the case when colleagues or supervisors observe erroneous acts and react to them.

Finally, error recovery may take place at the planning stage. This is when the operator recognises mismatches between his/her own intentions and formulated plans. Self-checking and vigilance for possible errors and planning for correction in advance are important error-recovery mechanisms. To be able to change plans is not unproblematic. Research shows that operators have a tendency to persist with the original plan and to disregard evidence that speaks in favour of changing the plan (so-called tunnel vision, see next Section). Other people with a fresh view of the situation are better at questioning existing plans.

8.2.4 The influence of emotion

Until now, we have concentrated on the 'cold' cognitive aspects of human information processing. The handling of a production disturbance and the recovery of a human error also has a 'hot' aspect of subjective feelings and emotions (see e.g. Lazarus and Folkman, 1984; Persson and Sjöberg, 1978). Emotions arise when the individual believes that highly valued wishes will either be frustrated or satisfied. We will here focus on emotions that come about when an operator faces embarrassing or threatening information. This may be about production disturbances, about the impact of his or her own errors or about an incident that may evolve into an accident.

Emotions are adaptive in the sense that they lead to arousal and increased and redirected attention and information seeking. The individual will interrupt the ongoing activity to prepare for flight or fight. When in an emotional state, the operator will focus his or her attention on information sources (cues) associated with the emotional state and disregard cues that are not associated with it. Increased cognitive activity towards possible ways of handling the situation takes place. There is a tendency to make fast and what may finally turn out to have been premature decisions.

Emotions often have a dysfunctional effect on the activities to recover errors and disturbances:

- They invoke 'tunnel vision', i.e. a reduction in attention capacity and a focus on central cues and on tasks requiring attention. This reduces the operator's ability to observe and comprehend peripheral cues and information that challenge the prevailing understanding of the situation. The operator may continue with one course of action and overlook conflicting information, although the situation requires a reassessment and redirection of the activities.
- A decreased tolerance of ambiguity takes place and the operator triggers a response as soon as he or she has a minimum of information to get out of the unpleasant situation. Decisions and actions may thus take place based on a premature understanding of the situation and the dangerous situation may escalate rather than being put under control.
- People behave differently under extreme stress. Certain people will freeze and avoid actions through repression and denial. Others will experience increased vigilance, speed up performance and take a large number of erroneous actions.

To reduce or eliminate the impairing effects of emotions, the personnel must be highly motivated and provided with anticipatory knowledge and coping strategies to be able to handle emotionally charged situations. There should be simple and crude procedures for handling the situation and well-defined decision criteria to reduce the need for cognitive capacity and subtle decision-making.

It is also important that the man–machine interface is adequately designed. The information required in handling emotionally charged situations should be presented on the central displays and the need for switching between information-scanning modes should be eliminated. The information should be presented in a clear and concise way and there should be a minimum of need for well-co-ordinated actions.

8.2.5 Preventing human errors and promoting error recovery

8.2.5.1 Personnel-related measures

Personnel-related measures to reduce the frequency of human errors and promote safe behaviour include personnel selection, education and training, and safety campaigns and performance feedback (Salvendy, 1987, Sanders and McCormick, 1992). There are three main objectives of the personnel-related counter-measures:

- To ensure an adequate *knowledge*, for example, about the technical design of the workplace and about the hazards involved in specific jobs,
- To ensure adequate *skills* in how to perform the work safely and efficiently, and
- To ensure adequate *attitudes* among the personnel so that they are able to put their knowledge and skills to use and make a commitment to behave safely.

The *selection of personnel* for hazardous work involves judgements regarding their physical ability (sight, hearing, motor skills, etc.), competence, motive structure, stress tolerance and experience. For certain types of hazardous work such as crane operation and lorry driving, there are formal requirements as to the testing and certification of the personnel. Personnel selection involves challenges regarding which selection criteria to use, the validity and reliability of the selection tests and the availability of an adequate supply of qualified personnel.

Knowledge, skills and attitudes to perform work correctly and safely are acquired through experience, but the learning process may be speeded up and controlled by formal *education and training*. Increasingly, government regulations require employers to provide adequate education and training of their personnel. Education and training is especially relevant when the operators routinely meet similar hazardous situations and errors and there is a need for developing adequate routines and ingrained behaviour to handle stress and to recover the system or bring it to a safe halt. Over-training may be necessary to make the response automatic. Education and training is also needed in situations where the operators routinely meet new hazardous situations that have to be resolved through knowledge and problem-solving skills. When interactions between team members play an important role in error detection and handling, each team member should be acquainted with the different skills of the other team members as well. Team training is especially important in this situation.

Safety campaigns involve the use of such different means as the distribution of posters and pamphlets, showing video films, safety meetings and direct talks about safety. The aims of such campaigns are to change behavioural patterns (e.g. by defining the 'ideal' safe behaviour), to affect attitudes, to put the focus on safety matters and to give warnings about negative consequences of unsafe behaviour, etc.

Performance feedback is based on behaviour theory. To improve the proportion of safe acts, it applies mutually agreed targets on safe behaviour and feedback to the workers to information about their actual performance. We will come back to this method in Section 18.2.

8.2.5.2 Workplace design

Workplace design is an engineering solution that aims at facilitating the use of safe and error-free work practices and at promoting error recovery. An advantage with this solution is that good design is permanent, whereas personnel selection, education and training and campaigns are measures that have to be repeated all the time to be efficient. There are different means of preventing human errors through design. One is to make it impossible to commit the error altogether, e.g. by eliminating the error-prone task. Design may also make errors less likely, e.g. by being compliant with people's ingrained behaviour or so-called population stereotypes (Salvendy, 1987; Sanders and McCormick, 1992). A good working environment will promote safe behaviour in general by reducing the likelihood of fatigue-induced errors.

It is also important to provide for an error-tolerant environment that makes it easy for the operator to observe, trace and correct errors before it is too late. There are certain design principles that promote error recovery (Kontogiannis, 1999):

- Observable and transparent, i.e. errors are easily detected through appropriate and immediate feedback;
- Traceable, the causes of inadequate systems performance are easily and promptly traced back to the erroneous act; and
- Reversible, the effects of errors are easily mitigated and the system is easily brought back to a safe state.

8.3 The role of the operators in major-accident prevention

8.3.1 Unscheduled manual interventions

In hazardous industries, human actions may determine whether a major accident can be avoided or not. In this Section, we will look closer into the role of the operators in the *control of major accidents*. We have in Chapter 7 identified the different barriers applied in the prevention of fires and explosions in a plant for processing of hydrocarbons. They represent an engineering solution, based on the defences-in-depth philosophy. The idea is to minimise the system's vulnerability to human and technical errors by providing independent and diverse layers of protection. Whereas the operator has a distinct role in maintaining production, the safety systems will take over in cases where the system transgresses certain defined limits. There are many examples, however, where these different barriers are dependent on and vulnerable to operator actions.

Examples: Cases where operator interventions into the production process of an oil and gas processing plant affect the barriers against fires and explosions are listed below:

- *The central control room (CCR) operator surveys the process from the operator control station in order to intervene in case of a process upset or in case a process parameter develops in an unsatisfactory direction. The aim is primarily to maintain production speed and avoid shut-down, but their actions may also affect safety in a negative way, for example, when alarm levels are altered or disconnected.*
- *An outdoor operator smells gas and traces the leak to a coupling to a valve. In agreement with the CCR operator, the outdoor operator tightens the coupling. This activity replaces the need for shut-down and maintenance but there is always a risk that the tightening operation will cause an increased leak or rupture.*
- *The CCR operator disconnects gas detectors from the operator control station. The aim is to avoid spurious shut-down due to gas detection during planned discharge of small quantities of gas from the process.*

These cases are examples of so-called **unscheduled manual interventions** (**UMIs**) into the automatic production process. This term was originally used to describe actions where the control-room operators assume control of production during periods when the system has been scheduled to be under automatic control (Hockey and Maule, 1995). The primary aim is to maintain production speed and quality. We will here use the term UMI to describe any action, where the control-room or outdoor operators strive to bring the production process back to normal operation or eventually to a safe shut-down after the occurrence of a process disturbance. This application of the concept of UMI is illustrated by Figure 8.5.

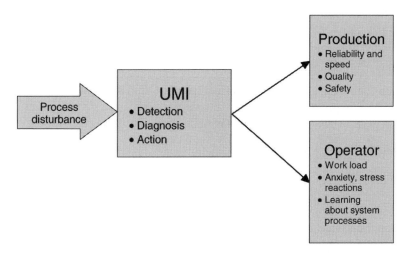

Figure 8.5 Unscheduled manual interventions (UMIs) following a process disturbance. UMIs are executed either from the control room alone or jointly by control room and outdoor operators.

The outcome of the UMI is dependent on the operators' ability to correctly detect and diagnose the process disturbance and to select and execute the necessary actions. The actions are executed from the operator control station or by an outdoor operator in communication with the control room operator. The design of the man–machine interfaces is crucial to the operators' ability to perform well. It is thus important in the design of such interfaces to take the handling of production disturbances into account.

Studies show that UMIs make up a considerable part of the work in a typical control room and have positive as well as negative effects on the psychological work contents. They contribute to the workload and may increase anxiety and stress (Hockey and Maule, 1995). They are at the same time essential to the operators' ability to learn about the process and to develop suitable mental models of it through their day-to-day interactions with the process. This is done both at the individual and work-team levels.

There are individual differences in the way the operators handle disturbances and learn from them (Döös, 1997). Whereas some operators take a proactive and problem-solving approach, others are more detached. The proactive operators will monitor the process closely and update their mental models continuously. They will be able to intervene at an early phase when the process shows signs of deteriorating, e.g. due to the fact that a process parameter develops in an unsatisfactory direction. The more detached operators will rely on the automatic systems to bring the system back to normal. There are safety pros and cons in both cases. The proactive operators may be able to maintain production and avoid spurious shut-downs more often.

They will be quicker in learning and in developing the necessary skills to handle the process than the more detached operators. There are risks, however, that the interventions of the proactive operators will be erroneous and cause hazardous situations.

There is a similar concern that has to do with design. Should the possibilities for the operators to make UMIs be arranged for and even encouraged? Or should the process be completely automatic in order to avoid any type of UMI? A common solution to this question in the design of process plants is to allow the operators to intervene in the production process. If the situation deteriorates further to the point where the emergency shutdown system takes over, this shutdown is pre-programmed, and the operators are not allowed to intervene to stop it.

8.3.2 Fallacy of the defences-in-depth philosophy

The UMIs and the associated learning that takes place are mainly concerned with the maintenance of production goals, where there are immediate feedback loops concerning production speed and quality. There are often no corresponding feedback loops concerning safety in well-defended systems such as plants for the processing of oil and gas. A problem is that failures may be countered by the system or concealed, without the operators being aware of their existence. This so-called *fallacy of the defences-in-depth philosophy* means that failures and mistakes may go undetected, and neither correction nor learning will take place (Rasmussen, 1993). Many barriers must fail before the system shows obvious signs of reduced safety. An accumulation of such latent errors can in combination with a sudden disturbance result in a catastrophic event.

Perrow, in his famous book 'Normal Accidents: Living with High-Risk Technologies', draws the attention to the contradictions in applying a multiple-barrier safety philosophy (Perrow, 1984; Reason, 1997). He draws his conclusions from experiences with high-technology systems involving major accident risks such as nuclear power. He argues that such systems have become more complex and opaque and thus more difficult to operate and maintain. This fact has some important implications:

- The assumptions of barrier independence may apply to foreseeable technical failures. There are, however, many case histories in the literature, where unforeseen common-mode failures introduced by the operators or maintenance personnel have caused multiple barrier breakdowns. Due to the complexity of the system, the operators have not been able to understand the situation and foresee the consequences of their actions.
- The increased complexity and opaqueness also causes another problem. Lack of feedback mechanisms to the operators on the status of safety barriers causes latent barrier failure to prevail and only become visible when it is too late.

- Well-defended systems give rise to a false sense of security among operators and management. They may use this experienced safety margin in a way that counteracts safety by disconnecting one or a few of the safety barriers to increase production, believing that the remaining safety barriers are intact.

8.3.3 High-reliability organisations

Perrow represents a rather pessimistic view of the operators' and maintenance personnel's ability to maintain a safe state in high-technology systems involving major accident risks. Experience shows, however, that some organisations are able to sustain failure-free operation of such systems. Studies of so-called *high-reliability organisations (HRO)* have identified three aspects characterising such systems in particular (LaPorte and Consolini, 1991; Reason, 1997; Rijpma, 1997).

The first strategy is based on similar principles to those discussed in Section 8.2.3 on error recovery. The HRO uses a strategy of *redundancy* in avoiding errors and/or recovering from them. Operators receive backup from colleagues who are ready to give advice and to correct errors, or to take over if an operator fails to execute an action correctly. A prerequisite is that the operators are able to listen to the reasoning of their colleagues, to observe their behaviour when performing critical tasks and to understand the consequences of their actions. The operators must also be able to exchange information and challenge each other's decisions and actions when there are conflicting views as to the correct course of action.

The HRO shows an ability to *change organisational structure* in a critical situation. In the day-to-day operation, the organisation is hierarchical, and decisions and actions are controlled by well-formulated and tested standard operating procedures. When the workload increases, for example during disturbance handling, the operators are protected from interference from the outside and are granted a high degree of autonomy in handling the situation. The organisation relies on the technical skills of the operators in this situation. The operators must have clear and well-agreed-upon operational goals and decision criteria to be able to handle the situation efficiently. In an emergency situation, the organisational structure changes again. There are clearly specified emergency events, and the organisation follows carefully assigned and practised operations to handle each type of situation.

Finally, the HRO maintains *conceptual slack* to be able to handle new situations and to facilitate learning from experience. Multiple theories on the technology and the production process are maintained simultaneously among the operators. In critical situations, they negotiate a course of action in order to avoid hasty inappropriate actions. This strategy becomes problematic if the production process deteriorates rapidly and time is too short to allow for discussions to agree on a proper course of action.

9 The occurrence of accidents over time

We will now change the perspective from the individual incident or accident to the occurrence of accidents over time. The occurrence of accidents follows the same basic laws of statistics as such events as telephone calls to a home, customers arriving at a department store or radioactive decay.

It is a commonly accepted assumption in SHE management that accident frequency, i.e. the number of accidents occurring at a company in the course of, for example, a year is a measure of the SHE performance of the company. In this Chapter, we will focus how we can use basic statistical theory in evaluating data on the occurrence of accidents. We will apply this theory in Part IV on SHE performance indicators.

Example: A company has a stable number of employees. Over a period of ten years, the average yearly number of accidents was 4, Figure 9.1. In the following year, there were 2 accidents. Does this number represent a significant improvement compared to the yearly average value or may it be explained by pure chance?

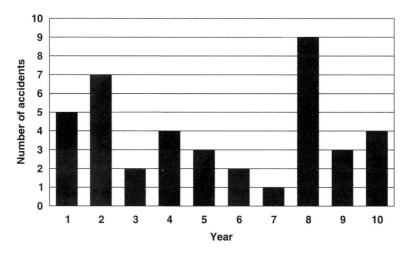

Figure 9.1 Recorded number of accidents per year.

To be able to answer this question, we must look into the statistical theory of *Poisson distributions*. We will here merely review the most important characteristics of such a statistical distribution. Standard textbooks on statistics should be referred to for further details.

Let us assume that accidents occur at random points in time. Let us call *c* the *intensity*, which is the average number of accidents per unit of time (for example one year). Let *x* be the number of accidents occurring during *t* time periods, then $f(x)$ is the frequency function of the Poisson distribution $(c \times t)$, i.e.:

$$f(x) = \frac{(c \times t)^x}{x!} \times e^{-(c \times t)}$$

where $f(x)$ is the probability that *x* accidents will occur during *t* time periods.

A Poisson distribution has the following characteristics:

- The frequency functions for the number of events (accidents) during time periods that are not overlapping are independent stochastic variables.
- The probability of one event during a short time period Δt is approximately equal to $(c \times \Delta t)$.

Example: Let us calculate the probability of 2 or less accidents at the company in the example above:

$$P(x < 3) = \sum_{k=0}^{2} \frac{4^k}{k!} \times e^{-4} = 0.0183 + 0.0733 + 0.1465 = 0.23$$

We see that this probability is 0.23. It means that in almost one out of every four years, we will expect from pure chance that there are two or less accidents. Figure 9.2 illustrates this particular Poisson distribution. From the figure we can see that the probability of exactly four accidents during one year is only about 0.2. Similarly, the probability of zero accidents during one year is 0.02, i.e. one in every 50 years.

The Poisson distribution has some other important characteristics. For 'large' values of $(c \times t)$, the Poisson distribution is approximately represented by a Normal distribution with:

- a mean value that equals $(c \times t)$ and
- a standard deviation that equals $\sqrt{(c \times t)}$.

The example above illustrated a typical example, where we were interested in knowing whether the number of accidents in one period (e.g. a year) significantly differs from the mean number of accidents during previous years. For 'large' numbers of accidents, where we can assume that the number

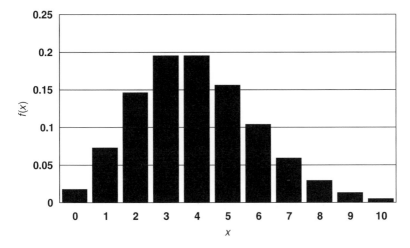

Figure 9.2 Graphical illustration of the frequency function for a Poisson distribution where $(c \times t)$ equals 4.

of accidents per period is Normally distributed, this is easily answered. For most practical circumstances, this means five or more accidents per period. For a normal frequency distribution, the probability of an outcome between plus and minus two standard deviations (μ +/– 2σ) is about 95 per cent. This level of confidence is often used when determining whether a change is significant or not.

Example: There were on average 100 accidents at a company per year. It follows that for 19 of 20 years we shall expect a yearly frequency of between (100 – 2 × √100) and (100 + 2 × √100), i.e. between 80 and 120 accidents per year. If the number of accidents for one year falls outside this range, we consider this as a significant change in the frequency rate.

The average number of accidents per period is our best *estimate* of the underlying or 'true' value of the accident intensity at a company. The 95 per cent confidence interval of such an estimate can easily be calculated by applying the same method as above.

10 Feedback and use of experiences in decision-making

10.1 Overview of feedback mechanisms

Industrial systems are designed and operated by people. Decisions at the different levels of the industrial organisation will determine the level of safety of the system. These decisions are usually made as part of a continuous flow of management actions in real time and are guided by a set of goals. Experiences show that decision-makers seldom focus on safety by habit in this process. Other factors such as productivity, quality and progress have their immediate attention. This has to do with the fact that production demands are continuously present and direct and attract the attention of operators and managers. Accidents, on the other hand, are rare events and the associated feedback information to the organisation is infrequent and often difficult to interpret.

The traditional SHE management approach to this problem has been to introduce formal decision-making routines that managers have to follow. The aim is to shape the behaviour of the decision-makers in a direction that gives safety sufficient attention. Heinrich pioneered this approach in his classical book on Industrial Safety Management (Heinrich, 1959). He introduced five steps that decision makers have to go through in order to prevent accidents in a systematic way: (1) collection of data about accidents and near-accidents; (2) analysis of these data; (3) selection of remedies; (4) implementation; and (5) evaluation of effects.

We will distinguish between two different feedback mechanisms; diagnosis and persistent feedback control. These processes represent two views of the traditional SHE management approach of pre-programmed activities. We will later also look into how decision-makers may query a database on accident-risk experience on an ad-hoc basis. This approach is rather new and aims at integrating SHE management aspects in the routine management actions and decision-making. It has been made possible through recent developments in information technology.

We associate *diagnosis* with the method used by medical doctors in identifying the relations between symptoms (e.g. body temperature above normal range, sore throat) and illness (e.g. flu) in order to prescribe the correct remedy. Heinrich's five steps have many similarities to the diagnostic process. We consider an accident investigation to be a diagnostic process where the accident is a symptom of underlying deficiencies in the design and/or management of the industrial system in question. The diagnostic process is thus not a continuously ongoing process but is started when the symptoms reach an awareness threshold (e.g. an accident or near-accident occurs). In this chapter, we will review some basic characteristics of the diagnostic process. In Part III, we will return to applications in accidents and near-accident investigations, workplace inspections and SHE audits, and in Part V in risk analyses.

Persistent feedback control, on the other hand, is an ongoing process where the SHE performance is monitored periodically and compared to pre-established SHE goals, compare Figure 3.2. Basic principles of feedback control are reviewed in this chapter. We will revert to applications in Part IV on SHE performance measurement.

It is not always clear in SHE management practice which principle applies. The diagnostic process and persistent feedback control are partially overlapping. In both cases, it is important to *close the loop*, i.e. to ensure that the necessary actions are identified following the diagnosis or measurement and that these are implemented and the effects monitored.

10.2 Uses of SHE-related information in decision-making

We must design SHE information systems with user needs in mind. Figure 10.1 illustrates examples of information uses in SHE-related decision-making in a simplified organisation with two line-management levels and a staff organisation. At the top management level, summary data on accidents is used mainly for monitoring purposes. The SHE information system will support such a feedback control process by providing:

• Input to the establishment of goals or norms, based on so-called SHE performance indicators.
• Measurement and follow-up of results in order to look for trends and to compare the results with pre-established goals.

Actions are taken to prevent the goals not being met. Top management also requests summary accident statistics for prioritisation purposes.

The diagnostic process represents many different activities in SHE management. They are typically found at the workplace level, where rich data on accidents is used locally for prevention. Here, supervisors and workers solve SHE-related problems by evaluating output from a SHE information system in relation to their own experiences. Below are some examples illustrating different needs of SHE related information at this level:

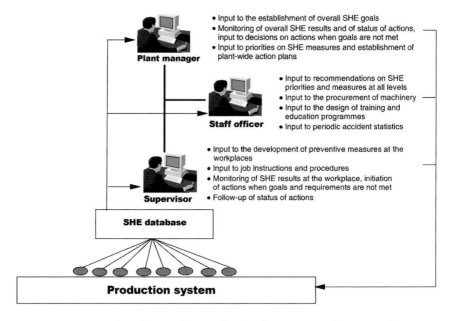

Plant manager
- Input to the establishment of overall SHE goals
- Monitoring of overall SHE results and of status of actions, input to decisions on actions when goals are not met
- Input to priorities on SHE measures and establishment of plant-wide action plans

Staff officer
- Input to recommendations on SHE priorities and measures at all levels
- Input to the procurement of machinery
- Input to the design of training and education programmes
- Input to periodic accident statistics

Supervisor
- Input to the development of preventive measures at the workplaces
- Input to job instructions and procedures
- Monitoring of SHE results at the workplace, initiation of actions when goals and requirements are not met
- Follow-up of status of actions

SHE database

Production system

Figure 10.1 Examples of uses of SHE-related information in decision-making.

- Input to priorities, selection and implementation of SHE measures.
  Example: The kitchen supervisor of a hotel was committed to a reduction in the number of cut injuries. The accident statistics were used as input to a decision about the selection of kitchen tools and equipment and personal protective equipment. The personnel participated in the development of solutions, and they helped in interpreting the accident statistics in relation to their experiences.
- Input to the development of job instructions, procedures, and technical specifications.
  Example: In developing job instructions for a rolling mill, the accident statistics were analysed in order to identify hazards during different activities (start-up and shut-down, operation, cleaning, maintenance and handling of disturbances). The hazards associated with these different activities were mentioned in the job instructions together with the necessary safety precautions.

We find application of persistent-feedback-control principles at this level. In workplace inspections, the conditions at the workplace are monitored and compared to pre-established norms (the regulations, internal procedures, accepted practice). Decisions are made during the inspections to correct deviations (e.g. replace faulty guards, improve housekeeping).

The SHE staff are often the most frequent users of the SHE information system. They support line management and union representatives by providing the requested reports. Periodic summary reports on accident statistics, for example, present the necessary information for monitoring and priority-making purposes at the plant and department levels. Risk analyses represent another area requiring input on accidents. Such analyses give overviews of the hazards at the workplaces and the associated risk, and are used as input to decisions on actions to reduce the risk. Line management makes the actual decisions.

We also find uses where line management and staff functions make ad-hoc queries, for example, about accident occurrences. This application does not necessarily follow the systematic diagnostic process. The SHE information system is used as a search tool rather than as a pipeline for transfer of information. Examples are:

- Needs of input to the design of machinery concerning accidents with similar types of machinery.
  Example: Management at a workshop had decided to buy a new rolling mill. Previous accidents with a similar mill were analysed and the safety measures that came out of this review were included as technical requirements in the purchase order.
- Needs of input to educational programmes.
  Example: The educational department was planning a first-aid course. They consulted the accident statistics for illustrative examples, showing the benefits of an immediate treatment of the victim.
- Identification of accident repeaters.
  Example: An accident-investigation team wanted to consult the SHE information system for similar occurrences. The results were used to determine the need for further analyses to identify root causes and remedial actions.

Unfortunately, experience shows that these uses rarely occur spontaneously. Management may find the accident database to be too diluted to allow for retrieval of significant information and may be sceptical about data quality and usefulness. We will have to address these concerns to avoid the SHE information becoming a graveyard for accident data rather then an instrument for SHE improvements.

10.3 The diagnostic process

Diagnosis is defined as the complete decision cycle consisting of: (1) identification of symptoms, (2) determining causes and (3) prescription of remedy. A *symptom* is a deviation of the system's behaviour from what is considered to be 'normal', a term we are familiar with from the process model of accidents (Section 5.3).

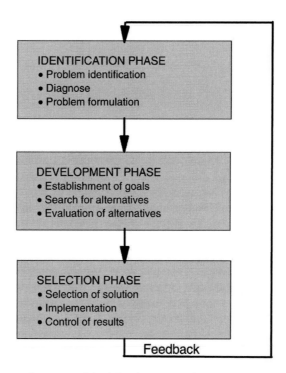

Figure 10.2 Mintzberg's model of the decision-making process.
Source: Mintzberg *et al.*, 1976.

A structured approach is represented by Mintzberg's conceptual model of the decision-making process, Figure 10.2. The first main phase, identification, involves the recognition that there is a SHE problem present. Mintzberg sees diagnosis as the inquiry into the available information in order to determine the causes. When these are known, the problem to be solved is defined.

In the development phase, the decision-maker establishes goals for the problem-solving activities, identifies alternative solutions to solve the problem and evaluates these in relation to the pre-established goals. Finally, a decision is made on the choice of solutions, these are implemented and the results are followed up.

Heinrich's and Mintzberg's models are examples of classical rational-choice models of decision-making. They do not describe how decisions actually are made. Instead, they prescribe how organisations should make decisions in order to arrive at satisfactory solutions. There are a number of assumptions behind these models (March and Simon, 1958):

- There is a whole set of alternative actions to choose between.
- Each action alternative is linked to a set of consequences. These may be certain or associated with an uncertainty.

- There exists a utility function that describes the relation between these consequences and the decision-maker's preferences.
- The decision-maker is rational in the sense that he or she is fully informed and selects the set of alternatives leading to the preferred set of consequences.

10.3.1 *Effects of limitations in human information-processing capacity*

In SHE decision-making in practice, the different assumptions behind the 'ideal' rational-choice models of decision-making are rarely present. One important aspect is the limitations in the information-processing capacity of the decision-makers. There are costs associated with the search for and processing of information due to the fundamental memory, decisional, response and attention limitations of human performance. These are treated in many textbooks on human factors (see e.g. Sanders and McCormick, 1992). We will here focus on how people handle these limitations by economising on their search for and processing of information. The information search and processing strategies that people apply will result in biases seen from the perspective of the rational-choice model of decision-making. They are exemplified below:

- Information that is received early in the search is given an unduly high weight.
- People do not utilise the information sources fully to look for alternative explanations or solutions.
- People have difficulties in taking uncertainty in the information into account.
- People are only able to consider a limited number of (three or four) alternative explanations to the problem at the same time.
- People are only able to evaluate a few (two to four) alternative solutions at a time and will only focus on a few critical characteristics of each solution.
- People tend to look for information that confirms the selected solution and to avoid information in support of an alternative solution.

In general, people tend to terminate the search when they have found a satisfying solution, rather than searching for more information to find an optimal solution. The subjective cost-benefit evaluation made by the decision-maker in searching for information is illustrated by Figure 10.3.

People will terminate the search for additional information when they experience that the subjective costs (time, attention, etc.) exceed the benefits. In practice, SHE-related information has to compete with other types of information to receive the decision-makers' attention. We have to design the SHE information systems in a way that minimises the costs of searching for and analysing information and maximises the usefulness of the information.

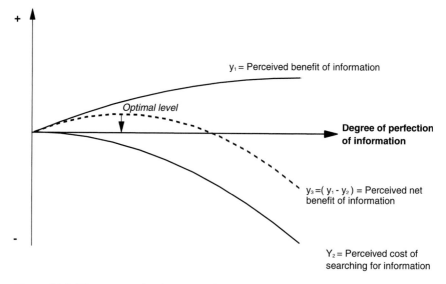

Figure 10.3 The marginal subjective value and cost of additional information.
Source: Adapted from Harrison, 1987.

It follows that the information has to be relevant to the decision-maker's needs, it must be comprehensive yet easy to survey. The decision-maker must have easy access to the information when it is needed.

There are other decision-making models that better describe how organisations actually arrive at decisions. Organisations that apply a *bureaucratic model* of decision-making will apply pre-defined rules and procedures rather than making trade-offs between different action alternatives. This approach will reduce the demands on the decision-maker's information-processing capabilities.

We also have to acknowledge that there are a diversity of goals and conflicting interests within an industrial organisation, not least in the area of safety, health and environment. The *political model* of decision-making analyses how decisions are made through bargaining and compromises between the different stakeholders (top management, line management, employees, owners, external interest groups, authorities). Information from a SHE information system may play an important role in the bargaining process by supplying the different stakeholders with the necessary arguments. We will later see how these motives will also affect the input side of the SHE information system through tactical reporting of SHE concerns and problems.

10.3.2 Hale's problem-solving cycle

Hale's problem-solving cycle is a typical example of a rational decision-making model that has been adapted to the needs of SHE management,

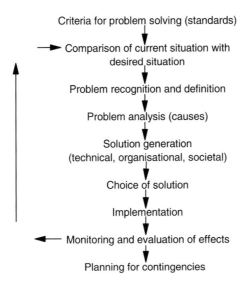

Figure 10.4 The problem-solving cycle according to Hale.
Source: Van der Schaaf *et al.*, 1991.

Figure 10.4. It is an idealised model for structured decision-making with many similarities to the Mintzberg model. Accident risks are recognised, defined and analysed with respect to causes. Decision rules (criteria for problem-solving) and priorities are established. Only then can the work to solve the problem (identified accident risk) start. Decisions concerning choice of solution are made and the selected solution is implemented. Ideally, there is also a monitoring and evaluation of effects.

The model has pedagogical merits by identifying important activities within a rational SHE management system. We may, for example, compare the way accidents are investigated in relation to this model in order to identify potential for improvement. It generates such questions as:

- Have we established SHE policy and goals to guide our decisions?
- Do we conduct accident investigations in a way that documents all relevant symptoms, i.e. deviations from the normal or accepted system state?
- Do we make a thorough analysis of causes rather than jumping directly to the conclusions, i.e. finding the easiest available solution?
- Do we review all possible remedial actions before we make a decision?
- Do we implement the decision and follow up on the results?

In a real situation, we will find many negative answers to such questions. The intention is to serve as a guide in improvements in a more rational direction.

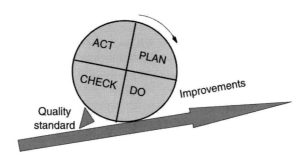

Figure 10.5 Deming's circle and the quality standards necessary to prevent fall-
back.
Source: Wig, 1996. Reproduced by permission from the author.

10.3.3 Deming's circle

The so-called Deming's circle named after its inventor is a central concept
in quality management (Wig, 1996). It shows the cycle in a learning process
from planning, through execution and check to correction, Figure 10.5.

Deming's four basic processes are further broken down in the following
way:

Plan
    Where are we?
    Where do we want to go (goal)?
    How do we get there?
Do
    Communicate and train
    Secure resources
    Execute
Check
    Do we follow the plan?
    Do we meet our goal?
Act
    Implement corrective actions
    Sum up experiences
    Standardise

The idea behind the circle is to ensure continuous improvements through
consecutive rotations of the circle or wheel. This concept has been widely
adopted by industry.

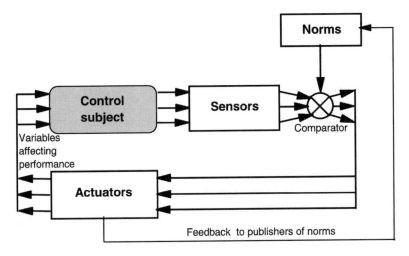

Figure 10.6 Feedback cycle for the control of anything.
Source: Adapted from Juran, 1989.

10.4 Persistent feedback control

Control by *negative feedback* is a regulating mechanism that produces corrective action. The system output is monitored continuously or sampled at discrete points in time and the results are compared to pre-established norms or goals. The measured difference between results and norm is used as input to control actions. It affects the system in a direction that gives a result closer to the norm.

These principles are applied in quality assurance and may be illustrated by Juran's model for the control of anything, Figure 10.6. Juran's model is applicable to different types of control systems such as the control of the quantity or quality of produced goods or the control of the risk of accidents.

In the feedback control of accidents, we use different types of standards. In workplace inspections, for example, we are concerned with deviations from technical requirements defined in safety regulations, etc. Another example is feedback control of SHE performance, as measured by one or more SHE performance indicators. We will present such indicators in Part IV.

An associated term is *feed forward*. Anticipation is the heart of the feed-forward mechanism. Here, the information used as input to control the system is not obtained directly by measuring system performance but indirectly through anticipation. One typical example is when we introduce control measures on the basis of results of risk analyses.

10.5 Ashby's law of requisite variety

When we analyse the types of safety measures that are implemented at a workplace following an accident, we often notice a lack of imagination in the development and selection of measures. We see repetitive use of a few and relatively simple types of measures such as instructions to the injured person to be more careful, repair of technical faults, mounting of guards and removing of litter. The simplicity of the selected measures is remarkable when we consider the complexity of the accident sequence and the conditions at the workplace which have influenced it. **Ashby's law of requisite variety** helps us in analysing the weaknesses in a feedback control system of this type. It tells us about the relation between the variety of the types of measure that need to be employed in order to exercise full control and the complexity of the phenomenon that we want to control. It reads (Ashby in Van Court Hare, 1967): 'For an analyst to gain control over a system, he must be able to take at least as many distinct actions, i.e. as great a variety of countermeasures, as the observed system can exhibit.'

In accident control, the requisite variety of countermeasures is dependent on the systems level. At the work-system level, the 'analyst' (the worker) faces accident risks that are determined by complex and varying changes in the environment and his/her effect on it. The worker must implement a great variety of measures in order to maintain control. Top management, on the other hand, is concerned with accidents at an aggregate level, where the different short-term variations and details even off. At this level, the performance data are much less complex and, as a consequence, the control actions at this level may be less complex. Experience shows that a simple high-level decision on changes in the SHE policy will influence the complex behaviour at lower management levels such that positive SHE effects are achieved.

The successive information filtering from the work-system level through different managerial levels to top management is necessary in order to avoid information overload. It is a concern, however, that this necessary filtering should not result in distorted or biased information. If this is the case, the feedback processes of the SHE management systems will actually be counterproductive. They may reinforce wrong accident perceptions and thus result in inefficient counter-measures.

Figure 10.7 shows the results of an analysis performed by the author of safety measures documented in accident reports at seven shipyards. There were no actions documented in more than a third of the accident reports. Instructions 'to take care' or to use prescribed personal protection or method of work were otherwise the most common types of measures. Only very few measures were of a preventive type, e.g. changes in method of work, procedure or organisation to prevent recurrence. The results from the yards are typical for many workplaces and illustrate that only very few types of measure are actually employed in accident prevention.

There are means of increasing the variety of the measures taken to prevent accidents. One possible way is to train the organisation to conduct more comprehensive accident investigations.

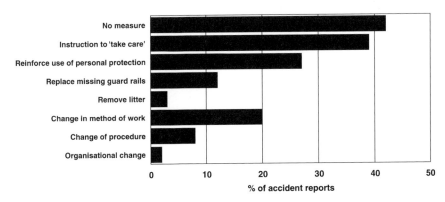

Figure 10.7 Percentage of the accident reports at seven yards containing information on safety measures by type of measure.
Source: Kjellén, 1997.

The 'defences-in-depth' philosophy presented in Section 7.1 can also be interpreted in terms of Ashby's law of requisite variety. Here, the system's designer has the clear ambition of preventing major accidents through design of barriers. This will make it less demanding for the operations organisation to manage the major accident risks.

There is another important aspect of Ashby's law of requisite variety. It has to do with the rate of change or variety in a system to be controlled. In this sense, Ashby's law of requisite variety reads: 'For an analyst to gain control over a system, he must be able to generate countermeasures at least at a rate corresponding to the rate of variety that the observed system can exhibit.'

Let us illustrate this principle with two examples. A construction site is characterised by rapid changes in technology, organisation and manning, methods of work, etc., as the construction project progresses. A process industry, such as a fertiliser plant, will look more or less the same from one day to the next. We expect changes with possible negative effects on safety to be introduced at a much higher rate at a construction site than at a process plant. It follows that we will put different requirements to the SHE information systems in these two cases, as to their ability to detect and process information on changes at the workplaces.

To sum up, Ashby's law of requisite variety implies that an analyst must show the following three abilities in order to exercise full control of a system:

1 He must be able to take at least as great a variety of actions as the observed system can show.
2 He must be able to take precisely the correct set of action alternatives to counter those changes generated by the system.
3 He must be able to collect and process information and decide on and implement measures at a rate at least equal to the rate of change of the system.

Table 10.1 Hierarchy of feedback systems arranged by order of feedback

System order	Characteristics	Traditional decision level	Examples from SHE management
0	Simple transformation without feedback	Workers	No follow-up of accidents with remedial actions, the loop is not closed
I	Simple machine with direct feedback but without selective memory	Foremen	Correction of deviations identified by accident investigations or safety inspections
II	Tactical system with memory organisation, conditional selection of pre-established plans and predictive feedback	Middle management	Starting a pre-planned eye-protection campaign following an increase in eye injuries
III	Strategic system, system that learns from experience and has ability to correct selection of plans and develop new plans	President and Staff	Change in routines, instructions, rules or design on the basis of accident experience
IV	Goal-changing system, system that learns and consciously develops, selects and implements new plans	Board of Directors	Change of safety policy and goals on the basis of accident experience

10.6 Van Court Hare's hierarchy of order of feedback

Van Court Hare distinguishes between different orders or levels of feedback control (Van Court Hare, 1967). Table 10.1 shows these different levels and gives examples from accident control. The order of feedback is an indicator of the degree of learning from previous experience. In zero and first order of feedback, there is no such learning. In the area of SHE management, this means that the same types of deviations, incidents and accidents will be able to re-occur. Long-term learning, which manifests itself in preventive actions that continuously reduce the risk of accidents at the workplaces, are here referred to as third or fourth order of feedback.

Van Court Hare's hierarchy was developed on the basis of experience in traditional industrial organisations and military organisations. His focus on the type of learning that is involved in the feedback control is, however, of interest also to safety in modern industry. In practice, we too often see easy fixes such as replacing a missing guard after a person has fallen from a scaffold or housekeeping after a person has tripped on debris. These are examples of first-order feedback. The opportunity of learning with more lasting effects has been lost.

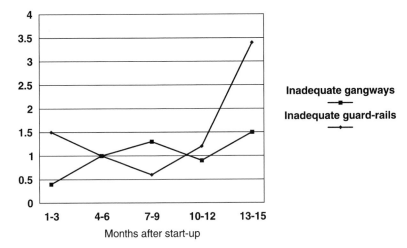

Figure 10.8 Number of actions per inspection at a construction site, by type of action.
Source: Kjellén, 1982.

Example: An analysis of the accident statistics from a construction site showed that there had been several falls on the same level and falls to a lower level during the last year. These types of accidents are typically prevented through adequate gangways and guard-rails. Workplace inspections were performed weekly at the site. An analysis of the results of the inspections showed that corrective action concerning missing guard-rails and inadequate gangways dominated the picture. The frequency of these actions (number of actions per inspection) had been relatively stable during the last year, Figure 10.8. Site management had failed to implement preventive measures in order to avoid these types of deviations reoccurring.

We regard the implementation of measures following an accident as a means for the industrial organisation to store experience from the event in its collective memory. We will here identify four different types of 'memory' for storing accident experience, corresponding roughly to Van Court Hare's hierarchy:

I.   Correction of deviations, i.e. only the 'short-term memory' is employed. The deviation may reoccur.

II. Long-term storing of experience by means of changes in design, work procedures, etc. at the workplace of the accident. The conclusions drawn from the experience will have lasting effects and may prevent a recurrence, but will be of a limited scope and will not affect accident risks at other workplaces.

III. Long-term storing of experience by means of changes in supervision of the personnel and in technical and administrative systems for production control at the functional department. These types of

Exercise: Identify type of preventive strategy according to Haddon and order of feedback according to Van Court Hare in the examples from accident reports from a yard in the table below. Discuss the efficiency of the measures (expected effect locally/at company level).

Event	Action	Type of preventive strategy	Order of feedback
A robot arm squeezed a repairman when he activated the local control by mistake.	Procedure to ensure that hydraulic equipment is depressurised and isolated before start of repair work.		
The welder took a step backwards and stepped into an opening in the deck. His back was injured.	Opening was protected by guardrail.		
The operator jumped from 0.6 m. He twisted his foot.	Common accident, no action.		
The operator stumbled on a piece of pipe lying on the deck.	The pipe was removed.		
The operator hit his head on scaffolding above the gangway. The clearance was only 1.6 m.	Warning sign was put up.		
The operator's arm got caught in a packing machine while removing a parcel that had got stuck.	Increased inspection frequency to reduce production disturbances.		
The operator was sandblasting in a narrow space. He hit his leg against a support.	Written instructions to use knee protection.		
Management was unsatisfied with the high accident-frequency rate.	Change in SHE policy to signal increased management attention to the prevention of accidents.		

change will also have lasting effects and will affect other workplaces as well.

IV. Long-term storing of experience by means of changes in the general and SHE management systems and norms (policy, goals, specifications, etc.). The changes will not only have lasting effects but will also have a wide scope and affect many workplaces all over the company.

The observant reader will note that the four types of accident experience have their counterparts in the different elements of the accident analysis framework of Figure 6.4. When we move upwards in Van Court Hare's hierarchy, we can also expect that the measures will involve a higher level in the causal-factor hierarchy.

10.7 Obstacles to an efficient learning from experience

10.7.1 Organisational defences

Deviations or errors, i.e. mismatches between our plans or intentions and the actual outcome, produce opportunities for learning. We distinguish between four different orders of feedback and the associated results of learning from experience. In practice, we notice that many opportunities for learning are lost. This is when the organisation is only able to accomplish first-order feedback, i.e. no long-term effects are produced as a result of the experience.

Argyris has analysed the organisational obstacles or defences that prevent efficient learning (Argyris, 1992). He speaks about cold and hot variables that distort human and organisational information processing and make feedback inefficient. The 'cold' variables are equal to the limitations in human information processing discussed in Section 10.3. By 'hot' variables, Argyris means individual and organisational defence strategies in order to avoid embarrassment and threat. At the individual level, we here find such psychological defence mechanisms as suppression, denial and displacement.

In exploring the organisational defences, Argyris makes the distinction between single-loop and double-loop learning, Figure 10.9.

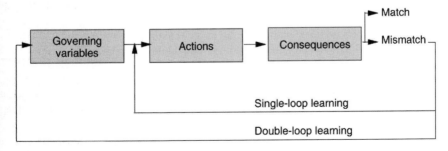

Figure 10.9 Single-loop and double-loop learning.
Source: Argyris, 1992. Reproduced by permission from the author.

Single-loop learning corresponds to the first order of feedback according to Van Court Hare. It is appropriate in routine issues, where it is necessary to get the job done. Double-loop learning, on the other hand, will affect the governing variables of the organisation. This type of learning is necessary for long-term effectiveness and ultimately for the survival of the organisation. It corresponds to the higher orders of feedback according to Van Court Hare.

Argyris' studies of organisations show that single-loop learning is dominant also in cases where double-loop learning would have been more appropriate. When the management of hierarchical organisations faces mismatches between intentions and results, they typically apply a strategy to achieve unilateral control and self-protection and discourage inquiry from others. Rationality is emphasised and feelings, especially negative ones, are suppressed. This creates an atmosphere where problems are not discussed and where errors escalate.

This type of condition will prevail as long as the costs for hiding errors that are difficult to solve or are experienced as embarrassing and threatening are felt to be lower than the costs for remedying them. This is typically the case for accidents. Management often prefers to explain the occurrence of accidents by referring to unique and situational causes rather than to system failures. They thereby reduce the immediate need to change governing variables that serve their purpose well in other instances. Accidents are rare events, and it is usually unlikely that the manager in question will be accountable for a similar type of accident again in the near future. Such a strategy may thus be rational from an individual point of view, even if the company as a whole has missed an opportunity for learning.

Processes to support double-loop learning have to come from the top of the organisation in order to create an atmosphere of trust. They involve supporting participation of all involved parties in defining purpose and in making inquiry, minimising unilateral control, creating win–win situations, and allowing feelings to be expressed.

10.7.2 Local information and the SHE information system

Argyris' analysis of organisational defences also has implications for the design of SHE information systems. A general dilemma in designing information systems has to do with our limited information processing capacity as human beings. The information provided by such systems should be comprehensive yet manageable (cf. Ashby's law of requisite variety).

At the workplace level, operators and supervisors have direct experience of accidents that are specific, concrete and rich in details. This type of information typically does not allow for generalisation and is unusable by anyone other than those who have generated it. Management at higher levels of the hierarchy, on the other hand, asks for coded summary data on accidents and quantified information on key performance indicators such as the accident frequency rate. This information represents the aggregated result of the complex processes at the local level.

Consequently, operators and first-line supervisors are exposed to direct information on accidents, whereas higher-level management receives filtered and interpreted information from the SHE information system. As a result, the different organisational levels hold different conceptions about the causes of accidents and about conditions for effective prevention. The feedback loops to each level will reinforce their respective accident conception.

When top management uses summary data for unilateral control purposes, such as when requiring unrealistic reductions in the accident frequency rate, the local level will according to Argyris respond in ways that may counteract management's intentions. They may withhold information or send distorted information through the SHE information system. They also tend to withdraw their commitment and feeling of responsibility. Typical responses from top management in such a situation are to require more detailed and tamper-proof information and to increase the orientation towards unilateral control. We thus face a situation where feedback mechanisms between the organisational levels escalate errors and maladaptive behaviour.

To counteract this 'Big brother is watching you' syndrome, we here emphasise the need to develop a spirit of co-operation and shared ownership. The SHE information system should not be used as an instrument for unilateral control but as a tool in a problem-solving process, where the different levels of the organisation participate. Top management's legitimate need for summary information for monitoring purposes has to be acknowledged. At the same time, the intrinsic limitations of this information have to be understood. It follows that the authority making decisions about remedial actions should be placed at as low a level of the organisation as possible.

10.7.3 Culpability and liability

In a police investigation of a severe accident, the investigators will look for breaches of statutory duties and other evidence of criminal negligence or wilful criminal acts. When addressing the question of culpability, the investigation will produce defensive reactions on behalf of the individuals and the organisation concerned. These reactions will counteract openness and trust and will obstruct the learning process.

Earlier, it was common to put the blame on the persons closest to the scene and with direct influence on the accidental event, i.e. workers and first-line supervisors. New legislation on requirements for companies' SHE management systems has made it possible to impose penalties on the company or its higher management (Hale *et al.*, 1997). The legislation on liability for compensatory payment in the area of product safety works in the same direction.

Police investigations and subsequent trials and negative publicity have deterring effects. They will under certain circumstances contribute to an improved safety level by focusing management's attention on its responsibilities, and may counteract complacency. It must be acknowledged, however, that organisational learning from accidents and deterrence are separate goals that often are in conflict. The industrial organisation has to address this conflict in its internal procedures for accident investigations. We will come back to this issue in Chapter 13.

At the same time, the authorities must be aware of the negative side-effects of police investigations. It has been proposed that the authorities' law-enforcement role can be played down in cases where the company demonstrates a clear willingness to learn from the accident. This principle should also apply when there is significant new learning involved as compared to a case when the same type of accident happens again.

10.8 A balanced approach

In Part II, we have been through a number of different accident-prevention principles described in the literature. Experience shows that the application of any single one of these principles alone will not lead to a substantial risk reduction. Figure 10.10 illustrates a balanced approach. The intention is to achieve a high-level SHE performance through synergetic effects by combining different accident-prevention principles.

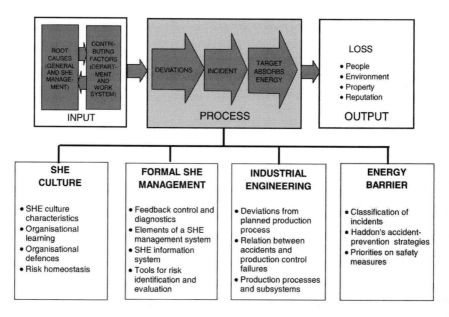

Figure 10.10 Balanced approach in the prevention of accidents.

There are four elements in this approach:

1 The *energy barrier element* has been selected because it represents a systematisation of the classical safety-engineering approach that is still valid and of high significance.
2 The *industrial engineering element* highlights the importance to SHE of controlling production and preventing disturbances. It has developed from F.W. Taylor's scientific management school of the early 1900s and, in its classical sense, is synonymous with the structural perspective of organisations. Different studies have shown that the 'control climate' determines more than many other factors the probability of all types of losses resulting from unwanted events, i.e. injuries, accidental environmental emissions, production disturbances and delays, reduced quality, etc.
3 The *formal SHE management element* is rooted in the scientific management school as well (Heinrich, 1959). It represents a necessary complement to ensure that the SHE aspects are given adequate attention by management. It involves the use of SHE information systems and different specialised tools for risk identification and assessment.
4 We will here treat the less formalistic and more human and symbolic-oriented aspects of accident prevention under the headline of the *SHE culture element*. It is the most basic of the four approaches. It helps to analyse the organisation's basic values and belief structure and how this affects behaviour and learning at the different levels of the organisation. Whereas the three previous approaches can be characterised as 'cold' in respect of rationality, this perspective also involves the analysis of 'hot' aspects. These are concerned with the organisational members' feelings of embarrassment and threat when facing danger. We will here also be concerned with the human aspects of involvement and empowerment through worker participation in the decision-making process. These are important provisions in order to accomplish effective organisational learning.

11 Requirements for a SHE information system

We will here summarise the issues dealt with in Part II by addressing a focal question: 'How to distinguish adequate SHE information system elements and total systems from those that are less than adequate?' To be able to answer this question, we need a set of criteria or requirements that the SHE information systems and their elements have to satisfy.

We will start by focusing on one important aspect of SHE information systems. Which criteria will make us able to distinguish good from poor SHE performance indicators or measures for use in feedback control? We will then shift the focus to the SHE information system in general. Here we define criteria for the evaluation of data collection, analysis and presentation of information and for an evaluation of the SHE information system as a whole.

In establishing the different criteria, we must acknowledge that one single theory or perspective will not give the complete answer. Instead, we have to understand the different needs that a SHE information system should satisfy and hence draw from the different theories and perspectives that best suits each need. We will in particular rely on the following theories and principles (compare Figure 10.10):

- Feedback control: We will here build on criteria derived from theory on measurement and on control systems and emphasise efficiency and cost-effectiveness.
- Organisational learning: Here we will emphasise aspects of the SHE information system that promote experience exchange and learning. We will also focus on provisions to minimise dysfunctional effects of organisational defences.
- Human information processing: We touched upon this area in Section 10.3. Here, we will focus on one specific aspect, the decision-maker's limited information-processing capacity (Sanders and McCormick, 1992).

An important aspect of SHE information systems is how we document and store data on accident risks. First we distinguish between two different **data types**, qualitative and quantitative data. By *qualitative data* we mean free-text descriptions. In an accident investigation, for example, this is similar to establishing a chronicle of the sequence of events.

We also have a need to code data. This is done to reduce its complexity and to allow for statistical analyses. We apply four different scales in coding or *quantifying data*:

- *Nominal scale*, i.e. a classification of the data by applying a classification scheme. A classification scheme is a **taxonomy**, when it consists of a complete set of mutually exclusive classes. Sex is an example of a taxonomy with two classes (male/female).
- *Ordinal scale*, i.e. ordering as larger than or smaller than.
- *Interval scale*, where there is the same distance between intervals. The Celsius scale for temperature is an example.
- *Ratio scale*, where there is an absolute zero (e.g. metre as a measurement of distance).

11.1 Requirements for SHE performance indicators

In Part IV, we will review different types of SHE performance indicators for use in feedback control. Table 11.1 shows the basic requirements that such indicators must satisfy (Tarrants, 1980).

The first four criteria are derived from feedback-control theory. A SHE performance indicator must be *observable and quantifiable*, i.e. it must be possible to observe and measure performance by applying a recognised data-collection method and scale of measurement. The nominal scale is the simplest type. This means that we must be able to tell whether the result represents a deviation from a norm or not. Usually, the SHE performance indicators are expressed on a ration scale of measurement. A typical example is the LTI-rate, i.e. the number of lost-time injuries per one million hours of work.

Table 11.1 Requirements for SHE performance indicators

1	*Observable and quantifiable*
2	*Valid indicator of the risk of loss*
3	*Sensitive to change*
4	*Compatible*
5	*Transparent and easily understood*
6	*Robust against manipulation*

The SHE performance indicator must also be a *valid indicator of the risk of loss*. We are especially concerned with so-called criterion-related validity. We have to ask whether the SHE performance indicator actually measures what we intend to measure, in our case the risk of losses due to accidents. Since accidents are rare events, we also look for other types of SHE performance indicators such as the frequency of unsafe acts and conditions. In risk analyses, we predict the frequency and consequences of accidents. Accident statistics are then used to validate the methods (Suokas, 1985).

The SHE performance measure must be *sensitive to change*. It must allow for early warning by capturing changes in an industrial system that have significant effects on the risk of losses due to accidents. The indicator must also be *compatible* with other performance indicators to prevent the decision-makers receiving contradictory control signals.

We now move to criteria derived from theories on human information processing and organisational learning. The SHE performance indicator must be *transparent and easily understood* in that its meaning is apparent and compatible with the users's theoretical understanding and unconscious mental models.

It must also be *robust against manipulation*. This is a variation of the validity requirement. Through SHE performance monitoring and feedback we want to achieve reductions in the risk of accidents. We expect the monitored organisation to change its behaviour in order to achieve improvements. The question is here whether the indicator allows the organisation to 'look good' by, for example, changing reporting behaviour, rather than making the necessary basic changes that reduce the risk of accidents.

11.2 Requirements for the SHE information system as a whole

We will here summarise some important requirements to a SHE information system to support feedback control and diagnosis processes, Table 11.2.

Table 11.2 Requirements for a SHE information system

A   Data collection:
1   *Reliability*
2   *Accuracy*
3   *Adequate coverage*

B   Distribution and presentation of information:
1   *Relevance*
2   *Comprehensible and easy to survey*
3   *Timeliness*
4   *Availability of the information when it is needed*

C   The SHE information system as a whole:
1   *Easily understood and acceptable methods*
2   *Promotion of involvement*
3   *Cost-efficient*

11.2.1 Data collection

We first focus on criteria developed from feedback-control theory. Efficient control requires a reliable measurement and feedback of performance. *Reliability* is here defined as the extent to which repeated measurements give the same results. A reliable reporting and counting of accidents is important if the frequency of accidents (number of accidents per one million hours of work) is used as a measure of the SHE performance of the company or department in question. Another example has to do with whether different investigators who look into the same accident will come up with the same results concerning causal factors (intra-observer reliability).

The documented facts about accident risks must be *accurate* in relation to the factual circumstances. Reliable documentation is not necessarily accurate. We may have systematic errors due to some common misconception of the nature of accidents. Or our data may be contaminated by extraneous factors such as rumours or direct manipulation.

The requirement for *adequate coverage* comes from Ashby's law of requisite variety. To exercise efficient control, we must receive data on the different technical, organisational and human factors that affect the risk of accidents and are controllable through management decisions. It also means that data collection must be flexible enough to handle unexpected threats and changes.

11.2.2 Distribution and presentation of information

We now shift the focus to the limitations in human information processing. We are here concerned with the limitations of the decision-maker as an information processor. In particular, we are concerned with the decision-maker's experience of the benefits of receiving additional information in relation to the perceived 'costs' (time, attention) in finding the information (Harrison, 1987). To avoid information overload, the information that is presented to the decision-maker must be *relevant* in relation the decision-making context. Data on, for example, the victim's domestic situation ('quarrel with husband/wife on the morning of the accident day') may be of little relevance to decision-making inside the company. Relevance is dependent on the types of use of the information and the associated needs of data (compare Figure 10.1):

- In SHE performance monitoring, the user is concerned with information on a few so-called *key performance indicators*. The SHE information system must be able to provide information on these indicators.
- In analyses of accident statistics and risk analyses, the user is concerned with a limited set of factual data on each accident and near-accident occurrence of interest. As long as the user only applies standard analysis methods, the type of data needed is rather limited and can often be determined in advance.

- When searching for answers to specific questions, the user may need to put a whole range of queries to the database, i.e. the 'memory' of the SHE information system. He/she may also be interested in results of earlier queries in order to build experience. Since it is not possible to know all types of queries in advance, the relevance will be decided by the coverage of the data.
- A common use of SHE information systems is in monitoring the status of accident counter-measures. In this case, relevant information has to do with responsibilities for actions, deadlines and the extent to which they have been met or not. The aim is to ensure that the feedback loop has been closed.

Also, to avoid information overload, the presented information must be *comprehensible and easy to survey*. Managers, especially at top level, must not be overwhelmed by detailed accident data, where it is impossible to 'see the wood for the trees'.

Excessive time delays will jeopardise the possibilities of efficient control. In general, the time it takes to detect and process data on changes and to implement corrective actions must not be larger than the rate of change of the control object (see Ashby's law of requisite variety). *Timeliness* is important in order to avoid hazardous deterioration of a system resulting from the non-detection of hazardous changes over a long period. In safety inspection, the maximum time lag is determined by the inspection frequency. This means that the inspection frequency must be higher at workplaces with rapid changes (such as a construction site) than at workplaces that remain unchanged during long periods (e.g. an office).

In using the accident frequency rate (i.e. the number of accidents per one million hours of work) as a key performance indicator, timeliness in detecting changes that affect the accident frequency rate is a concern. This has to do with the fact that accidents are a rare phenomenon. Often, there has to be a significant time lapse before an increase in the accident frequency rate is detectable for use in feedback control. We will come back to this issue in Part IV.

From the decision-maker's perspective, it is important that the *information is available* when it is needed. Computer support has significantly improved the possibilities of accessing experience data for use in decisions. We distinguish between periodic reporting, follow-up of actions and querying. In the first case, the user must have access to the SHE information system at periodic instances in time to get support in generating the necessary standard reports. Decision-makers need to access the information system easily at any time to get the status of outstanding actions. Users making queries have similar needs. When a company buys a new truck, for example, the person responsible for purchasing may be interested in reviewing earlier truck accidents for use as input to specifications.

11.2.3 The SHE information system as a whole

At this level, we apply criteria mainly derived from theory on organisational learning. To minimise extraneous influences, the methods for data collection, analysis and distribution of information must be *easily understood and acceptable* to the involved parties. It is important, for example, in the reporting of human errors, that the employees have trust in the system and that the reports are not used to punish the individual. The methods should also *promote involvement* of management and employees, and the development of a shared understanding on SHE goals, accident causes and conditions for efficient prevention.

Finally, we are concerned with *cost-efficiency* with respect to the resources necessary for the establishment and operation of the system, seen in relation to the benefits of the system. The costs are relatively easy to assess by standard methods and include investment and operations costs. In evaluating the benefits of the system, we have to consider two aspects. One is the administrative support the system gives in the handling, storing and distribution of SHE-related documentation (accident and near-accident reports, minutes for workplace inspections, meetings, etc.) and in the follow-up of actions. This support will result in savings in internal man-hours. The second benefit is the support the system yields in reducing the risk of accidents and of other types of unwanted events (production disturbances, reduced quality, etc.). It is more difficult to assess this benefit, since the SHE information system cannot be evaluated in isolation. It is recommended first to evaluate the visible costs and savings for hardware, software, internal hours, etc. In the next step, it is possible to proceed with a qualitative argumentation about the potential that the system offers in reducing the risk of accidents and other types of unwanted events. The benefits in terms of an improved quality of the working life should also be considered.

Exercise: Evaluate both the old and the new systems for feedback control of accidental emissions from a fertiliser plant presented in Chapter 4. Use the criteria presented in Tables 11.1 and 11.2 in the evaluation. Identify the most significant characteristics of the new system that explain the positive results.

Part III

Learning from incidents and deviations

In our model of a SHE information system in Chapter 1, we identified four different subsystems, i.e. data collection, data analysis or processing, a memory and distribution of information. We will here first focus on the data-collection subsystem. Chapter 12 presents an overview and Chapters 13 and 14 give details about different methods of data collection in accident and near-accident reporting and investigation, workplace inspections and SHE audits. We will then proceed in Chapter 15 to go through some basic principles for the establishment of a memory (database) on accident risks. After that, we will review different methods for data analysis based on data from accidents and near-accidents in particular. In practice, it is not always easy to separate data collection from data analysis or processing, because these activities often interact.

The principles and methods presented in Part III are reactive in the sense that they are used in an approach where we learn from incidents and deviations after the occurrence. In Part V we will look into a proactive approach, where we identify and analyse accident risks before they have been manifested in actual occurrences.

12 Sources of data on accident risks

12.1 The ideal scope of different data-collection methods

In Part III, we are concerned with the collection and analysis of data on accident risks. The term 'accident risk' is used as a collective name for factors at the workplace or in the management system that increase the risk of accidents. It includes hazards, deviations, contributing factors and root causes. There are different means of collecting data on accident risks. Accident and near-accident reporting and workplace inspections represent 'traditional' means, where data is collected 'after the fact', i.e. after the accident risk has emerged. SHE audits are now also established as a mandatory part of such a system in many companies. Figure 12.1 shows the 'theoretical' coverage of these different means of data collection. Risk analysis offers an opportunity of collecting data on accident risks before they have occurred. We will come back to this theme in Part V.

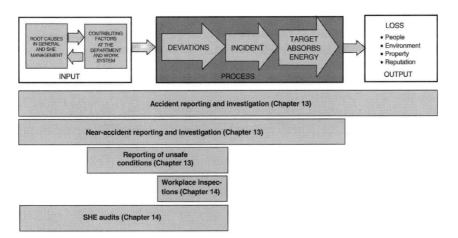

Figure 12.1 Overview of different means of collecting data on accident risks and their ideal scope seen in relation to the accident-analysis framework of Chapter 6.

12.2 Filters and barriers in data collection

At the beginning of the 1980s, the author conducted an evaluation of six large Swedish companies' accident and near-accident reporting and workplace-inspection routines (Kjellén, 1982). The study was performed before any of the companies had been affected by the recent trends within corporate SHE management. The data that actually were collected in these activities depart from the ideal scope according to Figure 12.1. There were barriers and 'filters' in the data collection at the six companies that suppressed certain types of data on accident risks.

Reporting on lost-time accidents frequently lacked data on deviations in the initial phase of the accident sequence and on contributing factors, although these types of data were available, though latent, at the workplaces. The accident investigators (usually the supervisors) lacked adequate accident perception and motivation to execute a thorough investigation. The supervisor's legal responsibility for safety, and lack of feedback, were identified as factors influencing motivation.

Only two of the six companies conducted near-accident reporting on a routine basis. At those two companies, the near-accident reports mainly contained data on technical incidents, deviations and causal factors. The lack of data on human errors was explained by the fact that the reporter's anonymity could not be guaranteed and by fear of disciplinary action.

Workplace inspections functioned as a means of tracing deviations representing unsafe conditions at the workplaces; i.e., technical deviations, poor housekeeping and faulty guards. Only exceptionally had the inspectors identified deviations relating to the method of work or contributing factors such as shortcomings in design or routines. SHE audits were not an issue at that time.

The different filters narrowed the scope of the companies' learning from their experiences. They limited the types of accident prevention and the remedial actions that were taken and their timing. The focus in accident investigations, for example, on the incident rather than on the precursory deviations will direct the investigator towards remedies to stop the energy flow, i.e. protective devices. Workplace inspections mainly resulted in corrections of deviations.

Let us analyse these findings in relation to Ashby's law of requisite variety. This law conveys an important message concerning the necessary conditions for us to be able to exercise full control of a system. If the available information on accident risks has a limited coverage, it will limit the variety of the action alternatives taken to control accident risks. The variety of the actions may well fall below the critical level; the accident risk will remain at the same level or even increase as a result. Many opportunities for accident control are lost if the investigator fails to identify how production disturbances affect the risk of accidents.

It further follows from Ashby's law that changes in production affecting the accident risk must be detected not only through accident reporting but through other means as well. This point does not only follow from the need to be proactive, i.e. to act before the occurrence of losses. Accidents are rare events and will only trigger infrequent data collection. In all practical cases, the rate of data collection and processing will be too low in comparison with the rate of change in production. We are thus dependent on near-accident reporting and workplace inspection in order to identify these changes quickly enough. It is important that the near-accident reporting is efficient in picking up significant incidents and that workplace inspections are carried out with a satisfactory scope and frequency. Changes in production may also be evaluated by risk analyses. We will come back to this issue in Part V.

It is obvious that there was significant improvement potential in many of the SHE information systems existing in industry at that time. The author conducted a similar study in the beginning of the 1990s (Kjellén, 1993). It involved an oil company that had implemented a SHE management system based on the ILCI principles (Section 5.7). It was interesting to study whether the application of modern SHE management principles would help the oil company to overcome some of the barriers and filters identified in the earlier study.

There was an increased focus on the identification of causal factors in accident investigations by use of checklists. The supervisors making the investigations had, however, a tendency to choose causal-factor alternatives that were not possible to verify and that brought about minimum obligations to carry out remedial actions. The same lack of variety in the selection of remedial actions was observed in this oil company as in the six Swedish companies a decade before. There was no obvious link between the types of causal factors that had been identified and the selection of remedial actions. It follows that the supervisors did not use the rational approach shown in Figure 6.6 in their decisions on accident prevention.

Near-accident reporting had improved dramatically, from in the order of one near accident per lost-time accident in the two Swedish companies reporting near accidents to in the order of 100 near accidents per lost-time accident in the oil company. The near-accident reports of the oil company were still dominated by technical events. The reporting of unsafe conditions was a new phenomenon. A closer analysis of the accident, near-accident and unsafe condition reports revealed the intentions of the reporters. The oil company had introduced an efficient system for follow-up of reports with remedial action. This fact was known among the employees, who used the near-accident and unsafe-condition reporting systems as a means of getting working-environment problems solved.

In designing SHE information systems, we must be aware of these filters and biases. We will here be concerned with methods for the collection and analysis of data on accident risks. How to accomplish reliable and comprehensive data collection and analysis, while taking the knowledge and motive structure of the reporters and investigators into account, will be a recurring theme.

13 Accident and near-accident reporting and investigation

13.1 Why report and investigate accidents and near accidents?

There are different reasons why an industrial organisation establishes and maintains routines for the reporting, investigation and documentation of accidents and near accidents:

- Meeting regulatory requirements. Most countries have mandatory provisions for accident reporting within the company and for reporting to the authorities, see further Section 2.2.
- Compensation to the victim. An occupational accident has to be reported to the national social security office or to the insurance company in order to ensure that victim receives compensation for lost wages, medical bills and for suffering, see Section 2.2. Also accidents resulting in material or production losses have to be reported to receive payment from the insurance company.
- Learning from the occurrences. Accidents and near accidents are unwanted occurrences. At the same time, they represent invitations to learn about the hazards at the workplace and about weaknesses in the systems for the control of such hazards (Kletz, 1994). Near accidents also give important experience about accidents that have been avoided by the operators through proper recovery actions. By using such experience properly, the organisation will be able to improve its performance in the area of accident prevention. This aim will be the focus here.
- Monitoring of the SHE performance. Accidents and near accidents are used as input in important measures of the SHE performance. We will revert to this aspect in Part IV.
- Creation of positive safety attitudes and alertness. Severe accident occurrences will shape the accident perceptions of those concerned. A well-functioning programme for accident and near-accident reporting, investigation and follow-up will support the development of a positive climate for safety-related behaviour and awareness.

There are also disciplinary and juridical reasons for the investigation of accidents. We are here considering rule violations and criminal negligence. It is the duty of the workers to follow the work and safety rules. An accident investigation will often reveal violations of such rules. We also have to consider the legal consequences of the fact that safety is a line-management responsibility. Supervisors and managers at higher levels may be held accountable for the accident in cases where the investigation reveals substandard practices or conditions for which management is responsible.

Accident investigations with the purpose of allocating accountability in order to undertake disciplinary or legal actions are conducted both internally in the company and by the police. Such purposes will influence the search for causes and restrict the range of viable remedial actions. They will create an atmosphere of blame and guilt that will induce self-protective behaviour on the part of the personnel involved, such as the hiding and distortion of information. It will also affect the personnel's willingness to report accidents and near-accidents in the future.

We will here de-emphasise the disciplinary and legal purposes of an accident investigation. Investigations with such purposes should, if possible be avoided altogether or conducted separately from the investigations with the purposes of learning, monitoring and motivation.

13.2 Investigations at three levels

An accident and near-accident investigation is a diagnostic process, involving: (1) reporting of the event and mapping the sequence of events (fact-finding), (2) identification of basic causes (3) development of remedial actions and (4) implementation and follow-up. The accident-investigation stairs in Figure 6.6 illustrated this process. All investigations have to go through these four phases in order to close the loop, i.e. to ensure that the experiences are utilised to reduce the risk of accidents.

In SHE practice, we will not be prepared to follow the same procedure and use the same amounts of resources every time an accident or near-accident occurs. Certain priorities have to be made in order to focus on the vital accidents and near accidents that offer the most significant opportunities of learning.

A comprehensive approach is here recommended, Figure 13.1.

1   All reported incidents (accidents and near accidents) are investigated immediately at this first level by the supervisor and safety representative.
2   A selection of serious incidents, i.e. frequently recurring types of incidents and incidents with high loss potential (actual or possible) are subsequently investigated by a problem-solving group.
3   On rare occasions, when the actual or potential loss is high, an accident-investigation commission carries out the investigation. This corresponds, for example, to incidents with Grade 4 or 5 according to Table 6.4. The

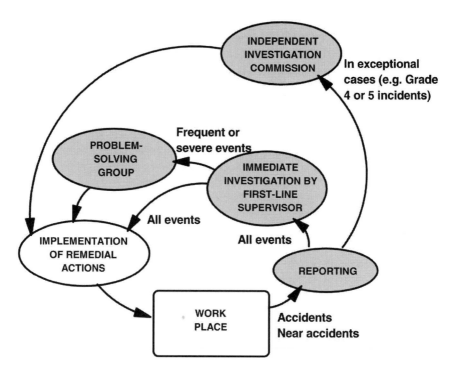

Figure 13.1 Comprehensive approach for accident investigations at three levels. For definition of Grade 4 and 5, see Table 6.4.

commission has an independent status in relation to the organisations that are responsible for the occurrence. Such investigations have many similarities to SHE audits.

We may expect different outcomes from the investigations at the three levels, roughly corresponding to the levels of feedback according to Van Court Hare. This means that the supervisor's first report (level 1) corresponds to the first order of feedback, i.e. correction of deviations. The problem-solving group (level 2) will be concerned with the change of contributing factors at the workplace and department, roughly corresponding to the second order of feedback. An independent investigation (level 3) will look at these types of measures as well, but will have the duty to evaluate root causes and to come up with recommendations corresponding to the third and fourth orders of feedback.

The judgement to proceed with the investigation at level 2 or 3 should not be left to the immediate supervisor alone. Rather, this is the duty of the middle or top management. An alternative approach is to appoint a screening group with representatives of management, employees and SHE staff which has the authority to decide about the level of investigation.

The rationale for this approach is:

- Ensuring adequate focus and use of resources in relation to the events that have a high learning value and, at the same time, ensuring that all accidents and near accidents are investigated. The three-level investigation routine meets this dual goal.
- Allowing for local use of accident and near-accident experience. Accidents and near accidents are rich in information and it is not possible to record all this information for use by higher-level management and staff officers in decision-making. Problem-solving has to take place in a group setting, where there are members with direct experience of the context in which the events have occurred.
- Ensuring an adequate investigation into sensitive issues. There are often circumstances around an accident that are sensitive, especially those related to the distribution of responsibilities and to the safety culture at the workplace and in the company. We cannot expect the involved parties to make a comprehensive and unbiased investigation into such factors. When it is important to go into depth with these factors, especially in cases of severe occurrences, there is a need to ensure that the investigation is carried out by an independent body.

Example: There was a severe accident at a yard during maintenance of a crane. A worker fell 6m from the crane boom. This accident triggered the setting up of both a local investigation team and an independent investigation commission. The local team focused on the immediate causes of the accident and the absence of adequate fall protection. They proposed changes in the work instructions to ensure that the necessary safety precautions were taken. The commission looked into root causes involving management acceptance of unsafe behaviour and lack of routines to review new jobs from a safety point of view. The recommendations by the commission were thus more far-reaching. Top management decided not only to request a follow-up of the commission's recommendations at the workplace. All other departments with similar types of job had to report on how they implemented the recommendations.

13.3 Reporting

Reporting aims to bring the incidents to the attention of the person(s) in the organisation responsible for investigation and documentation of the events. A person involved in the event (the victim or a witness) usually does the reporting. We will here be concerned about two important aspects concerning the reporting of incidents, i.e. the criteria used in the reporting and the reliability of the reporting.

The *reporting criteria* define what types of events are to be reported. Such criteria may be formalised but may also be informal and based on a shared understanding within the organisation. Reporting criteria for accidents are usually defined on a consequence scale of measurement such as whether the accident involves lost time after the day of the event or not (yes/no). This particular criterion is relatively simple to communicate and apply, since it also has implications as to the victim's right to compensation for sick leave. A problem is its limited coverage. Minor accidents with potentially severe consequences will pass undetected and we will miss an opportunity of learning from such experience.

Alternatively, we may base our reporting criteria for accidents and near-accidents on subjective judgements of potential losses. Such criteria involve problems of inter-subjectivity. Different persons may have varying opinions on what constitutes a potentially severe incident.

This brings us to the question of *reporting reliability*. This is expressed on a scale from zero to 100 per cent and is defined as the number of reported incidents in relation to the 'true' number of incidents (as defined by the reporting criterion). The reporting criteria and the reliability in the reporting are closely linked. To achieve a high degree of reliability in the reporting, well defined and easily understood criteria have to be applied.

The reporting reliability is also affected by the severity of the event. We may expect a high reporting reliability when applying reporting criteria corresponding to a 'high threshold' on the consequence scale. Reporting of fatalities is an obvious example. There will be few events that meet such a reporting criterion, and the number of opportunities for learning will be very low. At the other end of the consequence scale we find such criteria as 'to report all accidents and near accidents'. In this case, it is up to the individual employee to determine whether an event is significant enough to be reported or not. It follows that we here expect a very low and even undefined reporting reliability. When we apply a low reporting threshold, many 'trivial' events with a low learning value will be reported. We run the risk of information overload that ultimately will discredit the system. To counteract this problem, an efficient system for the classification of the reports with respect to severity is crucial.

13.3.1 Reporting to the authorities

Most countries have provisions for the reporting of occupational accidents to the authorities and/or to insurance companies, see Section 2.2. The aim is to ensure compensation to the victim for sickness leave, medical treatment, etc., and to provide the authorities with a basis for regulation of the conditions at the workplaces. As an example, the table below lists the requirement to report accidents to the authorities in Norway. Other countries have similar provisions.

Table 13.1 Accident-reporting requirements in Norway

Authority	*Occurrence*	*When?*	*By whom?*
Police	Fatality, serious injury	Immediate notification	Employer
	Serious pollution of the environment	Immediate notification	Principal enterprise
Directorate of labour inspection	Fatality, serious injury	Notification and report	Employer
Health insurance office	Injuries resulting in medical treatment or three days of absence or more	Report within three days	Employer
Environmental pollution authorities	Accidental contamination of the environment	Immediate notification, report within 24 hours	Principal enterprise
Directorate of fire and explosion prevention	Fire, explosion	Immediate notification, report on request	Employer

13.3.2 Problems of under-reporting

Earlier studies in the Scandinavian countries have shown that the reporting of accidents to the authorities is unreliable and that on average about 50 per cent of the reportable accidents never are reported (Kjellén *et al.*, 1986). Differences between the countries could be explained by varying benefit levels, as determined by the workmen's compensation system. There are reasons to assume that the accidents that are not reported to the authorities are not documented and followed up at company level either. This means that important experience is lost.

These conclusions are supported by a more recent study from the USA, indicating an under-reporting in the order of magnitude of 50 per cent (Weddle, 1996). The workers stated reasons for not reporting such as that the injury was too minor to be reported and that they were afraid of being judged as careless or accident-prone. Older workers and workers that had worked for a long time at the same job were less willing to report injuries.

The severity of the accident and the size of the company affect the reporting behaviour as well. Under-reporting decreases with increased severity, as measured e.g. by the number of days of absence. Senneck has shown a simple model of the relation between reporting reliability and the severity of accidents, Figure 13.2. This relationship offers a means of assessing a company's reporting system. We start by looking into the average severity of the accidents in the company's files. If this analysis reveals a high average severity, we must expect the company to be de-emphasising accident reporting with a low reporting reliability as a consequence.

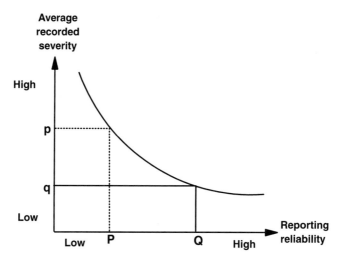

Figure 13.2 Relation between the reporting reliability and the average severity of accidents.

Source: Senneck, 1973.

Q = the reliability is high which results in a low average severity.

P = the reliability is low which results in a high average severity.

Research shows that large companies and the government sector have more reliable reporting than small companies (Kjellén *et al.*, 1986). The reporting reliability also varies between different branches of industry. Differences in reporting reliability are to a large extent explained by differences in the incentives.

Workers' and supervisors' inclination to report accidents can be interpreted in terms of **behavioural theory**. This theory has primarily been used in explaining why people deliberately violate safety rules (Saari, 1998). Behavioural theory focuses on the consequences of behaviour and how these affect people's judgements in relation to recurring situations and their ingrained action patterns in these situations. We can illustrate this theory by a simplified example, where a person has two action alternatives to choose between. One of the alternatives is considered 'safe' and the other 'unsafe', e.g. to use personal protective equipment or not. Experiences show that people usually choose the 'unsafe' alternative when this has positive foreseeable consequences such as savings in time and effort. The possible negative consequence of 'unsafe' behaviour, i.e. an injury, does not follow each time the 'unsafe' act is conducted. The expected positive consequences of 'unsafe' behaviour are much more frequent than the negative consequences and this feedback mechanism will affect people's behaviour. Deliberate violations will thus become an integral part of a skill.

Let us now turn to the present case. Workers' and supervisors' inclination to report accidents can be analysed in terms of action alternatives, the 'safe' act to report an accident and the 'unsafe' act to refrain from reporting the event. For each action alternative, there are positive as well as negative consequences. These are summarised below with focus on the workers (victims).

In order to achieve reliable reporting of accidents, the different incentives and disincentives need to be analysed. The following questions should be asked:

- *Are the incentives that promote the reporting of accidents sufficient?* The compensation to the victim is an important incentive. High reporting reliability will follow if the victim receives economic benefits from it. These benefits may be immediate (higher sickness pay) or delayed (injuries emerging at a later time can be attributed to the accident). Other examples of such incentives are signs of appreciation and an efficient follow-up of the reporting with remedial actions and feedback of the results to the reporter.
- *Have the disadvantages of reporting been reduced to an acceptable level?* Measures need to be taken to reduce the victim's feeling of guilt and blame. The extra time needed for reporting can be reduced through use of simple reporting forms.
- *Are there any negative consequences to the individual if accidents are not reported?* The employees should be informed about the negative consequences of not reporting accidents immediately, for example, the possibility of not receiving compensation in the future.
- *Have the advantages of not reporting accidents been reduced?* Incentive schemes in industry, such as safety prizes for 'zero accidents', may cause under-reporting. It is important to evaluate such schemes in relation to the need for reliable accident reporting.

Examples of practical measures to improve reliability in accident reporting:

- Criteria as to which accidents to report should be well defined and easily communicated. It is recommended that they include accidents resulting in lost-time, transfer to another job or restricted work and medical treatment other than first aid.
- Simple and well defined reporting responsibilities and routines. It must be made clear that all employees are responsible for reporting accidents to their immediate supervisor.
- There should be a written instruction on accident reporting. It should be available to the supervisors and employees and the form should be readily available at the supervisor's office. Employees

> should be informed about the routines when they start work and at regular intervals.
> - De-emphasis on blame and guilt in the subsequent investigations and careful handling of sensitive information.
> - Feedback of results of the investigation to those concerned.
> - Avoid using incentive schemes that may counteract the reporting of accidents.

Exercise: Perform the same analysis of pros and cons seen from the perspective of a supervisor who faces two action alternatives. One is to promote accident reporting among the employees and the other to discourage reporting. Discuss how an increased focus on safety from top management may affect the supervisor's behaviour. Such focus may, for example, find its expression in ambitious company goals concerning a reduced accident frequency and requirements to give top management feedback on actions taken after accidents to avoid a recurrence.

13.3.3 Near-accident reporting

13.3.3.1 Why report near accidents?

According to conventional wisdom among safety experts, accidents and near-accidents are attributed to the same types of causal factor. The consequences of an incident or loss of control of energy is largely fortuitous. Pure chance will dictate whether an incident will result in an accident or not and the degree of severity of the accidents. Due to the fact that near accidents are more frequent, these will reveal causal factors earlier and thus make accident prevention possible before the occurrence of an accident (Tarrants, 1980). These conceptions are rooted in the iceberg theory according to Heinrich, Figure 13.3. He showed that out of 100 incidents, 0.3 resulted in major lost-time injuries, 9 in minor injuries and the rest in no injury at all. These figures are averaged over different industries and different types of incidents and causes.

The results of more recent studies contradict this hypothesis. Studies of the consequences of incidents have shown that the severity is dependent on the energy involved (Shannon and Manning, 1980; Salminen *et al.*, 1992). Fatalities, for example, are more likely when the victim is struck by a falling object or hit by a moving vehicle. It follows that there is no generally valid ratio between the number of accidents and the number of near accidents at a workplace. This is because the distributions of incidents differ between different types of workplace, mainly depending on the types of energies involved in production.

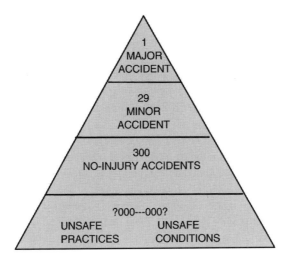

Figure 13.3 Iceberg theory.
Source: According to Heinrich, 1959.

Near accidents with the potential of resulting in severe injuries are important sources of information in accident control. Near-accident reporting also has other advantages (Carter and Menckel, 1985; Van der Schaaf *et al.*, 1991):

- Near accidents are less emotionally charged and are less likely to trigger organisational defences against openness and needs of change.
- Some near accidents provide important experience on how losses are avoided through recovery actions.
- Near-accident reporting creates an increased knowledge and awareness among the employees about accident risks.
- It also contributes to an improved co-operation between employees and management on safety matters.

Experiences from industry are used as 'evidence' on the positive effects of near-accident reporting. Figure 13.4 shows a typical example. In the course of a seven-year period, the yearly number of lost-time accidents went down by almost a factor of ten at an industrial plant. During the same period, the near-accident reporting frequency increased by about the same factor of ten.

These results have been interpreted as proof of the benefits of near-accident reporting. There are other possible explanations. A change, for example, in top management attention may explain both effects. There is also another concern. The plant did not experience a similar reduction in the frequency of severe accidents, indicating that near-accident reporting will not necessarily pick up the causes of these types of accidents.

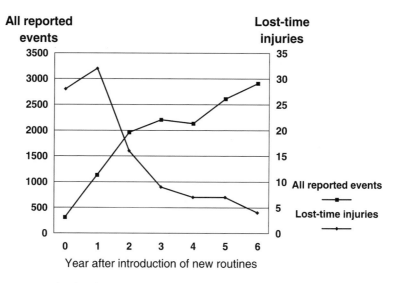

Figure 13.4 The development in the frequency of lost-time and first-aid accidents and near accidents at an industrial plant after the introduction of new near-accident reporting routines.

13.3.3.2 Provisions for reliable near-accident reporting

We have in Chapter 12 discussed research findings that show that near accidents involving human errors are reported to a lesser extent than near accidents with technical causes. The employees' feeling of guilt and the risk of blame that they run when reporting their own errors may explain these results. This leads us to the question of advantages and disadvantages to the employees in the reporting of near accidents.

In Section 13.3.2 we identified different incentives and disincentives in the reporting of accidents. In principle, the same types of incentives and disincentives influence the reporting of near accidents. This may be illustrated by an example from civil aviation.

Example 1: Figure 13.5 shows an analysis of the incentives that the different participants in the US Aviation Safety Reporting System (ASRS) receive. ASRS is a system for the reporting of incidents and human errors for use in accident prevention. To promote reporting, the following measures have been taken:

- *The research institute responsible for the reporting system (NASA) is separate from the enforcement agency (FAA);*
- *Confidentiality and de-identification of reporters;*
- *Rapid and concrete feedback to the reporting community (pilots and air traffic controllers); and*
- *Easy for reporters to complete and file reports.*

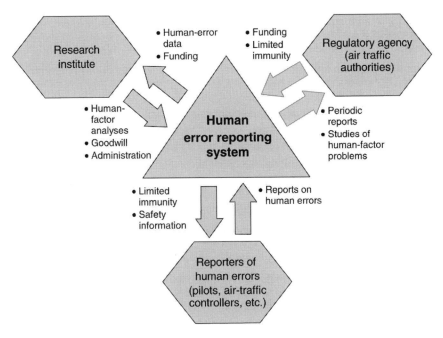

Figure 13.5 What different participants in a human-error reporting system give and receive in return.
Source: Reynard, 1986.

A successful reporting of human errors has been accomplished by giving pilots and air-traffic controllers benefits when reporting such events. Most important is the fact that they receive immunity from punitive actions as soon as they have filed a near-accident report. This limited immunity covers negligence but not criminal acts such as drug trafficking. They also receive feedback on the results of their reporting.

Example 2: A well functioning near-accident reporting system at a nuclear plant was jeopardised through changes in the reporting incentives (Ives in Van der Schaaf et al., 1991). Use of performance measures became a management focus and it was decided to measure safety performance by the number of reported near misses, especially those assessed as severe (i.e. grade 1 on a scale from 1 to 3). A reduction in the reporting frequency by 50 per cent immediately followed the introduction of this new monitoring system. A closer analysis showed that the situation at the plant was unchanged but that the under-reporting had increased. Time-consuming and unproductive discussions between the plant and head office on the classification of events also followed.

The lessons from these examples are that the different incentives and disincentives must be carefully analysed when designing near-accident reporting routines. The willingness of the workers to report near accidents is dependent on the extent to which there is openness and trust between workers and supervisors on these issues. The workers must feel a shared ownership to the reporting system and be confident that the results are used for the purpose of learning and improvements. The opposite, a 'Big brother is watching you' attitude, where the purpose is to monitor the organisation in order to allocate blame and punishment will jeopardise any near-accident (and accident) reporting system.

It is difficult to measure near-accident reporting reliability, because the definition of what is a reportable near accident is fuzzy. It is mainly left to the discretion of the reporter. Instead, we apply measures of the *reporting propensity*, i.e. the employees' willingness to report.

The most commonly used measures are:

* *The number of reported near accidents per reported lost-time injury.* According to Heinrich, this ratio should be of the order of 300:1. In practice, far lower ratios are acceptable. The total number of recordable incidents (lost-time and first-aid accidents and near accidents) per lost-time injury (TRI/LTI) is a variation of this measure.
* *The number of reported near accidents per employee and year.* This reporting frequency is often of the order of 1 to 2.

A high reporting ratio or frequency is not necessarily a goal in itself. Near-accident reporting may have counterproductive effects. This is the case when it directs the attention from important safety problems to minor one's and drowns the decision-makers with information that exceeds their capacity. It is important to combine an efficient near-accident reporting system with tools to help in making priorities and establishing focus on the vital few near accidents. Table 6.4 presents such a tool. The reporting of 'trivial' near accidents with a low learning value should not necessarily be promoted for the purpose of achieving a high near-accident/accident ratio.

The reliability of near-accident reporting may be improved by applying the same general principles as those for accident reporting. Measures directed especially at the improvement of near-accident reporting in industry include:

* Provide possibilities for anonymous reporting.
* Avoid blame and provide immunity against disciplinary action following the reporting of human errors.

- Simple and easily understood forms for self-reporting.
- Efficient follow-up of reports with remedial action and feedback to the reporters with visible results. The employees will use the near-accident reporting system for the purpose of achieving improvements in their working environment if they know that their reports are taken seriously.
- Communicate realistic goals on the reporting frequency to the employees and follow up and feedback on the results.

Exercise: FOUNDRY Inc. delivers aluminium wheels to the car industry and has 300 employees. The number of lost-time injuries per million hours of work (LTI-rate) has remained about the same during the last five years. Last year there were 30 lost-time injuries and 18 near accidents. The customer, a large car manufacturer, requires improvements in SHE results at the foundry. Management decides to work actively to improve the SHE results by reviewing the near-accident reporting routines. Your duty as SHE manager is to give recommendations to management on how to secure an adequate reporting and follow-up of near accidents. Your task is to plan and lead a quality-improvement project by applying Deming's circle:

1. *Choose the members of the project from the group of students and give them different relevant roles in relation to the purpose of the project (worker and supervisor representing different departments, etc.).*
2. *Go through the first three steps of Deming's circle (plan, do and check).*
3. *Analyse factors that affect near-accident reporting in a positive and negative way seen from the point of view of (a) the supervisors and (b) the operators.*
4. *Suggest realistic reporting goals during the current year.*
5. *Propose measures to improve near-accident reporting.*
6. *Propose an action plan for implementation and follow up.*

13.3.3.3 Reporting of unsafe conditions

The employees' self-reporting of unsafe conditions by filling in a form is complementary to the traditional informal notification of the immediate supervisor. This self-reporting follows the same basic principles as those of near-accident reporting, and usually the same form is employed. Information on deviations and shortcomings in design and routines at the workplace are documented in this way. As in near-accident reporting, an efficient follow-up of the reports on unsafe conditions will reinforce the reporting behaviour. It also represents a channel for communication on SHE problems, when the employees do not fully trust the informal system via their immediate supervisor.

13.4 Immediate investigation and follow-up

13.4.1 Quality of the supervisor's first report

The routine investigation into accidents and near accidents is usually carried out by the first-line supervisor and safety representative, or in case of near accidents, by the persons directly involved. In Chapter 12, we discussed problems with the quality of the supervisors' accident reports. The documentation in the reports focuses on the late phase of the sequence of events. There is often an arbitrary and incomplete documentation of deviations in the early phase of the sequence of events and of causal factors. This documentation thus suffers from problems of reliability and coverage.

If there is a substandard identification and documentation of deviations and contributing factors, we must expect the development of remedial actions to suffer as well. Figure 10.7 showed typical results from accident investigations in industry. As illustrated by this figure, the final result, i.e. the recommendations on remedial actions, are characterised by a poor variety and a low degree of learning (compare Ashby's law of requisite variety and Van Court Hare's feedback hierarchy). There is scope for large improvements in the quality of accident and near-accident investigations.

A number of factors affect the quality of accident and near-accident investigations in a negative direction (Kjellén *et al.*, 1986). The following factors are to a varying extent present, especially in smaller and medium-sized companies:

- *De-emphasising the use of the report in accident prevention.* The investigations of accidents are often confined to the filling in of the form for insurance purposes and are given minimum attention.
- *The investigators' knowledge about accident causal factors and investigation techniques.* The investigators usually have fragmentary knowledge about accident-investigation methods. Stereotypical attributions of accident causes also affect the quality of the investigation.
- *Punishments for making a good investigation.* A number of factors are often present at the workplaces that discourage the execution of a thorough investigation. One example is the fact that the first-line supervisor has a juridical responsibility for safety at the workplace. An in-depth investigation may reveal violations of regulations and procedures for which the supervisor is responsible. The supervisor will also be

responsible for implementing the results of the investigation. He may want to avoid safety measures that he experiences as too 'costly' in terms of working hours and money.

- *The time span between the occurrence of the event and the execution of the investigation.* Statistics from the Swedish Information System on Occupational Accidents show that 50 per cent of the investigations are carried out only after the workman's compensation board has filed a reminder. This means a delay in the investigation of at least two weeks. Such delays result in loss of information.

Research has explored different possibilities of improving the quality of the investigations. In this Section, we will focus on the support offered by analytic tools and checklists in improving the quality of the supervisor's first report on accidents and near accidents. The next Section will review findings concerning the use of group problem-solving techniques in accident and near-accident investigations.

13.4.2 Use of checklists and reporting forms

In Chapter 6, we looked into some of the checklists that have been developed for use in accident and near-accident investigations and in the coding of data from such investigations. Two alternative principles lie behind the design of checklists on deviations and causal factors. One principle is that the investigator makes an 'open' analysis by asking about the immediate and underlying causes of the event. The checklist is used to check the results for comprehensiveness and should support the investigator in considering all different circumstances of the event. The results are documented in free-text descriptions. The checklist may later be used for coding purposes. An alternative principle is to design a checklist with a multiple-choice format that is used during the interview to catch all the essential information in its detailed categories of deviations and causal factors. The circumstances of the event are hence documented by checking the correct alternative in the checklist.

The checklists on deviations and causal factors in Tables 6.8 and 6.11 have been designed according to the first principle. Different studies show that the use of tools designed according to this principle improve the reliability and coverage of data collection in accident and near-accident investigations (Kjellén, 1983).

The ILCI model has been applied in the design of accident investigation checklists according to the second principle. Such checklists are relatively easy to apply in accident investigations and in the subsequent analysis and coding of the results for statistical purposes. They suffer from a poor reliability (different investigators will come up with different results) and from an inadequate coverage of the detailed circumstances of the accidents and near accidents.

Experiences of the British Airway Safety Services' (BASIS) near-accident reporting system support the use of a form for self-reporting with open-ended questions according to the first principle (Reason, 1997). BASIS first tried a form with questions concerning types of human errors and contributing factors, where the answers were given in a multiple-choice format. The resulting data suffered from poor validity and reliability.

This was because the flight-crew personnel did not have a good understanding of the underlying concepts. The introduction of a new form with open-ended questions resulted in a dramatic improvement in the reliability of the data. It was also possible to collect data on a variety of issues that had not been covered before. The introduction of the new form was accompanied by a change in the organisation, where the causal-analysis work was moved from the flight-crew personnel to a team of human-factor analysts.

The use of detailed checklists is no panacea for improving the quality in accident and near-accident investigations. Such checklists may even be counterproductive if they replace thoughtful reflection on experience, and documentation of this in free-text descriptions.

Table 13.2 shows an example of an accident-investigation form which can be used in combination with the checklists in Tables 6.8 and 6.11 in documenting the sequence of event and immediate causes.

Good investigation tools such as simple and robust accident-investigation forms and checklists are important but not sufficient measures to ensure satisfactory investigations by supervisors and safety representatives. It is also important to provide for:

- Simple routines and clear responsibilities;
- Adequate education and training of the supervisors and safety representatives in accident-investigation methods;
- Quality control of reports and feedback of results; and
- Careful evaluation of the employees' and supervisors' incentives to produce a good investigation when designing and evaluating investigation routines. The questions presented in Section 13.3 are of relevance also when considering the conditions for reliable and comprehensive investigations.

Below are some guidelines for the immediate investigation:

1 The reporting routines should ensure that the supervisor gets an immediate notice about the occurrence.
2 The safety representative should participate in the investigation.
3 The site for the occurrence should remain unchanged until the immediate investigation has been carried out.

4 In interviewing the victim (if possible) and witnesses, explain that the purpose of the investigation is to learn and that the question of guilt is irrelevant.

    a Let the interviewee tell his story by describing the sequence of events in his own words.

    b Rehearse the explanation by putting control questions such as: 'What activities were the victim and his work fellows involved in immediately prior to the event?' 'Was everything normal or were there any irregularities or deviations?' 'What triggered the incident?' 'What were the consequences and why?'

    c Keep the checklist on deviations (Table 6.9) in mind when making sure jointly with the interviewee that the important circumstances have been identified.

    d Try to establish a mental picture of the sequence of events. Is it logical and complete?

    e Ask for the interviewee's opinion about the causes of the accident.

    f Ask for suggestions for remedial actions.

5 The actions from this first investigation are usually of limited scope and involve correction of deviations. Consult the checklist on contributing factors (Table 6.12) when developing more lasting measures.

6 Ensure to close the loop by prompt follow-up of the investigation with remedial actions.

Exercise: Decide on the different roles in an investigation into the accident in Table 13.2 including the supervisor and safety representative performing the investigation, the co-worker and the crane driver. Perform the investigation by carrying out interviews. Apply the checklist in Table 6.8 to secure information on deviations. Propose immediate remedial actions.

13.4.3 Displaying the sequence of events

Various analytic tools have been developed for the purpose of displaying the sequence of events in a graphical format. They have different aims:

- To help in checking that the relevant data about the sequence of events have been documented and that the data are logically coherent.
- To act as a communication tool, when groups discuss the sequence of events and causal factors.
- To provide coded data as input to accident databases.

Table 13.2 Example of an internal form for an immediate investigation. An accident at a construction site is shown

Report on unwanted occurrences					
Type of event	☐ First-aid injury ☒ Lost-time injury ☐ Uncontrolled emission		☐ Material damage ☐ Fire / explosion ☐ Near accident		
Time and place	Place *External gangway, house 4*				
	Date *20/12 1999*	Time *09.20*	Department *Lysaker construction site*		
Injured/ ill employee	Employee no. *12345*	Name *Anders Andersen*			
	Date of birth *22.03.69*	Occupation / position *Concrete worker*			
	Department / employer *Oslo Construction AS*				
Sequence of events	What was the employee doing at the time of the event? *Assembling concrete slabs*				
	How did the accident occur? *While Anders Andersen and Bjarne Boe were assembling concrete slabs on top of concrete beams, a slab got into the wrong position. Andersen went out on the beam to adjust the slab. The beam was icy, he slipped and fell to the floor below.*				
	Machinery, tools or materials involved? *Spit; Crane type 22B*				
Injury or damage	Describe injury / damage *Broken ribs, punctured lung*				
	☐ Eyes ☐ Head, face ☐ Back ☐ Trunk ☐ Arm	☐ Hand, wrist ☐ Leg ☐ Feet, ankles ☒ Internal ☐ Other	☐ Amputation ☐ Burn, scald ☐ Concussion ☐ Crushing ☒ Cut, puncture	☒ Fracture ☐ Hernia ☐ Bruise ☐ Sprain, strain ☐ Other	
Deviations	Describe deviations from regulations, instructions, common practice *No fall protection*				
Causal factors	What were the causes? *Time pressure, crane needed elsewhere. Fall protection not in place.*				
Actions to prevent recurrence	Immediate actions? *Temporary guard-rails*				
	Actions to prevent recurrence *Review planning of fall protection in new jobs* *Evaluate need for more cranes*		Responsible *Site manager* *Site manager*	Due date *15/1/00* *15/1/00*	
Signatures	Date *20/12/99*	Supervisor *Karl Karlsen*	Date *20/12/99*	Safety representative *Lars Larsen*	

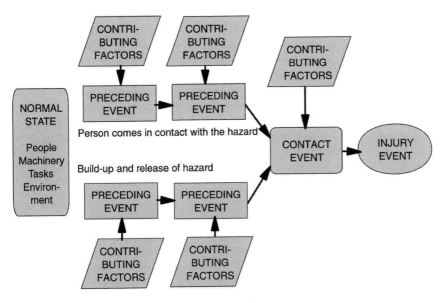

Figure 13.6 The Finnish accident-analysis model.
Source: Tuominen and Saari, 1982.

A typical example is the so-called **Finnish model for accident analysis,** Figure 13.6. It displays accidents in a standard format in terms of two separate event sequences. One of these considers the build-up of hazardous energy and the second the movements of the victim in relation to this energy source. The accident occurs when the person and the released hazard come in contact. We recognise the energy model presented in Section 5.4 in this accident-analysis model.

MAIM, the Merseyside Accident Information Model has many similarities to the Finnish accident-analysis model (Shannon and Davies, 1998). Both models are primarily designed for the purpose of recording and storing information from the supervisors' first reports in a database. The Finnish model has also been implemented in accident-investigation forms, where the investigator is requested to record the information directly into the different boxes of the model. Experiences from such an application are mixed, since it requires investigators with high analytic skills.

Tripod-Beta is a third variant based on the Tripod model of Section 5.2 (Reason, 1997). The associated PC software makes it possible for the analyst to document the findings of the investigation and to produce a graphic incident tree with the immediate circumstances of the accident (hazards and barrier failures) and the underlying active and latent failures and preconditions.

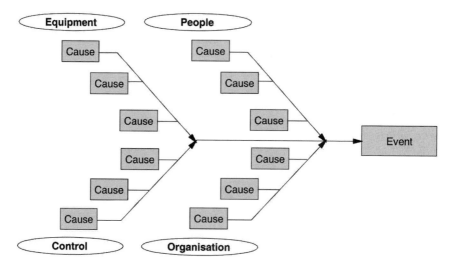

Figure 13.7 Basic outline of a cause–effect diagram.

The **logical tree diagrams** (or 'why-why' diagrams) also offer an opportunity for graphical display of the accident sequence. Figures 5.9, 6.5 and 13.10 show different examples. The so-called **fishbone** or **cause–effect diagram** serves this purpose as well. This standard quality-management tool was developed by Kauro Ishikawa in the 1950s to analyse causes of poor quality (Wig, 1996). Figure 13.7 shows a general outline of a fishbone diagram.

A **STEP-diagram** displays the sequence of events in a graphical format by applying the Multi-linear Events Chartering Method, see Section 5.3 (Hendrick and Benner, 1987). An example is shown in Figure 26.2.

The procedure to establish a STEP-diagram involves 14 activities:

1   A blank piece of paper (A3 or larger) is used.
2   All involved actors (persons, machines, materials, etc.) are identified and listed in a column on the left of the sheet.
3   Mark the start and the end of the sequence of events.
4   Put each event on a 'Post-it' slip. The events include deviations as well as expected events in accordance with plans, etc. Identify which actor is 'responsible' for the event.
5   Check that all relevant events have been documented.
6   Locate the event slips at correct position on the sheet in relation to the actors and the time sequence.

7 Describe the end state for the actors.
8 Extend with more actors and events if necessary.
9 A further breakdown of the events may prove necessary.
10 Identify missing information and fill in with new events. Consult witnesses if necessary.
11 Identify events that represent deviations.
12 Determine which effect each event has had on actors other than those responsible for the event.
13 Show relations by using arrows.
14 Review the sequence of events and identify conditions that have affected the events.

Exercise: Perform a STEP-analysis of the accident case in Table 13.2 by using results from the deviation analysis. Discuss advantages of performing a STEP-analysis. Will they balance the costs of doing the analysis?

We here recommend using graphical techniques in the display of accident sequences on an exceptional basis only. A cost–benefit consideration must be made, where the time and attention needed to establish and analyse the diagram are considered as well. Experience shows that the establishment of, for example, a STEP diagram, is merited in the investigation of accidents involving complex interaction and communication between different actors.

13.4.4 Computer-supported accident investigations

Table 13.2 showed an example of a traditional fixed form for the investigation and documentation of accidents and near accidents. In designing such a form, a trade-off has to be made between the need for comprehensive and detailed information on the one hand and the need for a form that is simple and easy to take in and use. Developments in information technology offer new possibilities for the collection of data on accidents and near accidents that represent a way out of this dilemma. Investigation results are successively fed into a computer, and an interactive program will choose the new questions depending on previous answers (Van der Schaaf *et al.*, 1991). Such programs are built up as a logical tree structure, for example, similar to the one shown in Figure 13.6.

Example: A transportation firm has developed an interactive program for the registration and analysis of accidents involving material damages. They are especially concerned with motor-vehicle and material-handling accidents. The interactive program first asks questions about the type of accident. If the answer is 'a motor vehicle accident', the program will proceed by asking questions about:

- *Date and time of occurrence*
- *Type of transportation task*
- *Place of occurrence, physical environment*
- *Drivers, age, experience*
- *Types of vehicles involved, condition of vehicles*
- *Extent of personal injury and material damages.*

Next, the program identifies typical traffic conflict situations and the investigator has to choose between these. Depending on the answer, the program generates detailed questions of relevance to the chosen situation.

In ISA (Intelligent Safety Assistant), the computer application has been developed even further (Hale *et al.*, 1997). ISA is based on MORT logic and provides the users with the following expert support in the investigation, diagnosis and selection of remedial actions:

- Check and verification of data that is fed into the computer.
- When to stop further investigations because enough data has been received for the generation of relevant diagnosis and recommendations.
- Identification of relevant weaknesses in the SHE management system.
- Recommendations concerning relevant management factors.

An advantage of these types of application is that the results will immediately become available in a database for statistical analyses and for follow-up. Computer support is best suited for companies where the accidents mainly fall within a few well-defined categories. The program must not be too closed in the sense that it makes the investigator overlook vital information and uncommon causal factors.

13.4.5 Registration of accident costs

Table 13.3 shows a simple scheme for the registration of accident costs. It is based on the market-pricing model and is intended for use by the investigators as a complement to the ordinary investigation form. In most practical circumstances, the lost working hours due to sickness leave will dominate the monetary losses. A simple alternative is to include the first line of the table in the form for routine accident investigations.

13.5 Group problem-solving

Various investigation methods have been developed that use a team-organisation and group-problem-solving techniques (Kjellén, 1983; Carter

Table 13.3 Form for registration of accident costs

1. Lost working hours for the victim

| 40 | × | 250 | | = | 10,000 |

Number of hours   Hourly salary

2. Lost working hours, other employees

| 3 | × | 250 | | = | 750 |

Number of hours   Hourly salary

3. Capital costs due to production stop

| 1 | × | 1000 | | = | 1000 |

Number of hours   Costs per hour

4. Material costs

| 0 | + | 0 | | = | 0 |

Scrap costs   Repair costs

5. Costs of medical treatment

| 150 | + | 0 | + | 0 | = | 150 |

Medical doctor   Nurse   Physiotherapist

6. Transportation

| 15 | × | 3.00 | + | 0 | = | 45 |

Number of km   Costs per km   Public transp.

7. Other costs

| 0 | | = | 0 |

Other costs

8. Total costs for the company   =   11,945

Source: Sklet and Mostue, 1993.

and Menckel, 1985, Hale *et al.*, 1997). The aim is to overcome some of the problems in accident investigations discussed above by ensuring:

- Access to the necessary knowledge and skills during the investigation by involving persons representing different areas of knowledge and competence.
- Organisational learning, whereby the team members' detailed knowledge about conditions at the workplace of relevance to the investigation are made available in the decision-making. Team discussions also contribute, first, to a transfer of tacit knowledge between team members in a process called socialisation, and second, to the development of a shared understanding of causal mechanisms.
- Exchange of opinions on alternative interpretations between team members in order to avoid bias and arrive at judgements of a satisfactory quality.
- A basis for influencing the decisions on the part of the employees.
- Ownership of and loyalty to decisions made by the team.
- Learning and attitude change among team members.

A typical team consists of 3–5 members representing the workers and supervisors from the workplace and SHE experts (e.g. a safety engineer or an occupational-health nurse). In cases where an accident involves more than one department, each department has to be represented in the team.

Figure 13.8 shows the results of an experiment where the **OARU accident-investigation method** was introduced. The method is based on principles similar to those of so-called quality-improvement projects (Wig, 1996). It applies checklists and a bi-level investigation routine. The immediate investigation of the accident sequence is carried out by the first-line super-

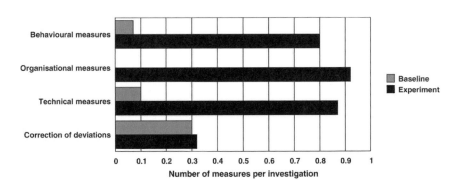

Figure 13.8 Results of an experiment involving training of supervisors and safety representatives at three construction sites in the use of the OARU accident-investigation method.
Source: Kjellén, 1983.

visor. He/she uses a checklist on deviations similar to that in Table 6.9 as an extra check to ensure comprehensiveness in this part of the investigation. Meetings in an investigation team follow. The team looks into contributing factors and long-term safety measures. The checklist in Table 6.12 is applied to support this part of the investigation.

Experience with the method shows that it results in more comprehensive and less biased investigations into causal factors. Experience also shows that the affected organisation becomes more willing to accept the solutions. We may expect changes in attitudes and accident perceptions among the participants in the problem-solving group.

Deming's circle defines the different steps in the planning, execution and follow-up of the activities of the investigation team:

1 *Establishment of criteria for selection of accidents and near accidents for team investigation.* Examples are incidents (accidents and near accidents) with a potential for severe consequences, frequently recurring incidents, and incidents in new production systems.

2 *Selection of team members.* These should include persons with detailed knowledge about the conditions at the workplace in question (operator, supervisor), persons who must accept the solutions, persons with special knowledge of importance to the analysis of basic causes and to the development of remedial actions, and persons with authority to make decisions.
   Example: An electrician fell off a scaffold at a construction site. An investigation team was set up, consisting of a fellow-electrician, the scaffolders' foreman, the general supervisor and the safety engineer.

3 *Resources at the team's disposal.* These include time during working hours, physical resources, training in investigation techniques, access to resource persons.

4 *Team meeting* with the following agenda:
   a *Review of available documentation.* The team reviews and checks the supervisor's first report on the accident and other relevant documentation such as work and safety instructions and safety regulations. Use the checklist in Table 6.9 to ensure that all deviations have been identified.
   b *Selections of deviations for further evaluation.* Discuss the significance of the different deviations and the incident with respect to how frequent they are at the workplace, their respective severity and causes. Use the checklist in Table 6.12 at the end of the discussion to ensure that all different possible causes have been considered.

> c *Development of remedial actions.* Propose remedial actions
> to prevent recurrence of deviations and incidents and/or to
> mitigate their consequences.
> 5 *Documentation of the team's meeting in a report.*
> 6 *Support from management* in follow-up of actions, informing the
> organisation, granting of authority, recognition of team members.

The recommendations above imply that the team uses a well-structured
problem-solving technique, for example, as in Figure 6.6, in its work. Ideally,
the accident-investigation process is sequential. In practice, there is some
overlapping and readjustment. As soon as the investigators have accumulated
enough information to have an adequate overview of the accident sequence
and the conditions at the workplace, they start to analyse the causes. This
analysis may lead back to a search for more facts to fill in gaps. They continue
by developing remedial actions. These should be of a long-term nature and
represent second to fourth order of feedback according to Van Court Hare's
hierarchy. The team's report should be kept simple and document the basic
recommendations, the sequence of events, and the analysis of basic causes.

---

*Exercise: A problem-solving group is given the task of finding measures
to reduce fall accidents at a construction site. The group starts by
looking into the accident in Table 13.2. Your tasks are:*

1 *To establish the group and to assign roles to the members of the
group. The group members should reflect the need for a vareity of
knowledge and experience of relevance to the accident.*
2 *To review the existing materials on the accident, the first report,
identified deviations, STEP diagram.*
3 *To analyse and identify causal factors. Use the checklist in Table
6.12 to ensure that all relevant factors have been considered.*
4 *Propose remedies.*

There is a need for evaluation of the proposed remedial actions. Different
schemes are offered for such evaluations.

Evaluation of remedial actions at the workplace level:

1 The following questions follow from the energy model:
a Has the hazard been eliminated or reduced? We here speak
about such measures as exchanging toxic chemicals with less

toxic substitutes or eliminating height differences to prevent falling.

b   What has been done to prevent the triggering of the hazard? These types of measure usually concern the prevention of incidents and of deviations such as human errors or mechanical failures.

c   What has been done to interrupt the uncontrolled energy flow? It is here a question of passive barriers (guards) or active safety systems such as fire detection and sprinkler systems.

d   Is it necessary to reduce harm by means of personal protection?

e   Is there a need for improved first-aid routines?

2   Additional considerations in deciding priorities:

a   Is it possible to implement the measure within a reasonable period of time? If the identified hazard needs immediate attention, short-term measures should be chosen pending a more permanent solution.

b   Are many people affected? A measure to reduce a hazard that currently puts many people at risk should be given high priority.

c   Does the measure have lasting effects?

d   Are no new hazards or other unwanted secondary effects created?

e   Is the measure feasible from an economic point of view?

f   Will those who are affected by it (workers and supervisors) accept the measure?

13.6 In-depth accident and near-accident investigations

According to the principles of accident and near-accident investigations at the three levels of Figure 13.1, an independent investigation commission should investigate severe accidents and near accidents. We will here describe a technique for this investigation similar to that of quality audits.

An immediate and necessary task for the commission is to establish the event sequence and an explanation of what happened. The commission will then be able to recommend actions to prevent a recurrence of this particular event in the future. Severe accidents, however, are usually the result of an unusual combination of circumstances, and it is unlikely that this particular combination will recur. A second and equally important task is therefore to make a systematic and independent examination of the organisational, technical and individual circumstances around the accident.

The commission has the duty to look into the root causes related to weaknesses in the general and SHE management systems of the company that are revealed by the event. In this way, the commission's recommendations will not only affect the probability of a recurrence of the particular event in question but the general safety level of the company. The commission will neutralise those interest groups inside the company that are benefited by a limited investigation into situational factors as discussed in Section 6.5. An independent commission will also be able to explore such aspects as management's commitment to SHE and the prevailing accident perceptions within the organisation.

The approach selected here for in-depth accident and near-accident investigation should not be confused with the police investigations of such events. It differs from the police investigations in that the latter aim to discover which laws or company procedures have been violated, who violated them and who is liable.

13.6.1 The steps in an in-depth investigation

We will here outline the typical steps in an in-depth investigation into a severe accident or near accident. They include:

1   Securing the scene
2   Appointing an investigation commission
3   Introductory meeting, planning the commission's work
4   Collection of information
5   Evaluations and organising of information
6   Preparing the commission's report
7   Follow-up meeting
8   Follow-up.

Whereas line management is responsible for steps 1, 2 and 8, the commission has principal responsibility for the other steps.

An important methodological issue in carrying out in-depth investigations is whether the investigation should be carried out on the scene or after the event. When the investigation is carried out *on the scene*, the investigation commission arrives at the scene immediately after the occurrence, when physical objects are unchanged and witnesses are available. The commission starts its work after the emergency response and health-care needs have been satisfied. There are obvious advantages with this immediate mobilisation of the commission:

- Time-dependent physical data such as weather and lighting conditions are directly observable.
- Physical evidence needed for reconstruction of the sequence of events is likely to be unchanged.
- Witness descriptions are largely unaffected by 'rationalisation'. Experience shows that the people involved have a wish to tell about their experiences, and that the statements given by witnesses and involved persons immediately after an accident are different from those given at a later stage.

When it is not feasible to mobilise an investigation commission at short notice, the investigation is carried out *after the event*, i.e. in practice one to a few days after the occurrence. For the in-depth investigation to be viable, parts of or the whole commission must visit the accident site and carry out investigations on the spot. There is a possible solution to the trade-off between the quality of the investigation and resources needed for investigations on the scene. This is to have one member of the commission available at short notice to carry out the initial investigation immediately after the occurrence. The whole commission will then assemble on the spot to review the findings and continue the investigation.

13.6.1.1 Securing the scene

Irrespective of whether the investigation is carried out on the scene or after the event, the scene of the accident has to be made safe and preserved for the forthcoming investigation. This is the responsibility of the immediate supervisor. The first task is to take immediate corrective actions to secure the site in order to avoid secondary accidents. It is then necessary to prevent physical items from being operated or moved. It is advisable at this stage to take photos of any physical evidence. When an investigation commission is appointed, the chairperson of the commission will take over responsibility for the site. It will not be released for clean-up and return to normal operation until all physical evidence has been secured.

13.6.1.2 Appointing an investigation commission

The appointment of the investigation commission lies with line management, usually at the department or plant level. If more than one department is involved in the incident, the decision to appoint a commission is made at a management level above these two department heads. Alternatively, the decision is made by the company's Working Environment Committee. The commission receives a mandate defining its members and describing its authority, the aim and scope of the investigation, resources and reporting schedule. The commission has to have access to any relevant documentation and personnel for interviews.

It is important to consider the following factors in designing the team:

- Credibility and competence. It is important to select team members who enjoy an adequate credibility in management's eyes. One possible solution is to choose one member of the commission from the responsible manager's peers (e.g. a manager of a similar department at another plant). The members of the team must represent competence both in the technical and organisational aspects of the event and in accident-investigation techniques. There is a need for formal training and certification of accident investigators in a way similar to that of accredited quality auditors (cf. ISO 10011 in ISO, 1990/91/92). The chairperson must come from a different unit from that affected by the event. The total number of members should not exceed five, but three members are a recommended size.

- Time away from competing duties. A serious incident will bring about a tremendous amount of attention and will also produce anxiety in the period immediately following the event. To be able to act on these circumstances, the commission should be freed from other duties immediately to focus on the investigation and come up with their report and recommendations promptly. This priority will also have a beneficial impact on the relations with the affected workplace. The commission will through their presence demonstrate that they take the problems of the local management and employees seriously.
- Power. The independent commission will not have the power to implement their recommendations. Authority and power are, however, important in another respect. The commission should have full access to all relevant sources of information, including the right to interview personnel and consult written and electronic files at their discretion.

13.6.1.3 Introductory-meeting, planning the commission's work

The commission starts its work by taking control of the scene of the event and by ensuring that it is preserved. After that, the chairperson calls a meeting and invites a person knowledgeable about the event and the affected organisation to give a briefing to the committee. The commission then decides about:

- Investigation of the scene of the event;
- Conducting interviews of participants, witnesses, line management, supporting staff;
- Obtaining documentation such as event logs, arrangement drawings and/or maps of the site, work and safety instructions; and
- Practical arrangements such as meeting room, scheduling of activities, contact person at the site, etc.

It is important that the investigation commission contributes to an atmosphere of openness and trust. The commission members' conduct during the introductory meeting, interviews and follow-up meeting is decisive. During the introductory and follow-up meetings, it is important to emphasise the learning perspective. Interviews have to be carried out in an atmosphere of mutual respect and recognition of the interviewees' legitimate self-interest. As in audits, the follow-up meeting represents an opportunity for the creation of ownership of the results among those affected by them.

13.6.1.4 Collection of information

Information-gathering involves inspecting the scene and other locations if relevant, interviewing participants and witnesses, expert examination of physical objects, document reviews, etc. It is important to make accurate notes of all gathered information concerning source, date, time, place, person responsible for data collection and type of information (factual or interpretive).

In inspecting the scene, care must be taken to note all relevant details such as debris, scratches, left tools, switch and handle positions, etc. The commission should establish sketches of the scene, showing the location of different objects. This should be accompanied by general and detailed photographs.

The interviews should start as soon as possible after the event and initially focus on the immediate sequence of events associated with the accident or near accident and the emergency response. Initially, the interviewee is asked to give a narrative of the event from his or her perspective. The interview is then used to examine detailed facts and to confirm earlier received information. First, participants and witnesses are interviewed in order to establish an accurate description of the sequence of events. Next, the line organisation and staff officers are interviewed to identify contributing factors and root causes.

13.6.1.5 Evaluations and organising of information

Evaluation and organising of the information starts after the initial inspection of the site and interviews with participants and witnesses. The aim is here to establish a description of the sequence of events and to compare it with existing information to determine whether there are any inconsistencies. One of the methods for displaying the sequence of events presented in Section 13.4.3 may support this work. It is here important to identify and evaluate deviations from normal practices and failures of existing barriers and controls. The commission will often find it necessary to return to the scene or to conduct additional interviews to check and complement already-gathered information.

In the subsequent analysis of contributing factors, the commission will look into specific aspects related to the workplace, see Section 6.5. The commission will evaluate such factors as:

- Hardware. It is here important to evaluate whether equipment has functioned as intended and whether the intended function was adequate. The commission will also determine if the design, inspection, testing and maintenance routines were adequate.
- Human performance. The commission first reviews the tasks performed by the operators to identify human errors and to evaluate the adequacy of recovery actions and emergency response. Influences from design, procedures and other circumstances (including emotions) should be evaluated, see Chapter 8. The commission should also look into errors and omissions made by management (latent failures).
- Procedures and accepted practice. The commission will evaluate the adequacy and coverage of procedures covering the work affected by the event. It is also important to review accepted practice and whether this

was followed or not. Identified discrepancies between procedures and accepted practice have to be addressed and evaluated.

Any changes from normal routines may contribute to an accident and have to be evaluated. Such changes may over-stress the organisation and make it more accident-prone. The commission may here apply a formal *change analysis* involving (Johnson, 1980):

1 Description of the accident (or near accident) and a comparable accident-free situation.
2 Comparison of the two situations and identification of any changes in the situation of the accident as compared to the accident-free situation.
3 Evaluation of the changes to determine their effects on the accident. It is here important to consider obscure and indirect relationships.

In the subsequent analysis, the commission will focus on root causes and on recommendations. Here different techniques are available to support the investigation, see Section 6.5. In the next Section, we will go into detail on one method, SMORT.

The commission members must be aware of the risk of getting into conflicts with management when pursuing underlying management and organisational factors. Company managers are usually overburdened with daily duties and routines and want to avoid adding more tasks to their agenda, especially those perceived as difficult to manage. They expect the commission to come up with immediate causes of the accident or near accident and tangible and effective counter-measures. Management also wants to avoid negative comments upon their area of responsibility.

The investigation commission, on the other hand, is responsible for a thorough analysis of the underlying root causes. The commission may also feel compelled to put forward many recommendations without considering the costs and benefits adequately. A gap between the expectations of management and the commission is a breeding ground for frustration, anxiety and self-censorship on the part of the commission. At the end of the day, the commission may find that management overlooks their recommendations. To circumvent such problems, the commission has to act in such a way that its integrity is maintained. The commission must at the same time strive for an open dialogue, based on mutual support and respect.

13.6.1.6 *Preparing the commission's report*

The results of the investigation are summarised in a report. The main results and conclusions in the report are presented at the follow-up meeting for quality check, see below. The report is then forwarded to the client (manager responsible for appointing the commission) for approval, implementation and follow-up. It should meet the general requirements of documentation of quality audits, see ISO 10011 (ISO, 1990/91/92). An accident investigation may reveal findings within different organisational units. The commission thus has to make recommendations to the client about the distribution of the report and responsibilities for follow-up.

A typical table of contents of an accident-investigation report is shown below:

1 *Scope of the investigation*, the accident that has been investigated, the mandate of the investigation commission.
2 *Executive summary*, presenting the essential facts and findings; probable causes and contributing factors; recommendations for managerial controls and safety measures that are necessary to prevent recurrence; and possibly, the need for further investigation.
3 *Method*, short description of inspections made, list of interviewees and documents reviewed, reference documentation, etc. Details can be given in an attachment.
4 *Facts*, i.e. a description of the workplace and the conditions at the time of the accident, and the sequence of events. A STEP-diagram may be used to present the results of the mapping of the sequence of events.
5 *Conclusions*, i.e. the commission's judgements regarding contributing factors and root causes of the accident.
6 *Recommendations* regarding remedial actions, further investigations, etc.

13.6.1.7 *Follow-up meeting*

The commission presents its conclusions and recommendations to the affected organisation in a follow-up meeting. The aim is to give the involved personnel a chance to clarify misunderstanding and give further explanations and comments. The follow-up meeting also fills the purpose of creating an atmosphere of trust and willingness to learn from the event.

13.6.1.8 *Follow-up*

The line manager who has appointed the commission will be responsible for distribution of the report and for the initiation of the required actions.

13.6.2 *Applying SMORT in in-depth investigations*

SMORT, the Safety Management and Organisation Review Technique, is a tool for use in in-depth investigations of accidents and near accidents

Table 13.4 SMORT checklist levels 1 and 2

	Level 1	**Sequence of events/risk situation**
☐	1.1	*Deviations in production*
☐	1.1.1	Personnel deviations
☐	1.1.2	Inadequate information
☐	1.1.3	Disturbance in material flow
☐	1.1.4	Human error
☐	1.1.5	Technical failure
☐	1.2	*Disturbances from the environment*
☐	1.2.1	Intersecting/parallel activities
☐	1.2.2	Poor housekeeping
☐	1.2.3	Poor environmental conditions
☐	1.3	*Incidents*
☐	1.3.1	Loss of control
☐	1.3.2	Failure in active safety systems
☐	1.3.3	Failure in fixed barriers
☐	1.3.4	Failure in personal protection
☐	1.3.5	Persons in danger zone
☐	1.4	*Development of injury/damage*
☐	1.4.1	Failure in alarm and mobilisation of emergency response team
☐	1.4.2	Failure in limiting injury/damage
☐	1.4.3	Failure in management of information

	Level 2	**Department and work system**
☐	2.1	*SHE culture and values*
☐	2.1.1	Line-management commitment
☐	2.1.2	Adherence to safety rules
☐	2.1.3	Communication on SHE
☐	2.1.4	Employee involvement
☐	2.2	*SHE management*
☐	2.2.1	SHE organisation
☐	2.2.2	SHE goals and action plans
☐	2.2.3	SHE routines and procedures
☐	2.2.4	Work and safety instructions
☐	2.2.5	Co-ordination of SHE work between subcontractors
☐	2.2.6	Emergency preparedness
☐	2.2.7	Documentation
☐	2.3	*Management responsibilities*
☐	2.3.1	Assignment of responsibilities
☐	2.3.2	Resources
☐	2.3.3	Supervision
☐	2.3.4	Production and activity plans, control of pace of work
☐	2.3.5	Co-ordination between activities
☐	2.3.6	Planning for disturbances
☐	2.4	*Human resources*
☐	2.4.1	Resources, manning
☐	2.4.2	Education, training
☐	2.4.3	Remuneration
☐	2.5	*Fixed assets*
☐	2.5.1	Plant layout, access
☐	2.5.2	Physical barriers, safety systems
☐	2.5.3	Physical working environment
☐	2.5.4	Design of machinery/equipment
☐	2.5.5	Man–machine interface
☐	2.5.6	Availability of machinery and equipment
☐	2.5.7	Documentation
☐	2.6	*Maintenance of fixed assets*
☐	2.6.1	Human resources
☐	2.6.2	Technical resources
☐	2.6.3	Routines, procedures
☐	2.6.4	Co-ordination with operations
☐	2.6.5	Permit-to-work system
☐	2.6.6	Modifications
☐	2.6.7	Documentation, change control
☐	2.7	*Material supply*
☐	2.7.1	Composition
☐	2.7.2	Material transportation, storage
☐	2.7.3	Waste handling
☐	2.7.4	Documentation
☐	2.7.5	Quality control

Table 13.5 SMORT checklist levels 3 and 4

Level 3	Project management
☐ 3.1	*SHE culture and values*
☐ 3.1.1	Project management commitment
☐ 3.1.2	Communication on SHE
☐ 3.2	*SHE management*
☐ 3.2.1	Organisation and responsibilities
☐ 3.2.2	Project's SHE programme
☐ 3.2.3	Suppliers' SHE programmes
☐ 3.2.4	Experience transfer
☐ 3.2.5	Regulations, standards, specifications
☐ 3.2.6	Analyses and verifications
☐ 3.2.7	Audits and management reviews
☐ 3.2.8	Nonconformity handling
☐ 3.3	*Resource management*
☐ 3.3.1	Human resources
☐ 3.3.2	Budget
☐ 3.3.3	Schedule
☐ 3.4	*Relation to stakeholders*
☐ 3.4.1	Co-ordination of stakeholders
☐ 3.4.2	Liaison with operations
☐ 3.4.3	Contact with authorities
☐ 3.5	*Study phase*
☐ 3.5.1	Concept exploration and definition
☐ 3.5.2	Demonstration and validation
☐ 3.5.3	Definition of scope, change control
☐ 3.6	*Project execution*
☐ 3.6.1	Engineering
☐ 3.6.2	Procurement
☐ 3.6.3	Mechanical completion and commissioning
☐ 3.6.4	Follow-up of SHE in fabrication and construction
☐ 3.7	*Take-over by operations*
☐ 3.7.1	Recruitment and training of operations personnel

☐ 3.7.2	Development of procedures and work instructions
☐ 3.7.3	Start-up
Level 4	Higher management
☐ 4.1	*SHE culture and values*
☐ 4.1.1	Leadership and commitment
☐ 4.1.2	Adherence to safety rules
☐ 4.1.3	Communication on SHE
☐ 4.1.4	Employee involvement
☐ 4.2	*Policy, goals and action plans*
☐ 4.2.1	Policy
☐ 4.2.2	Goals and acceptance criteria
☐ 4.2.3	Action plans
☐ 4.3	*Resource management*
☐ 4.3.1	Responsibilities
☐ 4.3.2	Human resources
☐ 4.3.3	Fixed assets and materials
☐ 4.3.4	SHE organisation
☐ 4.4	*Identification and evaluation of risks*
☐ 4.4.1	Routines
☐ 4.4.2	Data collection
☐ 4.4.3	Storage, analysis and distribution of information
☐ 4.5	*Handling of governing documents relating to SHE*
☐ 4.5.1	Routines
☐ 4.5.2	Regulations, codes, standards
☐ 4.5.3	SHE management programme
☐ 4.5.4	Procedures
☐ 4.6	*Performance monitoring and auditing*
☐ 4.6.1	Performance indicators
☐ 4.6.2	Follow-up of results
☐ 4.6.3	SHE auditing and management reviews

(Kjellén *et al.*, 1987). A SMORT accident investigation follows a diagnostic process, where the findings at one level are used as input to the analysis of causal factors at a higher level. This is illustrated by Figure 5.12. Tables 13.4 and 13.5 show updated versions of the SMORT checklists, and the questionnaires to match these are presented in Appendix II. These tools are based on many years of experience from accident inspections and from SHE management in general.

The SMORT checklists and questionnaires are open-ended and have the limited purpose of focusing the commission's attention on various problem areas. There does not exist any 'correct' answer independent of the examined company. In applying SMORT, the commission has to make a number of value judgements.

Examples of activities and remedial actions at each level of a SMORT analysis are shown below:

Level 1: Initial investigation of the accident sequence or potential accident scenario and identification of deficiencies and deviations at the workplace. This may involve a review of the accident investigation by the first-line supervisor, and interviews and observations at the workplace. The immediate use of the findings involves the elimination of deviations at the place of occurrence.

Example: A person inspects roofing tiles. He leans against a guard-rail at the end of the roof. The guard-rail is not adequately secured. It breaks and the person falls and is injured. An immediate cause of the accidents is a faulty guard-rail.

Level 2: Identification of risk factors at the workplace, i.e. in the work organisation and production system. The results are used to improve work organisation and production system.

Example (continued): A further analysis of the example above reveals weaknesses in the design of the guard-rails and in the routines for workplace inspections.

Level 3: Identification of how risk factors have occurred in production through a review of the organisation and routines for design, construction and start-up of new production systems. Remedial actions at this level include improved organisation and routines for project work.

Example (continued): A review of the design activities shows that the requirements for holders for guard-rail poles have not been adequately specified in the design of the roof.

Level 4: Identification of how risk factors have occurred through a review of supervision and management, including the safety-management system. Finally, remedial actions at this level involve improvements in supervision and safety, and in general-management organisation and routines.

Example, continued: An overall programme that defines the frequency and scope of workplace inspections to identify such deviations as inadequate guard-rails is missing.

13.6.2.1 Outline of a SMORT analysis

The stepwise procedure of a SMORT analysis follows the steps described in Section 13.6.1 and is similar to that of a quality audit (cf. ISO 10011 in ISO, 1990/91/92). We will here highlight those aspects of special relevance to SMORT.

- *Appointing an investigation commission*: One of the commission members must be acquainted with the SMORT method.
- *Introductory meeting and planning*: SMORT may be used to present the scope of the investigation in the introductory meeting. SMORT questions are selected and adapted to the specific circumstances of the event.
- *Collection of information*: Results are documented in the SMORT questionnaire in order to facilitate analysis and reporting.
- *Evaluation and organising information*: Evaluations of results from interviews, inspections, document reviews, etc. are summarised by use of the SMORT checklist and the following colour code:
  Red: Finding, i.e. nonconformity or observation
  Green: Checked and found acceptable
  Blue: Information is missing
  Black: Not relevant or not checked
- *Preparing the report and follow-up meeting*: SMORT may be applied to display the results in a summary way. Reference should be made to SMORT in the report to show that a systematic method has been applied.

SMORT has primarily been designed for use in investigation of severe accidents and near accidents. It complements existing routines for accident and near-accident investigations. SMORT provides a systematic and step-wise means of unfolding all relevant causal factors by starting with the specific and proceeding through the various managerial levels of the organisation.

An accident investigation is a search process and can only partly be planned in advance. It is thus recommended that the SMORT analysis of an accident is divided into different phases and that the procedure above is repeated for each phase. A full SMORT analysis may involve three phases: levels 1 plus 2; level 3; and finally level 4. The report from one phase is used as input to the SMORT analyses in subsequent phases.

The SMORT checklist and questionnaires are intended for use in planning the investigation and in interviews and document checks. The questions

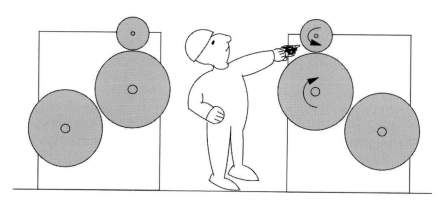

Figure 13.9 The victim's position at the time of the accident.

may need to be rephrased during the interview in order to cover the actual situation. Checklists and questionnaires help to ensure comprehensiveness and objectivity in the investigation. The investigation aims to reveal causal factors. It is here important to distinguish between facts (i.e. causal relations which may be objectively documented) and judgements about contributing causes. The accident investigation will also go far beyond specified requirements.

13.6.2.2 Case: A SMORT analysis of an injury in a rolling mill

A severe accident occurred in a levelling machine at an aluminium rolling mill. At the time of the accident, an operator was standing between two sets of rolls. He was cleaning a pair of rotating rolls with white spirit and a piece of cloth, Figure 13.9. He was working alone and had fixed the on/off button at the console to the machine in the 'on' position by means of the top of a pen and tape. This freed the second operator for housekeeping nearby. The two rolls caught the operator and his hand and arm got crushed between the rolls. The co-worker heard that something was wrong and ran to the accident site to activate the emergency stop.

On behalf of the Working Environment Committee, the plant manager appointed an investigation commission with two members. It consisted of a SHE expert and an engineer with a background in aluminium production. The mandate was to examine the formal, documented SHE management system in the light of the accident and the circumstances around it. The aim was to identify what improvements were necessary.

Initially, the commission met with plant and department management and safety representatives to review the event logs and results of the preliminary investigation by the plant personnel. The following facts were established concerning the conditions at the time of the accident (compare SMORT Level 1):

- *The two rolls were in contact with each other and were forced to rotate by the pushbutton being kept in an 'on'-position (1.1.4).*
- *The operator was working alone and was standing in front of the rolls at the suction side while cleaning the roll (1.1.4).*
- *Job instructions were incomplete and were not available at the machine (1.1.2).*

The commission verified these facts in interviews at the workplace. The subsequent investigation focused on SMORT level 2 and revealed the following conditions:

- *Requirements as to operator training had not been specified and the training did not ensure the establishment of safe practice (2.4.2).*
- *Safe work practices had not been defined in the work instructions. There were differences between individual operators and between shifts as to the methods of work (2.2.4).*
- *Unsafe practices were known at the supervisory level and were tacitly accepted (2.1.2).*
- *The supervisors had not received education on the plant's SHE policy and requirements and their SHE responsibility (2.3.2).*
- *There was no adequate work platform in between the sets of rolls, although this location was established as a workplace for cleaning. The operators had to balance on piping (2.5.1).*
- *There was inadequate protection against access to the danger zone (2.5.4).*

Figure 13.10 shows an arrangement of these findings in a diagnostic diagram, starting with the top event and tracing the underlying events and conditions back to the basic causes at the workplace level. The arrows do not represent causal relationships in a strict sense. Rather they represent the investigators' interpretations of the available facts and their arguments regarding the need for changes. In carrying out the investigation, evidence is collected in a stepwise fashion, starting with the top event and proceeding downwards as shown in the diagram.

The investigation report was handed over to the plant and department management, who initiated a number of actions. An immediate action was to review and update the work instructions. Training programmes were established for the supervisors and operators and for new employees. A full job-safety analysis of the machine was carried out. It was followed up by a number of technical measures to eliminate the need to work in the danger zone of rotating equipment.

13.6.3 Legal aspects of the commission's report

The in-depth accident report by the commission often reveals circumstances that are sensitive from a personal, legal or public-relations point of view. Examples of such circumstances are (Hale *et al.*, 1997):

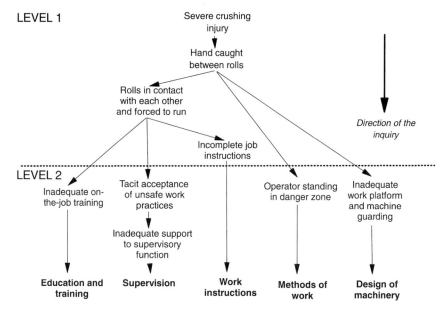

Figure 13.10 The results of the SMORT investigation displayed in a diagnostic diagram.

- Violations of statutory requirements.
- Evidence of hazards that have been known to management but have remained uncorrected.
- Failure to ensure compliance with company rules and procedures.
- Deliberate concealment of facts.

It is usually the client that decides on the status and distribution of the report. The client may decide to restrict the distribution of parts of the report or the full report.

Where the accident is under police investigation, the police have the right to request full access to the company's internal report. Knowledge of this fact will affect the commission's work and may hamper openness and trust. A careful balance must be kept between the need for openness, to take full advantage of the accident experience, and the involved persons' legitimate right of self-protection when they face the risk of legal action.

13.7 Computer-supported distribution of the investigation report

The widespread use of information technology allows for new means of distribution and follow-up of investigation reports. It is now possible to distribute reports conveniently and promptly to all concerned parties via electronic mail. It is also possible to store the reports in a shared database, where different persons can access the documents and comment on and edit them. These types of information technology solution provide several advantages:

- Enhanced quality of accident and near-accident reports and remedial actions by ensuring input and quality checks from people at different levels of the organisation and geographical locations.
- Transparency by making it possible for the reporter and other concerned parties (e.g. management and safety representative) to monitor the further handling of the report and the actions that have been identified and their status.
- Administrative simplification of distribution routines, approval of actions and close-out.

Although the developments in e-mail and group-ware for collaborative computing have many advantages from a SHE management perspective, there are also some concerns. It has to be ensured that the reports have passed through an adequate quality control before they are made accessible to all concerned parties through e-mail or a shared database. Commenting and editing has to be carefully organised in order to avoid contamination of the database with irrelevant or distorted information. Security routines to avoid misuse of information and corruption of the data are also concerns.

Accident reports may easily be distributed to a large number of people as e-mails. This distribution will add to the general information load within the organisation and may create problems of overload and inexpedient reactions. The problem may occur in large companies and in companies with an efficient reporting of near accidents. It is important to establish routines for the distribution of accident and near-accident reports inside the companies, based on the criticality of the events and the receiver's actual needs.

13.8 A procedure for accident and near-accident reporting and investigation

The routines for accident and near-accident reporting and investigation need to be documented in a procedure. Below is a proposed outline of such a procedure.

Proposed outline of a procedure for reporting and investigation into accidents and near accidents:

1 Scope and aim
2 Definition of types of events to be reported and investigated; definition of severity classes

3 Responsibilities for notification, investigation, documentation and follow-up of results
4 Description of routines for:
   a Immediate notification
   b Securing of the accident site
   c Supervisor's and safety representative's first report
   d Investigating team
   e Accident commission
   f Contractor accidents
   g Commenting on the report to ensure an adequate quality
   h Distribution of the report
   i Follow-up of remedial actions (assignment of responsibilities, deadlines, verification of implementation)
5 References
  Attachments: Forms for documentation of investigation results and follow-up of actions

14 SHE inspections and audits

14.1 Inspections

Inspections are continuous processes of vigilance. The type and timing of inspections are not primarily dictated by the occurrence of events. Rather, such activities have to be integrated as vital parts of the company's scheduled SHE management activities.

A main aim of the inspections is to check the technical standard of the workplaces by identifying and correcting deviations from regulatory requirements and company standards (i.e. first-order feedback according to Van Court Hare).

There are different types of inspections. Below are some examples:

- Plant operators inspect and monitor the equipment for which they are responsible on a day-to-day basis. It is, for example, a requirement that crane operators inspect and test certain safety critical functions such as brakes daily before they start to operate the crane.
- Supervisors and safety representatives co-operate in performing regular workplace inspections. These represent a tangible and repetitive SHE practice that stimulates co-operation.
- The technical department or a special plant-inspection department inspects and tests process safety systems such as pressure vessels, piping, safety valves and gas detectors at regular intervals.
- The fire department inspects buildings and fire-fighting equipment on a regular basis.
- The scaffolding department performs weekly checks on all scaffolding in use.
- The SHE staff inspects each plant department at regular intervals concerning such conditions as machine guarding, handling and documentation of chemical substances, waste handling.

The employer's duty to perform regular inspections of the workplaces is laid down in the legislation of many countries and in voluntary SHE management standards (Section 2.2). According to internal control legislation in e.g. the Scandinavian countries, the United Kingdom and the USA, the employer must set up SHE management programmes that include routine inspections of the workplaces. The aim of the inspections is to identify and assess hazards and to check compliance with the legislation.

14.1.1 Workplace inspections

We start by looking into the traditional workplace inspections that are carried out by the supervisor and safety representative, possibly with support from the SHE staff. There are documented routines for such inspections in companies that have implemented a formal SHE management system.

Inspection routines should be formalised in a procedure. The following outline is proposed:

1   Scope and aim of the procedure
2   Definitions of types of inspections with respect to scope (types of workplace/machinery to be covered *and* types of theme for the inspection)
3   Responsibilities for planning, execution and follow-up of inspections
4   Description of routines for each type of inspection with respect to frequency, type, participants, documentation and follow-up of results
5   References

Attachments: Inspection protocol forms

A traditional way of defining the *scope of the inspection* is to make a geographical delimitation. The company is partitioned into inspection areas, which usually coincide with the different departments. Each area is inspected at regular intervals by an assigned inspection team.

The theme of the inspection also has to do with its scope. Evaluation research shows that general inspections, where no theme has been pre-defined, only focus on a few types of deviations such as need for repair, missing guards and poor housekeeping, see Chapter 12. This narrow scope may be explained by our limited capacity as humans to be attentive to many different items simultaneously. The 'magic number seven plus or minus two' applies here. It means that we can not expect an inspector to cover more than a maximum of nine items in one inspection. In practice, five items seems to be the limit. Too limited a scope in this respect violates the criterion of coverage, compare Ashby's law of requisite variety. A common means of circumventing this problem is to select one or more themes for each inspection according to a pre-established plan and to apply theme-specific check-lists, Table 14.1.

Table 14.1 Example of a checklist on inspection themes

- Housekeeping;
- Traffic;
- Gangways, escape ways;
- Ladders, stairs;
- Fire prevention;
- Lifting equipment; cranes;
- Scaffolds, guard-rails, work at height;
- Storage and use of chemicals, waste handling;
- Machine guarding;
- Signs.

Theme-specific checklists are developed in order to secure a reliable and comprehensive mapping of deviations. It is recommended that the checklists be made company-specific. Typical input to the design of such checklists is regulatory requirements, company standards and experience from accidents, near accidents and previous inspections. An example of a checklist for inspection of fire protection is shown below:

Table 14.2 Example of a theme-specific inspection checklist

- Fire-extinguishing equipment (fire extinguishers, hydrants, hoses, etc.): according to safety plot plan, accessible, signed, fire extinguishers checked at regular intervals and sealed
- Manual call points: according to safety plot plan, accessible, signed, functioning
- Safety showers: according to safety plot plan, accessible, signed, functioning
- Emergency exits, escape ways: according to safety plot plan, accessible, signed
- Storage of flammable material: in assigned area, housekeeping, marking of material, fire protection

There are no generally valid recommendations as to an optimum *inspection frequency*. This will vary, dependent on the level of risk, the rate of change in production, the age of production and other circumstances. Weekly inspections may prove necessary, for example, at construction sites where there is a high risk of accidents and a high rate of change. At the other end of the spectrum are general offices, where yearly inspections may prove adequate.

Different conditions determine the *participation in the workplace inspection teams*:

- Workplace inspections are a means for the supervisors to follow up on their responsibility for safety.
- One of the participants must have authority to make decisions about remedial actions. Usually the supervisor fills this role.

- The participants must have the necessary knowledge about the specific hazards at the inspected places of work, about safety requirements and about available safety measures. Typically, the workplace inspection team is made up of the supervisor, the safety representative of the workers and a safety engineer. Experts may be called in when specific themes are covered such as fire protection, ergonomic or crane safety.
- The workplace inspections are part of the formal system for co-operation between management and unions on safety issues. The safety representative at the department thus shall participate.
- A recent development is to use workplace inspections as a means of demonstrating line management's concern for safety. In some companies, top management participates in the inspections at regular intervals (e.g. yearly).

So called 'one-on-one inspections' aim at strengthening line management's feeling of responsibility for safety. Management at different levels perform inspections of their area of responsibility together with the immediate superior. Here the focus is on the employees' work practices in order to identify and correct unsafe acts. Deviations related to work at height, use of personal protective equipment, use of equipment and tools, availability and use of safety instructions and housekeeping are typically identified in these inspections. An important aspect is how to communicate and follow up the findings. A report is written for each inspection and the results are followed up.

Another example of a practical tool with the purpose of reducing the frequency of unsafe acts is the so-called 'WOC' or 'Walk-Observe-Communicate' concept. Here, managers follow a specific agenda in observing workers and approaching them to discuss their behaviour. The aim is to arrive at a commitment to follow safe work practices. The WOC is announced in advance and accepted as legitimate by the personnel. At the workplace, the manager:

1   Stops and observes people.
2   Approaches a worker who has been observed committing unsafe work practices and explains his/her intentions in approaching the worker.
3   Asks about the job and how it is performed. Safe work practices are praised.
4   Asks about possible accidents in the job and the most severe consequences that may follow.
5   Asks why the worker used unsafe work practices and about corrective measures.
6   Gets commitment to act safely.

The workplace inspection is *documented* to ensure a timely implementation of measures. As a minimum, this report should include a description of identified deviations and hazards and the measures that have been decided, who is responsible for implementation of measures and due date for action. The reports are distributed to those participating in the inspection and to those responsible for safety measures. *Follow-up* should preferably utilise established management routines. In some companies it is common to take action on the spot and to omit to document the measures. The formal reports from such inspections will not give an accurate picture of the effects of the inspections.

In workplace inspection systems that function according to the principles of first-order feed back, the same types of deviations will recur. By periodic (e.g. yearly) analysis of the inspection reports, systematic recurrences of deviations will be identified. The group problem-solving technique that was presented in Section 13.5 is also applicable to the development of measures to prevent such recurrence.

Exercise: A SHE manager of a construction company summarises the results of the workplace inspection at one of the company's sites. He finds similar results to those illustrated by Figure 10.8. When the site manager is confronted with the results, he decides to set up a quality-improvement project. Your tasks as responsible for this project are:

1 *To establish the group and to assign appropriate roles to the group members;*
2 *To plan the work in accordance with the basic steps of Deming's circle;*
3 *To carry out the first part (Plan) and to propose solutions regarding inspection scope, frequency, follow-up and participants.*

14.1.2 Inspecting and testing barrier integrity

Regular inspections and tests are necessary tools to ensure that such barriers as those discussed in Chapter 7 against fires and explosions are maintained. Legislation in Europe and the USA make it a responsibility of the employer to ensure that the barriers against major accidents involving chemical hazards are inspected regularly (European Council, 1996; OSHA, 1994). The inspections have to be carried out by experts from the technical department of the plant or from a dedicated plant inspection department.

Below is a list of typical equipment of a process plant that is subject to inspections:

- Process vessels and storage tanks
- Process piping, valves and pumps
- Relief and ventilation system
- Emergency shutdown system
- Monitoring devices, interlocks and alarms.

The inspection or testing frequency has to be determined. Traditionally this has been based on manufacturer recommendations or on good engineering practice. A recent trend is to establish inspection intervals for individual types of equipment based on results of reliability analyses. Reliability data serve as necessary data inputs to such analyses. In order to establish a database on barrier reliability, it is necessary to tag each piece of equipment and to record its inspection/test, repair and maintenance history.

Inspection and testing procedures have to be documented for the different types of equipment and systems. It is important to ensure that overall barrier efficiency is tested. For gas alarms, for example, the complete loop has to be tested from the gas detector to the alarm presentation in the central control room. The results of each inspection/test have to be recorded, showing date, name of inspector, equipment tag number, description of inspection/test, results and actions. In cases where deficiencies are identified, the inspector has to evaluate whether they are of a nature requiring immediate shutdown and replacement of the failing unit. Less critical failure may wait for correction until the next scheduled revision stop.

14.2 SHE audits

A SHE audit is here defined as a systematic and independent examination of a company's SHE management system. The aim is 'to determine that the elements within a SHE management system have been established and are effective and suitable for achieving stated SHE requirements and goals' (compare ISO 8402 in ISO, 1994). Audits are a management tool to create confidence in the client of the audit that SHE is taken care of adequately. The client gives the audit team its legitimacy and authority to make the necessary enquiries. Legislation on internal control and voluntary SHE management standards identify SHE audits as an important element in the company's SHE management system (Section 2.2).

There are three different types of SHE audits:

1 *Internal audits* aim at evaluating whether the company's own SHE management system meets regulatory requirements, recognised national or international standards, company requirements, etc. The client is often top management or the Working Environment committee.
2 *Customer audits* are part of a scheme for pre-qualification of suppliers, where there is a desire to establish a contractual relationship.

Alternatively, they are carried out after order placement, when the customer wants to verify that the supplier's SHE management system continues to meet the contractual requirements. The customer's procurement department is often the client of the audit.

3   *Third-party audits* are conducted by an independent and recognised organisation on behalf of the management of the audited organisation. They aim at demonstrating to customers that the company's SHE management system meets specified requirements. The audits are often part of a certification process, where the audit is based on one or more standards such as ISO 14000 standards on environmental management systems. Following an audit with acceptable results, the third party is a guarantor for the quality of the audited company's SHE management system.

According to Norwegian Internal control regulations, companies that act as the principal enterprise shall ensure that sub-contractors and suppliers comply with the regulations (Section 2.2). The execution of SHE audits is a means for the principal enterprise to meet these regulations in this respect.

SHE audits apply the same basic principles as those of quality audits (ISO 10011 in ISO, 1990/91/92):

1   The following basic principles apply:
   a   Establishment of an audit plan for each audit, defining objective and scope, reference documents, organisational unit to be audited, schedule, audit team, reporting, etc.
   b   Competent and independent audit team.
   c   Preparation and use of checklists, based on previous experience and review of governing documents.
   d   Focus on conditions essential to the purpose of the SHE management system and on improvements. Avoid faultfinding.
   e   Building of trust through planned activities, including pre- and follow-up meetings with auditee.
   f   Conclusions (nonconformities and observations) derived from verifiable facts.
   g   Line-management responsibility for follow-up on actions.
2   The reference documents define the requirements against which the SHE management system is audited. In general, there are three different types of such documents:
   a   Laws and regulations, such as the European Council Regulations on eco-management and audits (EMAS, see European Council, 1993) and the Norwegian Internal control regulations (Arbeidstilsynet, 1996).

> b   Recognised international standards such as ISO/DIS 14001 'Environmental management system' and British standard BS 8800 on occupational health-and-safety management systems (ISO, 1996; British Standards Institution, 1996).
> c   The audit's internal governing documents such as the SHE policy and goals, action plans, job descriptions, safety rules, etc.
> 3   The *audit report* conveys the results and recommendations of the audit (including observations on nonconformities), the scope, objective and plan of the audit, reference documentation, and the audit team's judgement of the audited SHE management system. It is an important record for experience transfer and shall meet requirements to thoroughness, accuracy and objectivity.

14.2.1 Application of SMORT in audits

SMORT is also suitable for use as a tool in the planning and execution of SHE audits. The steps of a SMORT investigation, as described in Section 13.6, follow the basic procedure of an audit. The SMORT checklists and questionnaire give support in the development of detailed lists of questions for the audit.

> Principles for application of SMORT checklists in the planning of SHE audits:
>
> • Start by using the SMORT checklist in order to identify and delimit the areas to be covered according to the scope of the audit.
> • Identify the necessary reference documents (regulatory requirements, standards, specifications, procedures, etc.).
> • Identify the persons to be interviewed and documents to be reviewed.
> • Develop detailed questions for each area on the basis of SMORT questions and reference documents.

It is usually possible to plan the SMORT audit in detail in advance and no subdivision into phases is required. There is a clear difference between use of SMORT in accident investigation and in SHE audits when it comes to the presentation of results. SHE audits are closely linked to requirements as stated in various documents. When presenting *nonconformities*, the reference documentation should always be cited. The audit team also has the freedom to present other types of *observations* where their own judgements go beyond what is clearly specified.

Exercise: Planning the audit of a Permit-to-Work System

Permit-to-work systems should ensure that work is planned and carried out under adequate safety precautions in cases where the exposed persons do not have personal control of the involved hazards (Section 7.6). At the Ymer offshore platform, modifications of one of the compressor trains must take place while the platform is producing. This work has been subcontracted to an installation firm, BCC Installations. At the start of this work, the platform manager wants to be confident that the established work-permit-system is adequately designed and implemented. Your group is assigned the responsibility for the planning and execution of an audit. Its scope is the implementation of the work-permit-system for this particular modification work. Your tasks are to:

1 Establish the audit team and organise the responsibilities within the team.
2 Review the description above and to use the SMORT checklist in an identification of the areas to be covered by the audit.
3 Identify the necessary governing documents and requirements that must be reviewed.
4 Develop detailed questions and plan the interviews, document reviews and inspections.
5 Develop the information to be presented at the pre-audit meeting.

Remember that verification by interviews often requires the interviewing of representatives of the two parties, i.e. the Oil Company and the Contractor.

15 Accumulated accident experience

In this chapter, we will look into the establishment and use of accident and near-accident databases. Experience stored in such databases will help us in answering questions like:

- What are the characteristics of our accidents? What can we learn from them?
- Has this type of accident occurred before?
- Are there any particular accidents with this machine that we need to consider when buying a new machine?
- Where shall we allocate resources to prevent accidents?
- What are the causes of our accidents?
- What has been done to prevent the accidents? What is the status of the remedial actions that have been decided?

The database contains accumulated experiences on accidents and near accidents. *Answering questions* is one of the different uses of this type of database. In this application, the user queries the database in order to retrieve information about accident and near-accident cases. Experience shows that the users are mainly concerned with finding individual accident cases (Kjellén, 1987). There are other means as well:

- Establishment of *periodic summary reports* on accident statistics. Such summaries are distributed monthly, quarterly or yearly, dependent on the size of the company and the level of risk. Their aim is to provide feedback on SHE performance and to inform about accident cases and accident distributions.
- *Follow-up of actions.* This is a simple feedback to decision-makers on responsibilities, target dates and status in order to ensure timely implementation. A more advanced application is to study the long-term effects of implemented measures, e.g. by monitoring SHE performance.
- *Risk analysis.* Here we are concerned with the identification and evaluation of hazards at the workplaces and the estimation of risk. Data on accidents serve as an important input.

The second and third uses illustrate the support and rationalisation that computerisation offers the *SHE administrator*. We will present examples of these applications in Part VI. A more basic administrative function is to keep track of the accident and near-accident records and to help distribute these records within the company and to the insurer and the authorities.

We will here focus on another important advantage of computerisation, i.e. the support it offers the *SHE analyst* in making queries and in the application of different expert tools. We will start by looking into the principles for establishment of a database on accidents and near accidents. We will then proceed by reviewing various methods and tools for retrieving, analysis and presentation of data from such a database.

Computer support is not necessary when analysing accident data. For companies with only a few accidents and near accidents per year, paper-based archives will be adequate.

15.1 Database on accidents and near accidents

A database is a collection of related facts or data that has been designed, built and populated with data for a specific purpose (Elmasri and Navathe, 1994). We are here concerned with computerised databases on accidents and near accidents. These are electronic archives, containing a record of the essential facts about each unwanted event at the company. The purpose of storing the data electronically rather than in a paper-based archive is to make it easier to retrieve, analyse and present the data. Computer support also offers new possibilities for data analysis and presentation.

The users of the database have to handle the following two activities: (1) updating the database with new facts on accidents and near accidents; and (2) querying the database by retrieving relevant data and generating reports that present the results of the queries.

When we design a database on accidents and near accidents, we also have to consider what other types of data we may want to store in later developments of the database. In the end, we may, for example, be interested in integrating data from accident and near-accident investigations with reports on unsafe conditions, nonconformities, workplace inspections, risk analysis and SHE audits in a common database.

15.1.1 Database definition

In order to define an accident database, we have to specify the types of facts or *data elements* to be included in each accident or near-accident record. We also have to specify a *data type* for each element. This tells us whether the data is coded according to a nominal, ordinal, interval or ratio scale or stored in free text. Both these specifications are included in the so-called *database definition*. Each data element and its associated data-type specification are linked to a *field* in the database definition. In establishing a database definition, we have to decide what accident model to apply. This is a crucial decision. We have to make a careful balance between two concerns:

1   The amount and quality of the required information, as defined by the accident model; and
2   The time and attention needed for investigations and quality assurance of the results before storage in the database.

In Chapter 6, we discussed problems that the investigator faces when applying a complex accident causal model in the collection of data on accidents. A careful balance has to be made between the amount of information required, the quality of the information, and the time needed for the investigation. This balance is problematic, since the benefits of detailed information of a high quality are usually experienced by another part of the organisation than those responsible for the collection of the data. They have first-hand experience anyway.

Table 15.1 presents a proposed smallest efficient data set on accidents and near accidents. It is based on the framework for accident analysis presented in Chapter 6 and takes the author's experiences from evaluations of SHE information systems into consideration. In this proposal, no additional information is requested from the supervisor other than that collected on a traditional accident-investigation form. The reason why a SHE expert is needed to feed additional data is to secure the reliability of the data. In addition, the SHE expert will be responsible for checking the quality of the data from the supervisors.

Figure 15.1 shows an overall data model for storage of the information on accidents and near accidents in Table 15.1. It shall be possible to expand this model to make it applicable to data on unsafe conditions, nonconformities, workplace inspections, risk analysis and SHE audits as well.

15.1.2 Accessing the database

Figure 15.2 shows a simplified model of the environment of a database system, seen from the user's point of view. We recognise the stored data model for the actual application and the stored accident and near-accident data.

The user queries the database in two steps. These are repeated until a satisfactory result has been achieved. They include:

Step 1:   searches in the database to identify the subset of accidents and near accidents that meet certain search criteria; and
Step 2:   presentations of the results in the form of event descriptions, statistical summaries etc.

Table 15.1 Proposed 'smallest efficient set' of data elements in accident and near-accident records for computer storage

Category	Data elements and data types	
	Information from the supervisor's first investigation	*Information added by SHE expert*
Administrative conditions	Date, time Department (free-text description or fixed alternatives) Place (free-text description)	Registration number (added by the system) Person to contact for more details about the event
Injured persons	Information about injured persons (name, age, occupation)	
Work situation	Activity (free-text description) System/equipment involved (free-text description, eventually tag number)	
Sequence of events	Free-text description of sequence of events Free-text description of deviations	Type of occurrence (fixed alternatives, i.e. accident, near accident, material damage, fire) Accident type classification (fixed alternatives, see Table 6.6)
Consequences/losses	Description of injury/damage Type of injury/damage (fixed alternatives) Part of body (fixed alternatives)	Actual and potential loss (fixed alternatives, see Table 6.4)
Causal factors	Free-text description	
Remedial actions	Free-text description Responsible (name/position) Due date	

Example: An accident database of an aluminium plant includes about 500 records. The foundry manager is interested in the number and types of lost-time accidents at his department during the last year. He performs a search for events that fulfil three conditions, i.e. foundry, lost-time accidents, last year. The result is 58 hits. He then displays the events in three histograms showing the distribution of accidents by nature of injury, part of body affected and accident type.

The search criteria are determined by the database definition. There are three basically different principles for searches in the database:

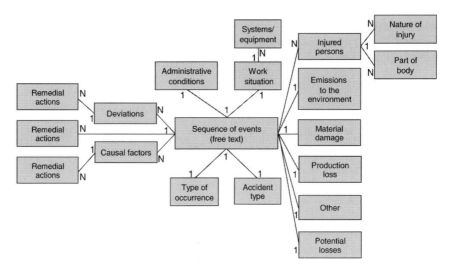

Figure 15.1 Overall data model for the proposed 'smallest efficient set' of data elements. (1–1) means a one-to-one relation; (1–N) a one-to-many relation.

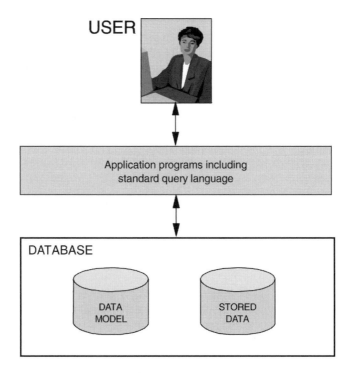

Figure 15.2 Simplified database environment.

1   Searches for events that meet criteria defined on an ordinal, interval or
    ratio scale of measurement.
    Example: Find all accidents for a particular year.
    Example: Find all accidents where the victim is younger than 25 years
    of age.
    Example: Find all accidents where the injured person has been absent
    for more than 30 days.
2   Searches for events that meet criteria on a nominal scale of measurement.
    Example: Find all accidents where women have been the victim.
    Example: Find all 'falling to a lower level' accidents.
    Example: Find all accidents caused by the personal factor 'improper
    motivation'.
3   Searches for events that include certain words in the free-text descrip-
    tions of, for example, the sequence of events.
    Example: Find all accidents where 'industrial trucks' are mentioned in
    the description of the sequence of events.
    Example: Find all accidents where the victim has been hit by falling
    scaffolding materials.

There are two types of error associated with these searches, Figure 15.3.
Type I-errors concern the *degree of retrieval*, i.e. wanted data that are not
found. Type II-errors concern the *degree of precision*, i.e. data identified as
hits that are not wanted. The way the database has been defined will affect
the size of these two errors. Experience shows that a skilled user will achieve
a higher degree of retrieval in free-text searches than in searches based on
fixed alternatives (Kjellén, 1987). This high degree of retrieval may, however,
be achieved at the cost of a low degree of precision.

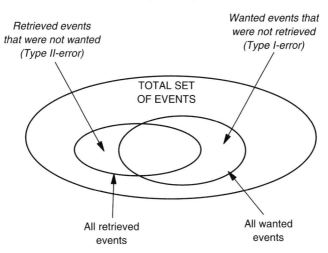

Figure 15.3 Illustration of errors relating to degree of retrieval and degree of
         precision.

In Figure 15.2 we identify the interface between the database system and the user and the application of specific programs developed to meet the user's needs. This interface is of great importance in relation to the prospects for an efficient utilisation of the database. We will here use the so-called 'Cake-bake analogy' in the identification of four different categories of users with different needs:

A   *The consumer buys a ready-made cake at the bakery.* This is equivalent to the user of accident statistics who receives a periodic report from the SHE manager.

B   *The consumer buys a cake mix and adds water before he bakes it in his own oven.* We here recognise the user who generates pre-defined displays of accident statistics by clicking on a menu on the computer screen. It is at this level possible to select the subset of data on accidents from pre-defined categories but not to customise the presentation.

C   *The consumer makes his own cake from a recipe.* The analogue is the SHE expert, who compiles accident statistics by using general application statistical packages and spreadsheets. It is at this level possible to generate a great variety of analyses and presentations of different subsets of the accident database.

D   *The consumer invents his own recipe.* At this level, the SHE (and computer) expert will develop special application programs, such as those needed for certain types of risk analyses.

In principle, the user program should satisfy all those different needs. We often see, however, that the systems designers have focused on category B users, making the database less flexible to meet the needs of category C and D users.

Figure 10.1 illustrated how different users within a company may access a SHE database for different types of information. A recent trend is to apply Intranet or Internet solutions by installing a common database at a server and making it accessible through the company or public computer network. We shall not go into the technicalities of this solution here, since it is common to information-technology applications in general. This solution has some definite advantages:

•   Simplified routines for feeding the database with new information and for quality assurance of this information;
•   Simplified routines for distribution of accident and near-accident reports and periodic reports;
•   Past accident and near-accident experience will be available throughout the organisation;
•   Simplified routines for updating and revision of the database structure and for the introduction of new tools for analysis and presentation; and

- A server-based database will be able to communicate with other databases such as personnel records and equipment lists. This makes it possible to validate the data and to simplify updating.

In the design of accident and near-accident databases, access control to avoid misuse has to be addressed. It is necessary to define the authority for access to the database for information retrieval (reading authority) and for updating of the database (writing authority). This has to be done for each category of user. By designing the database properly, it will be possible to specify read and write authority according to individual need.

Exercise: A construction firm runs five construction sites. The firm's accident database contains the following types of data about each accident: site of occurrence; name and personal data about the victim (age, sex, occupation, etc.); description of the sequence of events; description of causes; description of remedial actions. Propose read and write authority for the following categories of users: (a) managing director; (b) personnel manager; (c) SHE manager; (d) site manager; (e) safety representative at the site.

A common solution is to allow for widespread permission to read the accident and near-accident reports with the exception of those fields containing personal data on the victim. The possibility of commenting on or changing the document should be restricted to persons with special authority. A dedicated document owner should be given the responsibility to approve all changes in the report itself.

15.1.3 Coding of accident and near-accident data

In the previous Section, we distinguished between two different types of data on accidents and near accidents, i.e. coded data (applying a nominal, ordinal, interval or ratio scale) and free-text descriptions. We will here discuss the principal difference between qualitative or free-text data on the one hand, and coded data on the other hand. We will focus on the trade-off between using free-text descriptions and coding the data according to a coding schedule (i.e. a nominal scale).

Free-text data is by its nature unstructured and thus less suitable for statistical analysis of the type found in standard periodic reports on accident statistics. By coding the data, we will structure and standardise it for easy retrieval and presentation in tables and diagrams, showing the distribution of accidents by part of body, type of event, type of cause, etc. These advantages are acquired at the cost of loss of richness. We also run the risk of distorting the data in the coding process and concealing uncertainties in the data as discussed in Section 6.5.

In SHE practice, we find two different 'schools', as represented by the following extremes:

1   All relevant data about the accidents and near accidents are stored in a coded format. This coding may take place during data collection, see Section 13.4. Nominal scales of measurement are applied in coding the descriptions of losses, the sequence of event and causal factors, etc. The ISA, ILCI and MAIM accident models presented in Chapters 5 and 13 have typically been developed for this purpose. Table 15.2 shows a coding schedule based on the ILCI model.

2   All relevant data are documented and stored in free-text descriptions and no interpretation and coding takes place. In the completely unstructured format, the accidents and near accidents are presented as a chronicle, where the description follows the historical development. A semi-structured alternative is represented by free-text descriptions of the sequence of events, losses and causes under separate headings.

In coding accidents or near accidents, we run the risk of introducing errors and losing information. As an example, we will here use the ILCI model for classification of accidents and near accidents, Table 15.2. In applications of this model, the analyst checks the alternative in the second column that suits the actual circumstances. The reliability of this classification and the biases introduced are concerns. The analyst does not document these uncertainties, and the resulting accident statistics will appear as objective facts to the end user.

Exercise: Each student classifies three to five accident reports by applying the classification scheme of Table 15.2. Present the results to the group of students and answer the following questions:

1   *What is the interpersonal reliability in the classification? This is defined as the largest share of the classifications that agree as a percentage of the total number of classifications made.*
2   *Does the interpersonal reliability differ for the different types of data on accidents?*
3   *Discuss reasons why there is such a difference.*

Experience shows that reliability decreases and biases increase as we move from the right to the left in the 'causal chain', Figure 15.4. The errors that are introduced through poor reliability and biases in the coding process will propagate to the analysis and presentation of accident statistics to

Table 15.2 Overview of a coding schedule for accidents and near accidents

Type of data	Sub-categories
Nature of injury	Burn or scald (heat); concussion; crushing injury; cut, laceration; puncture, abrasion; fracture; hernia; bruise, contusion; sprain, strain; other.
Part of body	Head, face, neck; back; trunk (except back), arm; hand and wrist; fingers; legs; feet and ankles; toes; internal.
Accident type	Struck against; struck by; fall to lower level; fall on same level; caught in; caught on; caught between; contact with; overexertion/overload of muscle-skeletal system.
Substandard practices	Work without permission; failure to secure; carelessness or lack of attention; making safety devices inoperable or removing safety devices; wrong use of equipment; failing to use personal protective equipment properly; improper lifting technique or work posture; servicing equipment in operation; wrong operation; inadequate follow-up or control of safety provisions.
Substandard conditions	Inadequate safety precautions; inadequate access; inadequate ergonomic design; poor housekeeping; weather conditions; control system failure; influence from other work; inadequate work instructions; defective tools or equipment; inadequate planning; inadequate tools; inadequate warning system; fire or explosion hazards; lighting or viewing conditions; dropped object hazard; exposure to chemicals; radiation; hygienic conditions; psycho-social conditions; inadequate climate and ventilation; noise.
Basic causes related to personal factors	Inadequate information; inadequate experience; excessive work pressure; inadequate motivation; lack of rest; inadequate training; reluctance to ask.
Basic causes related to job factors	Inadequate supervision; inadequate design; inadequate purchasing; inadequate maintenance; inadequate work procedures; wear and tear; lack of communication; wrong function; inadequate planning; inadequate leadership.
Lack of control	Inadequate SHE management programme; inadequate programme standard; inadequate compliance with standard.

Source: Adapted from Bird and German, 1985. Copyright Det Norske Veritas and reproduced by permission.

management. We thereby run the risk of reinforcing biased and false accident perceptions.

Below we summarise the previous discussions on the conditions for a reliable and unbiased classification of data on accidents and near accidents. These are favourable when:

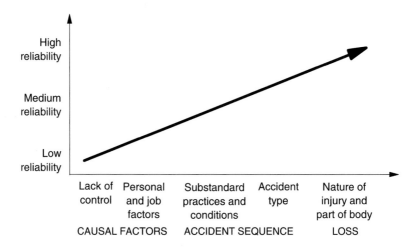

Figure 15.4 Tendency of coding reliability to decrease as the analysis moves backwards in the 'causal chain' of Table 15.2.

1   The classification scheme is represented by a taxonomy, i.e. the classes of the scheme are complete and mutually exclusive. It means that each accident or near accident can only belong to one class of the taxonomy. The victim's sex (male or female) is a straightforward example.
2   The data consist of observable facts rather than interpretations. Examples of observable facts are age and sex of the victim, physical conditions at the workplace and observable behaviour of people. When we move to the interpretations of causes or rather reasons for people's behaviour, we run the risk of introducing biases and errors.

In practice, these two conditions are rarely fulfilled other than for a few types of data. A compromise must be made. The proposed 'smallest efficient data set' of accidents and near-accident data in Table 15.1 represents such a compromise.

There is another problem connected to the classification of data. Such classification will result in a loss of information about the detailed circumstances of the accident. These details are often essential to our understanding of why the actual accident happened, especially when we study complex human behaviour. Fortunately, the problems with biases and with loss of information run in parallel. The losses and the incident at the right side of the ILCI model of Figure 15.4, for example, are usually adequately represented by coded data. The level of details in the precursory events and conditions and in the causal factors on the left side of the model is often too high to allow for a meaningful coding. We here face a situation where rich information has to be forced into too small a frame. The person responsible for coding will then make a more or less arbitrary selection. It follows that statistical summaries of accident causes too often represent arbitrary selections of facts taken out of their context. Decisions on remedial actions based on such statistics run the risk of being inefficient or even counterproductive.

It is recommended to store the unedited description of the accident sequence and the accident causes from the original report together with an accident type classification, see Table 15.1. The qualitative data are well suited for use in combination with coded data in statistical analyses. The different data-analysis techniques presented in Sections 15.2 to 15.4 below use coded data for data retrieval and presentation purposes. When an interesting subset of accidents has been identified and retrieved, the free-text descriptions of the events can be presented in a tabular format that is easy to survey. Such free-text summaries will help in interpreting the data in a meaningful way.

A recent development involves the use of software that automatically codes narrative or free-text data (Stout in Feyer and Williamson, 1998). This type of software has the potential to make coding more reliable and consistent, but is best suited for epidemiological analyses of large databases, e.g. on a nation-wide basis.

15.2 Analysis of accident and near-accident data

15.2.1 *Finding accident repeaters*

An important question in connection with accident investigations is whether the same type of accident has occurred before. This question is easily answered if a computerised accident database is available. The most efficient means of searching for similar occurrences is to utilise the free-text search facilities of the information retrieval programme. There are computerised tools for pattern recognition that may help identify accident repeaters. It is important to note that different persons may use different words to describe the same work operations, equipment, etc. There are also spelling errors and misconceptions that hinder efficient free-text search. The analyst must be well acquainted with the workplace in question and the different dialects in use.

Example 1: There had been a truck accident at an airport. The SHE manager wanted to know whether there had been similar accidents with trucks before. He first looked in the coded field 'injury agent' for vehicles and found two reports, one of which was relevant. Next, he made a free-text search on 'truck' in the sequence-of-events description and found these two reports and another two reports. One of these was also relevant.

There is a general need to query the database for accident and near-accident cases as input to decision-making. A computerised accident and near-accident database can easily help satisfy such a need. Below are some examples:

Example 1: A team on an offshore installation had performed a job-safety analysis of the entering of a tank. The safety engineer wanted to check whether any important hazards had been omitted. In a free-text search, an accident with electrical equipment was identified that had not been identified in the JSA.

Example 2: The catering manager on an offshore platform was planning a first-aid course for kitchen personnel. He searched the database for accident cases illustrating the importance of first aid.

Example 3: One of the compressor trains on an offshore installation was due for replacement. The technical manager searched the database for previous accidents on compressors in order to take experience from these into account in the purchase of the new compressor.

Case-based reasoning (CBS) is a relatively new branch of computer science and stems from research into artificial intelligence (Watson, 1997). A CBS system stores previously solved problems as 'cases' in a database. Each case is a record set of events or data such as an accident and is usually accompanied by decisions on solutions to the problem. When a user experiences a new problem, the CBS system gives support in matching this problem against the cases stored in the case database and in retrieving the most similar past cases. Among the retrieved cases, the system helps in selecting the most similar case. Different methods are applied in making the match and in finding the most relevant past case such as key-term, causal model and parameter matching. The old solution is then modified to conform to the new situation. If the new solution is successful, the new case and the solution are stored in the case database.

15.2.2 Uni- and bi-variate distribution analyses

These analyses employ coded data. Results are displayed in tables, histograms, etc., showing the distribution of the accidents/near accidents in absolute number or percentage. Uni-variate analyses show the distribution of one variable and bi-variate analyses display cross-tables of two variables. This is a well-established data-analysis method employed in companies' periodical statistical summaries of accidents and near accidents.

Example: The yearly accident and near-accident summary report of an aluminium plant displays results of various uni- and bi-variate distribution analyses, including:

- *A bar chart of the distribution of reported events by type of event, i.e. a uni-variate distribution. This graph showed that there had been 9 lost-time injuries (LTIs), 274 first-aid injuries, 450 material-damage incidents and 709 near accidents.*
- *A bar chart showing the distribution of events by type of event and year for the last three years, i.e. a bi-variate distribution. It showed a decrease in the number of LTIs during the last year (9), as compared to the previous two years (14 and 17 respectively). The near-accident reporting frequency developed in a positive direction.*

- *A pie chart of the distribution of LTIs by department. It showed that there had been five LTIs in the technical department, two in the electrolysis department and one in both the rolling mill and the carbon section.*
- *Separate tables of the distribution of lost-time and first-aid injuries by type of injury and part of body injured. Cuts in the finger were the most frequent combination.*
- *A pie chart of the distribution of injuries by unsafe act. 'Lack of attention' was the dominating category and occurred in 22 per cent of the accidents.*

Such summary reports are important tools for communication. The presentation method is a concern. Graphs are easier to read than tables but may lack precision.

Uni- and bi-variate distributions give an overview of the accidents and help prioritisation in relation to in-depth analyses and in selecting remedial actions. Figure 15.5 shows another illustrative example. Here, the analysis identifies the energy types involved in the accidents in different departments of an offshore installation.

15.2.3 Accident-concentration analysis

Accident-concentration analysis was first developed within traffic-safety research and was here denoted '*black-spot analysis*'. By analysing the geographical distribution of traffic accidents, instances of poor design of the road network could be identified. The accident black spots were the geographical locations (e.g. crossings), where many of the accidents occurred. The remedial actions were channelled to these locations.

In industrial safety, accident-concentration analysis is carried out in several 'dimensions', place of occurrence being one. The basic assumption is that each industrial system has its own distinguishable accident-risk characteristics. These reveal themselves through the accident occurrences. Analyses of accident statistics support this assumption. Figure 15.5 showed examples of distributions of accidents in different offshore activities. The associated hazards are easily traced back to the types of activities carried out in each type of system. The cut injuries in the living quarters, for example, occur during cutting of meat and vegetables in the kitchen.

The application of accident-concentration analysis is not meaningful for small data sets. There must be at least in the region of 50 accident cases. Useful types of data in accident-concentration analysis are location, activity, equipment, accident type, type of injury and part of body affected. The analysis is facilitated if some of the data is coded (i.e. presented on a nominal scale of measurement), especially if large quantities of data are handled. The coding should, however, not be done at the cost of the details in the information. The free-text description of the sequence of events should always be available.

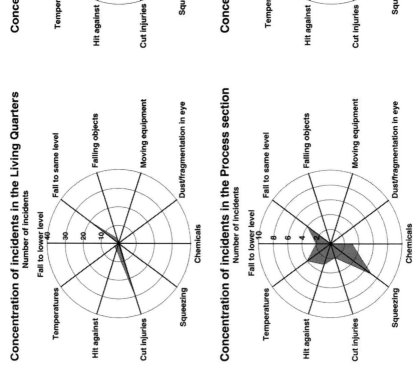

Figure 15.5 Distribution of lost-time and first-aid injuries at two offshore installations by type of event and department. In the living quarters, cut injuries predominate. A further analysis shows that they have mainly occurred in the kitchen. In the drilling section, squeezing accidents predominate, indicating safety problems associated with the handling of heavy drilling equipment.

In an accident-concentration analysis, the analyst must be familiar with the types of work studied. The analysis is carried out in steps:

1 Establish bi-variate distributions of the total set of accidents.
2 Identify cells with a high number of cases.
3 Repeat the procedure for cells with a high number of cases until meaningful accident concentrations have been identified. At the detailed level, it is often necessary to study the description of the sequence of events in the original reports and to arrange these in new, meaningful groups.

Example: A database of about 250 accidents in offshore drilling was analysed. A cross-tabulation of type of event and place of occurrence showed a concentration of accidents on the drill floor involving contact with moving objects (about 10 per cent of the total number of cases). A further analysis by type of activity showed that 17 of the drill-floor accidents had to do with the operation of tongs. Mechanisation of the activity was judged to be a feasible preventive measure.

Figure 15.6 shows another example from offshore yards. Accident-concentration analysis of this type helps prioritise remedial actions. Detailed

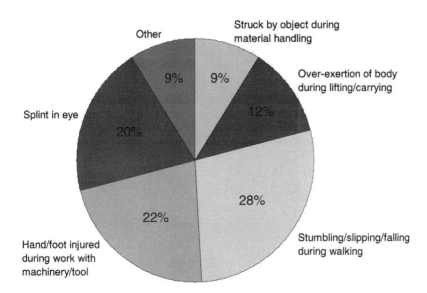

Figure 15.6 Results of an accident-concentration analysis of construction work in an offshore project. The results were used in planning the SHE activities in a subsequent project.
Source: Kjellén, 1997.

analyses of the accident concentrations are required in order to develop the actual measures. One of the remedies that followed from the analysis shown in Figure 15.6 was to introduce rules requiring the workers to use safety glasses.

Accident-concentration analysis is dependent on the analyst's analytic skills and knowledge about the studied industrial system. There are today statistical tools such as FAC (Factorial Analysis of Correspondence) and HAC (Hierarchical Ascendant Classification) that may help identify accident concentrations and thus make the analysis more objective. Such analyses do not replace but are complementary to the more intuitive analysis made by SHE experts. In a study of accidents among female automobile assembly workers, Laflamme (1996) identified four such concentrations by applying FAC and HAC:

- Overexertion causing sprain or strain of shoulder, arm or back;
- Fall or trip injuring lower limbs;
- Contusion or crushing injury involving lifting or transportation device;
- Hand and finger injury from handled objects.

Experience from accident-concentration analyses in different industries suggests that 50 per cent or more of the accidents belong to typical accident concentrations of the specific branch of industry. The rest of the accidents are more evenly distributed. When we introduce measures directed at the accidents of the concentrations, we will initially experience a positive development through relatively moderate efforts. The remaining accidents will, however, be more evenly distributed and hence more difficult to grasp and remedy through focused measures.

When we apply accident-concentration analysis, we must be aware of a common methodological problem due to the so-called *regression-to-the-mean* effect. Let us illustrate this effect by an example. There are on average four accidents a year at a particular workplace. The distribution of accidents at this workplace is shown by Figure 9.2 in Chapter 9. One year, when an accident-concentration analysis was carried out, there happened to be eight accidents at this particular workplace. This is not an unlikely figure, since the probability of eight or more accidents is about 0.05. Eight accidents was considered to be a high number and a safety campaign was initiated. Next year, there were three accidents, and the safety campaign was judged as successful. From statistics we know, however, that the probability of three or fewer accidents is about 0.4, so a reduction by this magnitude was not unlikely from pure chance.

The regression-to-the mean effect means that high or low accident levels tend to approach the mean value in the subsequent period by pure chance, and without any measures being taken. We run the risk of making erroneous conclusions about the positive effects of safety measures if we do not take this effect into account.

15.2.4 Analysis of accident causes

Results of uni- and bi-variate distribution analyses are often used as 'proof' of causal relationships. Let us look at two examples. We have earlier quoted a study that shows that 88 per cent of the accidents are caused primarily by unsafe acts (Heinrich, 1959). This statement is based on an uni-variate analysis of accidents by immediate cause. In the example in Section 15.2, an uni-variate distribution analysis 'showed' that inattention was the cause of almost a quarter of the accidents. These examples bring us to the basic questions about what we mean by an accident cause.

It is, for example, justifiable to speak about 'causes' of an accident if there is a logical relationship between the causes (specific events or conditions in the chain-of-events) and the loss. We thus may say, '*25 per cent of the injuries in the foundry were caused by contact with hot metal*' or that '*a third of the injuries in the slaughterhouse were caused by contact with knives*'.

'Unsafe acts' are examples of more distant factors from the injury event, where the logical relations are more uncertain. So-called **ex-post facto analyses** are employed to study whether a causal relationship may exist. In these analyses, we compare statistics on accidents with similar statistics for accident-free situations. The aim is to identify factors which are more common in the accident material than what is expected because of pure chance. In the next step, physical, physiological and psychological theories are brought in to explain the actual causal relationships.

Example 1: The effect on the risk of accident of using a mobile telephone while driving can be studied in this way. Such a causal analysis requires access to statistics on the percentage of drivers involved in accidents that use mobile telephones and statistics on the percentage of time that drivers in general use such telephones. A significantly higher percentage in the first case indicates that the use of mobile telephones increases the risk of accidents. The analysis does not, however, say anything about the causal mechanisms leading to an increased risk of accidents when using a mobile telephone.

Example 2: Let us consider a hypothetical example of injuries at a hamburger chain. An analysis shows the following distribution of injuries by the victim's job experience:

Table 15.3a Distribution of burn injuries at a hamburger chain by job experience

Job experience	Number of injuries
< 6 months	33
6–12 months	18
1–2 years	9
3–5 years	5
> 5 years	2

The conclusion was reached that inexperience was a primary 'cause' of accidents. A more detailed analysis revealed the following picture that rejects this hypothesis:

Table 15.3b Distribution of injuries at a hamburger chain by job experience

Job experience	Number of injuries	Number of employees	No. of injuries per employee and year
< 6 months	33	412	0.08
6–12 months	18	180	0.1
1–2 years	9	124	0.07
3–5 years	5	98	0.05
> 5 years	2	37	0.06

Statistical hypothesis tests such as χ^2 are used to study the significance of statistical relationships. Basic textbooks on statistical hypothesis testing are referred to. Even if such tests show strong statistical correlation, the causal explanatory value of the results can be questioned. A famous safety expert has commented on Heinrich's results showing that 88 per cent of the accidents are caused by unsafe acts in the following way: 'To say that most of the accidents are caused by human error is like saying that falls are caused by gravity'.

Statistical analyses of accident causes do not necessarily lead us closer to the goal of reducing the risk of accidents. Their value lies in the possibility of predicting the effects of measures to reduce the risk of accidents.

15.2.5 Severity-distribution analysis

Severity-distribution analysis is used in estimations of the probability of severe accidents at a workplace and in comparing different workplaces with respect to the expected severity of the accidents (Briscoe, 1982).

A severity-distribution analysis is based on the accidents for a specified period in time and follows a stepwise procedure:

1   Arrange the accidents by consequence (e.g. the number of days of absence) in an ascending order.
2   Divide the highest registered consequence value into intervals such that each interval has approximately the same size on a logarithmic scale. Table 15.4 shows an example.
3   Tally the number of accidents in each interval and the cumulative number (see Table 15.4).
4   Calculate the cumulative percentage of accidents for each interval and use a log-normal paper to plot the results (see Table 15.4 and Figure 15.7).

Table 15.4 Severity-distribution analysis of accidents at a steel mill. N is the total number of accidents

Days of absence	Number of accidents	Cumulative number of accidents (N_1)	% of the accidents $(N_1/(N+1))*100$
< 1	55	55	24.6
1–3	50	105	46.9
4–7	48	153	68.3
8–30	57	210	93.8
31–90	9	219	97.8
91–180	2	221	98.7
181–365	1	222	99.1
> 365	1	223	99.6

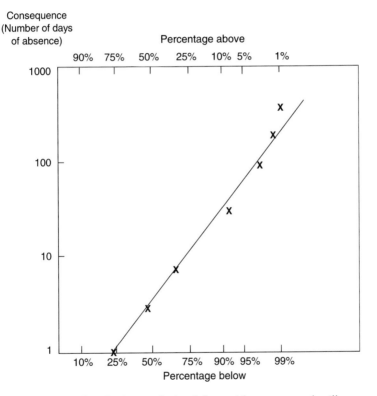

Figure 15.7 Severity-distribution analysis of the accidents at a steel mill.

Figure 15.7 shows an example of a distribution analysis of the accidents at a steel mill. The slope of the straight regression line is an indicator of how probable it is that an accident will result in severe consequences. The higher the slope, the higher is this probability. In the case illustrated by the figure, there is a probability of about 0.02 that an accident will result in more than 100 days of absence.

The points with the highest severity lie above the straight line. This is an indication of the fact that accidents with severe consequences result from a different set of causes than those of the less severe accidents. It follows that these causes are not subject to the same strict management control. This is a warning signal that there is a need to look more closely into the severe accidents.

15.2.6 Extreme-value projection

Extreme-value projection has some similarities to severity distribution analysis but utilises the most severe case for each period only (Briscoe, 1982). This can be a month, quarter or year. The value of this method lies in the possibility of anticipating how long a time we may expect it to take until the company experiences a really severe accident, e.g. a fatality or an accident involving major monetary loss. This is done through extrapolation based on data from less severe accidents. The method also gives input to an evaluation of whether a company has adequate control of its accident risks or not. Extreme-value projection utilises advanced statistics but is relatively simple to use in practice by applying special plotting paper.

An extreme-value projection is carried out according to a stepwise procedure:

1   Collect data on severity (e.g. number of days of absence for occupational accidents or monetary loss in euros for material damages) and date of occurrence for the accidents of the organisation for a specific period in time. There must be at least about 30 accidents for the analysis to be meaningful. The studied time period should not be longer than five years, since we have to assume that the conditions are stable during the period in question. Divide the total time period into sub-periods (e.g. months, quarters or years). There should be at least one accident in each sub-period. The number of sub-periods should not be less than ten and not more than 30.
2   Produce a table illustrating the number of accidents in each sub-period and the severity of the most severe accident of the sub-period. Arrange the sub-periods by severity in ascending order. Table 15.5 shows an example from a steel mill. The severity is measured as the number of lost workdays.
3   Calculate the cumulative probability for each sub-period by dividing N_i by $(N + 1)$, where N_i is the succession order for each sub-period and N is the number of sub-periods. This formula is approximate. Briscoe (1982) presents formulas and tables for calculations of more exact values. A differentiation is here made

between Type I and Type II extreme-value projections. These more exact values have been used in Table 15.5.

4 The severity and the cumulative probability of each sub-period are plotted in a special probability diagram, Figures 15.8 and 15.9. In extreme-value projection Type I, the Y-axis (severity) is linear and in Type II it is logarithmic. The vertical scale in the diagram is selected in a way that allows for extrapolations to two to three times the maximum severity of the material that was analysed.

Table 15.5 Data for extreme-value projection of the consequences of occupational accidents within a steel mill for 4.5 years (18 quarters). One fatality in period 12 was given a loss corresponding to 7500 days of absence

Period (quarter)	Loss (number of days of absence)	Type I	Type II
14	9	0.031	0.28
16	10	0.086	0.83
19	14	0.141	0.139
5	15	0.197	0.194
10	20	0.252	0.250
11	20	0.307	0.306
3	22	0.362	0.361
7	28	0.417	0.417
8	29	0.473	0.472
4	36	0.528	0.528
17	36	0.583	0.583
1	37	0.638	0.639
18	42	0.693	0.694
15	82	0.749	0.750
2	108	0.804	0.806
13	115	0.859	0.861
6	190	0.914	0.917
12	7500	0.969	0.972

Below follow interpretations of the results, Figures 15.8 and 15.9. They show that the Type II-analysis gives the best fit of a straight line to the points in the diagram. The interpretation is that the causes of the individual accident are related to each other. This means that there are deficiencies in the SHE management system resulting in inadequate control of the hazards. To reduce losses, there is a need to implement measures directed at this system. If the Type I-analysis had given the best fit, on the other hand, the interpretation would have been that the causes of the accidents were independent of each other. Accident prevention should in this case be directed at ameliorating the individual causes.

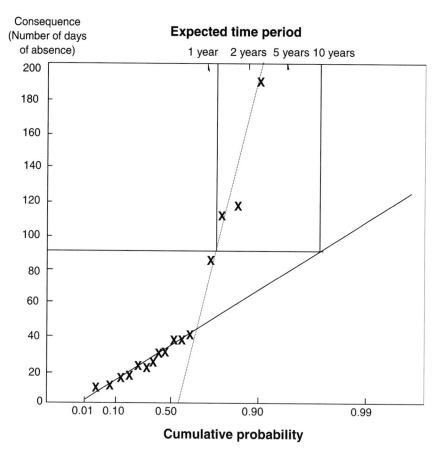

Figure 15.8 Extreme-value projection Type I of the accidents at the steel mill.

Two straight lines give the best fit to the points in the Type I-diagram (Figure 15.8). The interpretation is that there are two different distributions for the accidents with a short and long period of absence respectively. This type of diagram is called 'dog leg' and indicates that the accidents with a short and long period of absence have different causes.

In the Type II-diagram (Figure 15.9), one straight line fits the points. The implications are that there is no upper limit on how severe the accidents may become. If the diagram, on the other hand, had shown an asymptotic limit, there would have been an upper maximum severity of the accidents.

Figure 15.9 shows that an accident resulting in more than one year of absence is expected to occur every five years (50 per cent probability). In both analyses (Type I and II), the fatality (when given a severity corresponding to 7500 days of absence) falls outside the distributions represented by a straight line. These results imply that the fatality is an outlying

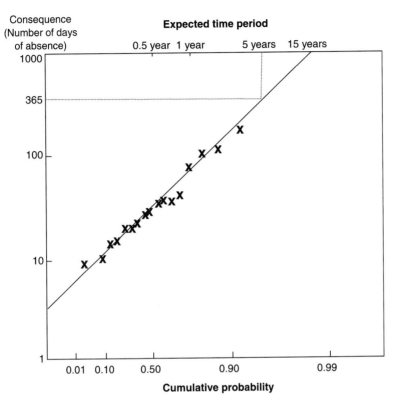

Figure 15.9 Extreme-value projection Type II of the accidents at the steel mill.

case that could not have been anticipated from the distributions of the other accidents.

15.3 Experience carriers

Experience shows that there are many obstacles to a systematic use of accident databases in preventive work. We often find a widespread scepticism within a company's organisations as to the quality and value of the information in the database. The information is found to be too diluted and incomprehensible for practical day-to-day use. Some of these problems may be overcome by the use of adequate analytic methods.

There is another possible way to ensure a systematic use of the accumulated accident experiences within a company. Figure 15.10 shows the application of experience carriers as an intermediate link between the accident database and the end use. This involves improvements of safety in production systems in operation or in the design of new production systems. An *experience carrier* here expresses a company's collective experiences in a format that

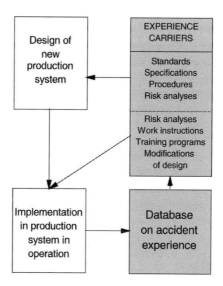

Figure 15.10 Examples of experience carriers used to transfer experiences from an accident database to existing and new production systems.

has meaning for those responsible for the design of safety measures. There are different types of experience carriers, dependent on the tasks and responsibilities of the users.

In the design and construction of new production systems, technical specifications play an important role. They represent a company's collective experiences from their production systems in operation. The company's project organisation instructs machinery and equipment vendors and engineering contractors to comply with the requirements in the specifications. We have to interpret the experiences from individual accidents and near accidents and from analyses of accident concentrations and translate them into specific requirements for use in such specifications. It is recommended that group problem-solving techniques be applied in this process. Regulatory requirements and standards on safety are similar experience carriers at the national or branch-of-industry levels. We will come back to this issue in Chapter 27.

A traditional means of implementing accident experiences in production systems in operation is to update the work instructions. These documents will successively accumulate accident experiences and become experience carriers for use in the training of new employees and in the follow-up of safe behaviour. Modifications of design are also a means of storing the accident experiences in the company's 'collective memory' as long as the changes represent the company's best practice and are used in subsequent design work.

The application of risk analysis offers another means of storing accident experiences in a systematic and meaningful way. Risk analysis is applied in design as well as during the operations phase. In Part V, we will look into different risk-analysis methods. Each method uses a specific means of breaking down complex production systems into 'units' that are meaningful and possible to handle. Systematic use of job-safety analysis, for example, implies that all hazardous jobs at a plant or a department are identified and that each job is broken down into steps. Next, accident risks are identified for each step. If a workplace has been subject to job-safety analyses, new accident experiences may be linked to the applicable step in the analyses. This will ensure that the analyses are kept updated with new experiences. The risk analyses thus provide a living tool for easy retrieval of accident experience for use, for example, in the development of work instructions and in the planning of new work places.

If we want to use experience carriers in a systematic way in accident prevention, we first need to identify all experience carriers of significance to safety. Next, an 'owner' has to be identified for each experience carrier and assigned the responsibility of implementing the new accident experiences. The methods for extraction of experiences from the accident database also have to be defined.

Part IV
Monitoring of SHE performance

In this Part, we will present SHE performance measures or indicators for use in the types of SHE management activities that were presented in Chapter 10 and named persistent feedback control. These activities involve the establishment of SHE performance goals and the follow-up of such goals through measurement of actual performance. The principles for establishing goals may differ:

- A fixed goal is established for a specific time period, e.g. the following year. This type of goal is usually based on previously achieved results at the company or on results of other companies that management wants to compare its company with.
- It is stated that the SHE performance indicator must show continuous improvements from one period to the next.

SHE performance indicators play an important motivational role, especially at the higher management levels. They make SHE visible in a summary way that is suited for communication, comparison and competition.

We will review different measures or indicators of SHE performance and their merits and shortcomings when applied in SHE management. Part IV starts in Chapter 16 with an overview of SHE performance indicators. The following chapter presents different indicators using data about accidents and losses as the main input. Chapter 18 presents various indirect indicators based on data about the process that may result in loss. In Chapter 19, finally, we will look into SHE performance indicators, where the focus is on the underlying contributing factors and basic causes.

16 Overview of SHE performance indicators

Figure 16.1 presents an overview of different SHE performance indicators. This overview is based on the framework for accident analysis in Chapter 6. Loss-based SHE performance indicators will be our starting point. Among these we find the most commonly used indicator, the lost-time injury frequency rate. We proceed by reviewing process-based indicators, similar to those developed in the fertiliser-plant case in Chapter 4. Finally, we will look into indicators relating to causal factors, i.e. indicators based on information about the organisation and SHE management system.

SHE performance measurement utilises all four sub-systems of a SHE information system. The performance indicators apply data from the different sources presented in Chapter 12 such as accident and near-accident reporting, workplace inspections and SHE audits. The data is analysed and presented in a specific way, enabling comparison with goals, previous results, results of other companies, etc.

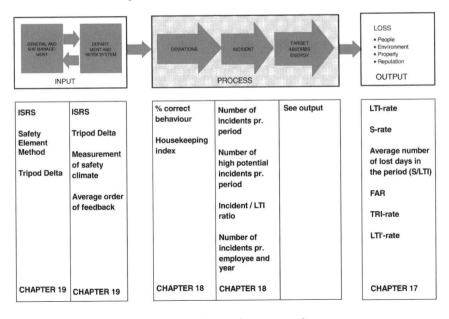

Figure 16.1 Overview of different SHE performance indicators.

17 Loss-based SHE performance indicators

17.1 The lost-time injury frequency rate

The lost-time injury frequency rate (LTI-rate) is the most commonly used indicator of SHE performance.

The LTI-rate is defined as the number of lost-time injuries per one million hours of work. A lost-time injury is an injury due to an accident at work, where the injured person does not return to work on the next shift.

In applications of the LTI-rate in SHE management, we calculate the LTI-rate from statistics on accidents and working hours and compare the results with the pre-established safety goals. We are also concerned with the development of the LTI-rate over time. We will here look more closely into the principles of analysing time series of the LTI-rate before we discuss the merits and shortcomings of the LTI-rate in general.

17.1.1 The control chart

In time-series analysis, we are interested in the development of the SHE performance indicators over time. This is done in so-called **control charts**, where the SHE performance indicator is calculated for consecutive periods and displayed in a diagram. These charts allow us to study changes in performance from one period to the next and trends over several periods.

We will here look into the control chart for the LTI-rate. Figure 17.1 shows an example.

The measured LTI-rate for a period is an estimate of the underlying or 'true' LTI-rate. From the statistical theory presented in Chapter 9 we know, however, that this estimate is uncertain and that we must expect the rate to fluctuate from period to period due to pure chance. The upper and lower control limits define the range in between which the LTI-rate will fall on average in 19 out of 20 periods, provided that the underlying accident risk at the company is unchanged. These limits are calculated from the mean LTI-rate during the studied periods, since this is considered to be the best estimate of the 'true' LTI-rate.

LTI-rate

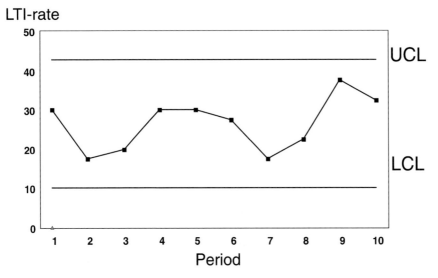

Figure 17.1 Control chart showing the development of the lost-time injury frequency
rate for a steel mill for ten consecutive periods.
UCL = Upper Control Limit; LCL = Lower Control Limit.

We consider a change in the LTI-rate for the latest period to be significant,
if its value falls outside the control limits for earlier periods.

Let us consider the example shown in Figure 17.1. The lower and
upper control limits (LCL and UCL) have been calculated on the basis of
the LTI-rates for ten consecutive years. There is a 5 per cent probability
that the value for the eleventh year will fall outside these limits by pure
chance (95 per cent confidence interval). If the LTI-rate for this year is
45, for example, we conclude that the risk of accidents has increased
significantly.

The upper and lower control limits are calculated by applying the following formulas (1), (2) and (3):

$$p = X_{tot} * 10^6/(N * e) \qquad (1)$$

$$UCL = p + 2 * \sqrt{(p * 10^6/e)} \qquad (2)$$

$$LCL = p - 2 * \sqrt{(p * 10^6/e)} \qquad (3)$$

'p' is the mean value of the LTI-rate for all periods that are included in
the control chart, 'X_{tot}' is the total number of lost-time accidents, 'N' is the

number of periods, and '*e*' is the mean number of working hours per period. Upper and lower control limits are applied in evaluating the significance in changes in the LTI-rate.

Example: An aluminium plant has 2000 employees and each employee works 1800 hours a year. During the last ten years, an average of 35 lost-time accidents per year was registered. For a control chart with a periodicity of one year, we calculate the following:

$$Average\ LTI\text{-}rate = 350 * 1000000/(10 * 2000 * 1800) = 9.7$$

$$UCL = 9.7 + \sqrt{(9.7 * 1000000/(2000 * 1800))} = 9.7 + 3.3 = 13.0$$

$$LCL = 9.7 - \sqrt{(9.7 * 1000000/(2000 * 1800))} = 9.7 - 3.3 = 6.4$$

The procedure for the establishment and use of control charts is as follows:

1 Start establishing the control chart only when at least 100 accidents have been registered. Do not use data of more than five years of age.
2 Decide about the period length: there must not be less than five accidents per period on average. An average of at least ten accidents per period is preferred. Period lengths of more than one year are not meaningful.
3 Calculate the LTI-rate for each period.
4 Calculate the mean LTI-rate for all periods and the upper and lower control limits according to formulas 1 – 3 above.
5 Draw the control chart.
6 If one point falls outside the control limits, calculate new control limits without this point. If still one or more points fall outside the control limits, the SHE performance is not stable. Causes of the unstable SHE performance have to be identified.
7 Use the control chart to determine whether the LTI-rate for the next period falls inside the control limits or not.
8 Use the control chart on a routine basis to evaluate the SHE performance for new periods. Keep the control limits updated with accident and exposure data until the period before the last period. Avoid using data for more than five years back in time, or from earlier periods that for some specific reason are not representative of the current situation.

Exercise: Ship Yard Inc. has 5000 employees and the number of work-ing hours per employee and year equals 1800 on average. There were the following number of lost-time injuries (LTI) per month during last year:

January:	35
February:	30
March:	25
April:	30
May:	29
June:	25
July:	16
August:	20
September:	24
October:	34
November:	30
December:	17

Draw a Control chart showing the LTI-rate for each month during last year and the upper and lower control limits for the whole year. In January this year there were 10 LTIs. Is the LTI-rate for this month within the control limits? What is your conclusion from the develop-ment of the LTI-rate in January?

In SHE practice, we rarely detect statistically significant changes in the LTI-rate. Let us look at an example where management establishes a goal for the LTI-rate for one year, which represents a 25 per cent reduction as compared to the average LTI-rate of previous years. When we apply the formula for the standard deviation of the Poisson distribution in Chapter 9, we can show that the mean number of lost-time injuries per period must be at least 60 to be able to say that a 25 per cent reduction in the LTI-rate is significant. Only the largest companies are able to base their accident statistics on that number of accidents per year. It is common practice to apply control charts of the LTI-rate in this way also when the goal does not represent a statistically significant change.

In quality control of production processes, control charts are applied to display results of statistical sampling tests of products. Basic textbooks propose the use of a confidence interval of +/− three standard deviations in this application (see e.g. Wig, 1996). For a stable distribution, 99.7 per cent of the results of the tests will fall within this interval. The underlying normal distribution for the production process will determine the width of the area between the control limits. When the production series is large enough, the sampling size and thus the degree of precision in the estimates of the mean value and the control limits are at the discretion of the quality-control engineer. In our application of control charts, the underlying number of accidents per period determines the width of the area between the control limits. We will often struggle with the problem that the frequency of accidents is too low for us to be able to establish meaningful control charts. We have therefore relaxed the requirements to a confidence interval of +/– two standard deviations.

17.1.1.1 Trends

We are not only interested in absolute changes in the LTI-rate but also in changes over several consecutive periods.

We speak about a positive or negative trend if the LTI-rate increases or decreases from one period to the next during five or more consecutive periods.

This figure is arrived at through simple probability calculus. The probability of an increase or decrease in two consecutive periods by pure chance is 0.5. Similarly, the probability for an increase or decrease during five consecutive periods is 0.06, which approximately coincides with our requirement as to significance.

Figure 17.2 shows an example of a trend. In this case it is not meaningful to plot the control limits.

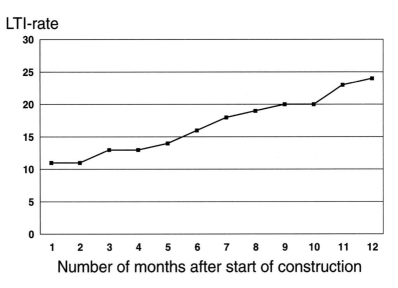

Figure 17.2 Chart displaying a trend in the development of the LTI-rate.

17.1.2 *The problems of SHE performance measurement*

Let us evaluate the LTI-rate in relation to the criteria that SHE performance indicators have to satisfy, see Table 11.1. Such an evaluation reveals a number of problems with the LTI-rate that are well known and have been discussed in the literature (see e.g. Rockwell, 1959).

First we ask whether the LTI-rate is quantifiable and measurable. This criterion is satisfied, since the LTI-rate is expressed on a ratio scale of measurement and draws from readily available statistics.

Next, we ask whether the LTI-rate is a valid indicator of the risk of losses due to accidents. This criterion is more problematic, since the LTI-rate is insensitive to the severity of the injuries. An eye injury resulting in a few days of absence and a severe fall injury with many months of sick leave count equally when calculating the LTI-rate. It is questionable whether the LTI-rate is a valid indicator of the risk of losses due to accidents. Other SHE performance indicators that we will look into in the next Section are better suited as to validity, since they account for the degree of harm (e.g. number of days of absence).

Is the LTI-rate robust against manipulation? Let us discuss this criterion by looking at an example.

Example: Seven yards received contracts for the manufacturing of different modules for an offshore installation. The client (an oil company) monitored the LTI-rates of the yards closely. Five of these ended up with an LTI-rate below 20; the LTI-rate was 55 and 70 for the two remaining yards. A closer analysis revealed that these latter yards had been more inclined to report minor accidents with only a few days of absence, Figure 17.3. There were reasons to believe that the five 'low LTI-rate yards' had established schemes for avoiding reporting minor injuries such as offering the injured employees alternative work.

Experience shows that it is possible to manipulate the registration and classification of injuries and thus to affect the LTI-rate in a desired direction. This phenomenon may take place, for example, when top management unilaterally establishes ambitious SHE goals for the workplaces. The effects occur when the LTI-rate is closely monitored and negative results have consequences to the supervisors and the work force. Examples of such consequences are top management attention and withdrawal of bonuses. A common means of reducing the number of reported lost-time injuries is to offer the injured person an alternative job that is less demanding. This phenomenon is an illustration of the risk-homeostasis theory of Chapter 3. We cannot expect to accomplish accident-risk reduction unless top management's ambitions to improve are shared by the whole organisation and affect the actual behaviour at the workplace level.

Is the LTI-rate sensitive to change? To put the question in another way, will a change in the risk level at a workplace show up in the control chart as a significant jump in the LTI-rate? The answer is most often 'no'. This

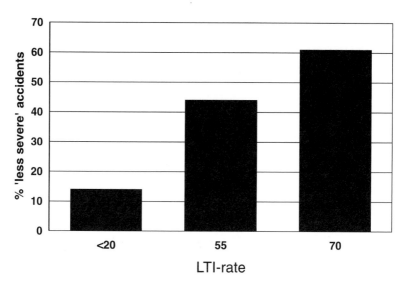

Figure 17.3 Percentage of the lost-time accidents at seven yards that were of a less severe type as a function of LTI-rate. This shows the average percentage for five yards with an LTI-rate below 20 and the actual values for the two remaining yards. 'Less severe' accident types include 'fall on same level', 'tripping' and 'fragment in eye'. Experience shows that these accident types often result in injuries with only a few days of absence.

has to do with the fact that lost-time injuries are rare events. Five lost-time injuries per period on average, for example, give statistical fluctuations in the LTI-rate in the order of +/– 100 per cent.

We will regard five accidents per period as a minimum number in order to produce meaningful control charts. Figure 17.4 shows the minimum period length for a given number of employees that follows from this figure.

For small companies, the period length will be too long in relation to our need for timely feedback. From Figure 17.4 we see that a company with 50 employees will have to use a period length of three years in their control chart. Two methods are applied to increase the basis for the LTI-rate calculations for small companies:

1   In the *moving average* method, the mean LTI-rate is calculated for a fixed number of periods ending with the last period. This method reduces the fluctuations but also the sensitivity to changes.
2   The *accumulated LTI-rate* is calculated from one point in time and for the following periods. In project work, where the activities have a defined beginning and end, this method has its advantages.

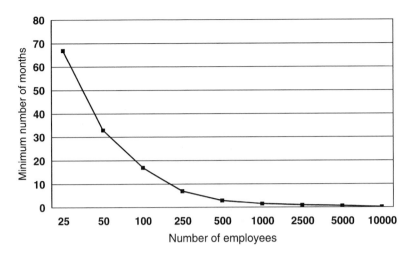

Figure 17.4 Relation between the minimum period length in months and the number of employees in order to produce meaningful control charts. A mean LTI-rate of 20 is assumed.

LTI-rate

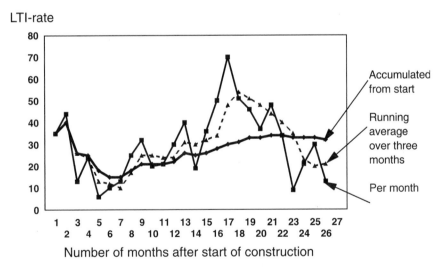

Figure 17.5 Chart showing the parallel developments of the LTI-rate, the accumulated LTI-rate from the first period and the moving 3-months average of the LTI-rate.

Figure 17.5 displays these two methods together with a traditional display of the LTI-rate.

First-aid accidents and near accidents are more frequent events. It is not, however, meaningful to use the frequency of such events as an indicator of the likely risk of injuries. On the contrary, experience shows that there often is a negative correlation between the frequency of near accidents and the frequency of lost-time injuries; i.e. when the near-accident frequency rate increases, a reduction in the LTI-rate follows (compare Figure 13.4).

Finally we ask whether the LTI-rate is easily understood and congruent with other performance indicators. Here we see the two criteria that explain the success of the LTI-rate in spite of its shortcomings. Experience shows that the LTI-rate is easily communicated to the layman and that the merits of a low LTI-rate are by-and-large accepted.

To summarise, we have discussed the following problems associated with the use of the LTI-rate in control charts and in trend analyses:

1   Insensitivity to the severity of the accidents.
2   Reporting inaccuracies.
3   Statistical fluctuations.

Exercise: Display the change in the yearly LTI-rate as a function of the number of employees for a company when there is one additional accident. Assume 1800 hours per employee per year. Discuss the results.

17.1.3 Zero-goal mindset

Many large companies, especially within major hazardous and construction industries, have adopted the so-called 'zero-goal mindset or philosophy' (Levitt and Samelson, 1993). This philosophy is based on the belief that all occupational injuries and illnesses and process and environmental incidents are preventable and management's goal for them all is zero. It is an easily communicated goal with many implications that are not always fully understood.

The zero-goal mindset or philosophy is rooted in the 'continuous improvements' programmes of the total-quality-management movement. The performance is expected to improve by a certain percentage each year (e.g. 25 per cent) as compared to the previous year. At the department level, where there may be only one or a few accidents each year, we often find goals of zero accidents. This type of goal is more problematic, since chance plays an important role. In practice, the counting of accidents is set at null after an accident and the goal is to have no accidents thereafter.

When dealing with occupational injuries, the focus is on performance indicators such as the LTI-rate and the goal is to reduce the measured values to zero. Management focuses its attention on minor incidents and injuries, motivated by the belief that a reduction of the LTI-rate to virtually zero will be followed by a similar reduction in the risk of major accidents.

Unfortunately, empirical evidence does not fully support this hypothesis. When the focus is on eliminating the causes of the few minor accidents that still occur, we cannot expect this to have any significant effect on the risk of major accidents. This is because there are different immediate causes of minor and major accidents.

17.2 Other loss-based SHE performance indicators

17.2.1 Measures of risk

The shortcomings of the LTI-rate have generated work to develop alternative SHE performance indicators. There is also a need for SHE performance indicators in areas other than occupational accidents. The LTI-rate belongs to the important category of loss-based SHE performance indicators. They all represent measures of the risk of accidents. This risk is related to an identified activity. We use SHE performance indicators to calculate the historical risk associated with activities that have taken place in the past. In Part V we will present methods to estimate the expected risk associated with future activities by applications of risk analysis.

We define the **risk of accidents** as a combination of the probability or frequency of accidents involving losses in a specified activity, and the extent of the losses (consequences). The measure of risk has three components:

1   A measure of the *exposure* to the activity involving accident risks.
    Example: In the LTI-rate, the exposure is measured in terms of the number of man-hours in the activity in question.
2   A measure of the *frequency or probability* of accidents in the activity per unit of exposure.
    Example, continued: In the LTI-rate, the number of accidents per million work-hours in the activity in question determines the frequency.
3   A measure of the *extent of loss (consequence)*.
    Example, continued: The consequence in the LTI-rate is measured on a dichotomous scale related to whether the accident has resulted in lost time or not.

There are two ways of reducing the risk of accidents in an activity. One way is to reduce the frequency of accidents per unit of exposure to the activity. The second way is to reduce the consequences. If we look at an organisation as a whole, there is also a third way of reducing the overall risk of accidents. The risk of accidents is not the same for all activities in the organisation. We may hence reduce the overall risk of accidents by reducing the exposure to high-risk activities. Transferring personnel to less hazardous work does the job. It is important, however, that the risk of accidents does not increase for the remaining personnel performing the high-risk activity.

17.2.2 *Standard loss-based SHE performance indicators*

Table 17.1 gives an overview of standard loss-based SHE performance indicators. In calculating the frequency of accidents, the size of the activity for which performance is assessed has to be considered. We must expect more accidents to occur in a large company than in a small one, even if the activities are similar. It is necessary to standardise the indicators in relation to the *exposure* to the risk of accidents. For occupational accidents, the most common exposure measure is the number of employee-hours. In traffic safety, for example, we use the number of vehicle kilometres or passenger kilometres as measures of exposure. By combining the frequency measures with the consequence measures, a measure of the risk is arrived at.

The **S-rate** (severity rate) is less sensitive to reporting inaccuracies than the LTI-rate. One single accident resulting in a long period of sick leave may, however, dominate the statistics. It follows that the S-rate may vary considerable from period to period, especially in small companies. Another problem with the S-rate has to do with the fact that the sick leave for an injury may extend into the following periods. The true S-rate is thus not available until all injuries from a period have been closed. Due to the low frequency of fatalities, **FAR** (fatal accident rate) is rarely useful as a SHE performance indicator other than for very large companies in hazardous branches of industry.

The types of performance indicators based on environmental and material losses vary from company to company dependent on type of production. Table 17.1 shows typical examples.

Table 17.1 Loss-related SHE performance indicators

Type	Definition
Occupational accidents:	
Lost-time injury frequency rate (LTI-rate)	Number of lost-time injuries per 10^6 employee-hours. For definition of lost-time injury, see Table 6.3.
Severity rate (S-Rate)	Number of working days lost due to lost-time injuries per 10^6 employee-hours. Fatalities and 100 per cent permanent disability account for 7500 days.
Average number of days lost	S-rate/LTI-rate
Fatal Accident Rate (FAR)	Number of fatalities per 10^8 working hours
Environmental pollution:	
Rate of emissions	Emissions due to accidents in kg or m^3 per ton production (e.g. emissions of fluor to the air in kg per ton produced primary aluminium)
Material losses:	
Loss rate	Number of accidents or loss in kNOK per produced unit (e.g. number of traffic accidents per 10^5 passenger-km)

Table 17.2 Example of untraditional SHE performance indicators

Type	Definition
Total recordable injury rate (TRI-rate)	Total number of recordable injuries per 10^6 employee-hours. The recordable injuries include fatalities, lost-time injuries, medical treatment injuries other than first aid, and injuries resulting in loss of consciousness, transfer to another job or restricted work (compare Table 6.3).
LTI'-rate	Lost-time injuries with the potential for severe harm per 10^6 employee-hours.
Days since last LTI	Number of calendar days since last lost-time injury.
Accumulated number of lost-time injuries	Accumulated number of lost-time injuries since the start of the period.

17.2.3 Untraditional SHE performance indicators

Above we discussed some problems with the traditional loss-based SHE performance indicators. A number of alternatives have been developed to compensate for some of these problems, Table 17.2.

The **TRI-rate** (total recordable injury rate) has been developed for two purposes. One purpose is to increase the statistical basis by including other types of injuries than those resulting in lost time. A second purpose is to make the indicator more robust against the type of 'manipulation' where the injured employee is transferred to another job rather than being sent home for sick leave. The TRI-rate shares the problem with the LTI-rate that it is insensitive to the degree of harm.

Figure 17.6 shows an example of TRI-rates and their components for a company's different plants. From this type of chart, not only the TRI-rate but also the LTI-rate and the rate of lost-time injuries plus restricted work injuries are easily distinguished. We can see that plant C does not practise alternative job assignment in case of injury. The LTI-rate from this plant is thus comparable to the rate of lost-time injuries plus restricted work injuries at the other plants.

The **LTI'-rate** only includes those lost-time injuries that have resulted in severe harm or have such a potential (Boe, 1996). In determining the potential of the events, the involved energy is considered. Accident types included in the definition of high-potential injuries are large falling object, fall to a lower level and being squeezed by large objects. A higher degree of validity in relation to the risk of accidents is thus achieved at the expense of a reduced statistical basis.

Two indicators, the **Number of days since last LTI** and the **accumulated number of lost-time injuries** both have the advantage of being readily understood. They are, however, insensitive to the size of the company and are

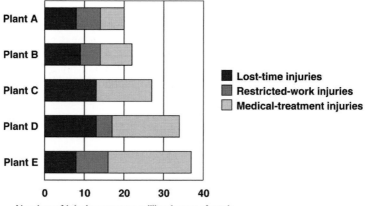

Figure 17.6 Total recordable injury rates and their components for five plants producing car components.

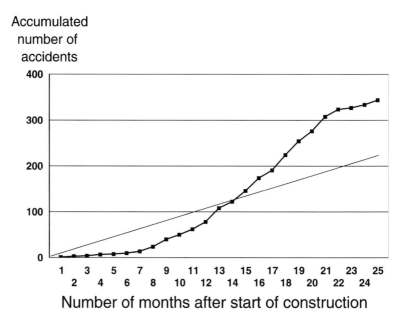

Figure 17.7 Nelson Aalen plot of accidents in a large construction project. The
straight line shows the accepted development, when the safety goal is
to end up at an accumulated LTI-rate below 20. This corresponds to a
maximum of 214 accidents.

thus not well suited for companies with significant changes in the number of employees.

Figure 17.7 shows a so-called **Nelson Aalen plot** of the accumulated number of accidents. The higher the inclination, the higher the frequency of accidents. A related SHE performance goal is to achieve a reduced number of accidents at the end of the period. This shows up as a straight line in the plot. When the actual performance lies above this line, corrective actions need to be taken.

18 Process-based SHE performance indicators

In the example in Chapter 4 we saw the advantages of feedback control through measurement of the process rather then the losses. This distinction between process and loss measurement relates to the basic accident framework of Chapter 6. We will here further explore the potential of the process-based SHE performance indicators, Table 18.1. An underlying assumption in using process-based SHE performance indicators is that these are valid indicators of the risk of accidents. We will discuss this issue further in connection with the individual measures.

18.1 SHE performance indicators based on near-accident reporting

In Table 18.1, we find different indicators of the near-accident or **incident rate**. As opposed to the loss-based SHE performance indicators, we are here in most cases interested in attaining high rates or ratios, since it is assumed to reflect the organisation's safety awareness and willingness to learn from the incidents. Goal-setting is a tool in use to acquire high reporting rates/ratios. A superficially high reporting rate may, however, be counterproductive when the learning value of the reported events is low.

Figure 13.4 showed an example where a plant experienced a negative correlation between the developments in the LTI-rate and the incident rate. The interpretation was that near-accident reporting had demonstrated its efficiency in promoting safety. The robustness against manipulation is not a concern here, since it is the actual reporting behaviour that is our measure. 'Manipulation' is a means of acquiring a high reporting rate/ratio.

A high or increased **rate of incidents with high potential** is an alarm bell, indicating the possibility of an increased risk of large losses. High-potential incidents include those incidents with the potential of resulting in severe injury or fatality. They are usually more difficult to hide, which means that we can expect an adequate reporting reliability in most cases.

The high-potential incident rate may be used directly as a SHE performance indicator. The control chart approach is here applicable, where the number of incidents per period is plotted for consecutive periods. The control limits are determined as $+/-2 \times \sqrt{(mean\ number\ of\ incidents\ per\ period)}$.

Table 18.1 Example of process-related SHE performance indicators

Type	Definition
Incidents:	
Incident rate	Number of incidents (near accidents) per period
High-potential incident rate	Number of incidents with grade 3–5 per period, see Table 6.4
Incident/LTI ratio	Ratio between the number of reported incidents in a period and the number of reported lost-time injuries in the same period
Number of incidents per employee and year	
Deviations:	
Percentage of unsafe acts/conditions	Ratio between the number of observed deviations and total number of observed opportunities of deviations per inspection

This approach is valid only when the activity level is relatively stable from period to period. It is highly relevant in high-risk industries, where specific types of accidents dominate the risk of fatalities and major material damage. In a yard, for example, incidents involving falling objects or risk of falling to a lower level are of special concern. In refineries, gas leakage is a precursor of explosions. Leakage frequency should be monitored carefully in order to identify increasing trends that are indications of reduced control.

The use of goal-setting is here applicable, but must be carefully evaluated, since this approach may have negative impact on the employees' willingness to report incidents, see example in Section 13.3.

18.2 Behavioural sampling

Behavioural sampling was developed in the 1950s to overcome some of the problems associated with the LTI-rate. It applies statistical sampling techniques in observing deviations from accepted safe work practices and conditions (Rockwell, 1959; Tarrants, 1980). The aim is to prevent losses by reducing the frequency of such deviations. The basic principle of accident prevention is the same as in the fertiliser plant case of Chapter 4. In behavioural sampling, we are concerned with the behaviour of people rather than of technical processes.

The measurement of SHE performance through behavioural sampling includes the following steps:

1 Identification of critical behaviour by analysing accident reports, safety instructions, inspection reports etc.

2 Selection of behaviour to be included in the performance indicator and establishment of a checklist with operational definitions of each item. The selected items have to be easily observable and the distinction between safe and unsafe has to be clear.
3 Inspections of the workplaces at randomly selected intervals to observe the items and whether the performance is correct or not.
4 Plotting the safety performance index on a control chart. The safety performance index is defined as the percentage of the observed items that are judged as correct.

The next step in the development of behavioural sampling took place in the 1970s (Komaki *et al.*, 1978). They based their work on behaviour theory (see Section 13.3) and introduced feedback to the workers as a consequence. (See Figure 18.1.) The rationale behind this scheme was to increase the immediate and positive consequences of safe behaviour. After a first 'secret' measurement of the baseline, the results were presented to the

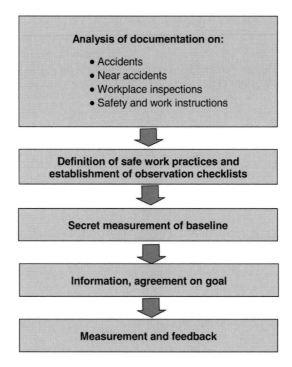

Figure 18.1 Application of behavioural sampling in controlling performance.
Source: Komaki *et al.*, 1978.

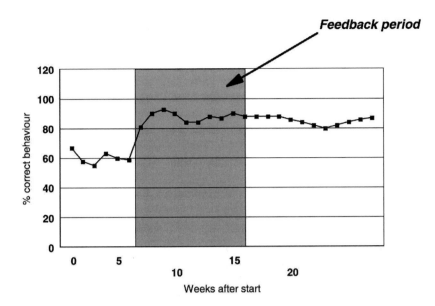

Figure 18.2 Development of the so-called housekeeping index at a shipyard. Neither the control limits nor the performance target are shown.
Source: Saari and Näsänen, 1989, page 207. Copyright 1989 by Elsevier Science and reproduced by permission.

workforce at the department. This baseline should preferably lie around 50–60 per cent to indicate potential for improvements without making the task to accomplish acceptably error-free performance seem insurmountable. An agreement on improvements was arrived at such as 90 per cent safe performance. Next, the safety performance was measured and the results were displayed together with the mutually agreed goal. Positive developments in safety performance as well as in the LTI-rate have been reported following the introduction of behavioural sampling (Sulzer-Azaroff *et al.*, 1994; Krause *et al.*, 1999).

Figure 18.2 shows an example of a chart where the development of the so-called 'housekeeping index' is displayed. The observation items in this application of behavioural sampling were observable conditions at the workplaces related to housekeeping, i.e. items that represent results of behaviour rather than the actual behaviour itself.

Behavioural sampling in its original version has been criticised because it represents a view of workers as objects for control rather than as active participants in the work processes to improve safety. It is thus not well suited for application where a more participatory tradition in SHE practice prevails. In Finland, a version of behavioural sampling called 'Tuttava' has been developed to make it more suited for these conditions. Tuttava addresses such problems as (Saari, 1998):

1   The dichotomy between 'safe' and 'unsafe'. Originally, this has been defined through analysis of historical documents on accident and near-accident record, safety and work instructions, etc. and may be problematic. This is because the investigator may not be fully aware of the different consequences of the 'safe' and 'unsafe' acts. In Tuttava, the definition is based on a thorough work analysis with participation of the workers concerned.

2   Outside investigators observing people. This may lead to blame and a feeling of 'big brother sees you'. In Tuttava, the focus is on physical conditions such as housekeeping rather than on the behaviour creating such conditions. This is illustrated by the example in Figure 18.2. The workers and supervisors make the observations themselves. By this focus on housekeeping, people at the workplaces can observe the positive results of better order directly.

3   Behavioural sampling being based on Van Court Hare's principles of first-order feedback. No lasting effects other than those concerned with workers' behaviour and cognition are expected. In Tuttava, so-called implementation teams of workers and supervisors analyse the causes of deviations in order to come up with technical, organisational and procedural measures to prevent recurrence and to facilitate correct behaviour. These teams represent the equivalent of the so-called quality improvement projects.

4   A narrow focus on safety. In Tuttava, the performance items will affect not only safety but also productivity and quality. The aim is to make the programme more interesting to management as well and to secure the necessary funding for improvements.

Exercise: Evaluate both the original behavioural sampling technique and Tuttava in relation to the criteria in Chapter 11. Discuss how behavioural sampling may affect the behaviour of the employees and the supervisors.

Behavioural sampling is a rather elaborate method. Below is an example of a simplified method that has been developed in industry.

Example: An aluminium plant has introduced a so-called housekeeping index. The plant is subdivided into housekeeping areas, and four times a year the safety engineer and the supervisor and safety representative from the area perform a safety inspection. A checklist with well-defined check items is used. Examples of such check items are:

- *Are walkways free from obstacles and adequately marked?*
- *Are floors and stairs kept free from litter?*
- *Are waste containers available and adequately used?*

- *Are tools and equipment kept in the right place?*
- *Are goods and products kept in designated areas?*
- *Is the area around machinery kept tidy?*
- *Are signs in place?*
- *Are windows and lighting appliances cleaned and in order?*
- *Are toilets and coffee bars kept tidy?*

Each area is evaluated by applying this checklist and the results are given a mark from 1 to 5, where 1 means that there are abundant deviations and 5 that no deviations have been found. The average value is calculated for each area and displayed in a diagram.

19 Causal factor-based SHE performance indicators

We now move further to the left in the accident-analysis framework (Figure 16.1). We will look into SHE performance indicators based on information about contributing factors and root causes. The organisation and SHE management system are in focus. These indicators have many similarities to the audit methods described in Section 14.2.

19.1 Rating the elements of a company's SHE management system

Recent developments in the development of standards on corporate SHE management systems have made it possible to introduce indicators that measure the degree of compliance with these standards. The different performance indicators or rating systems in this area are built up around the following principles:

1 An ideal SHE management model defines the elements of the corporate SHE management system and its contents.
2 A scale is established for measurement of each element with respect to degree of adequacy or degree of compliance with the standard.
3 A set of criteria is used in evaluating actual performance in relation to the ideal model.
4 Different data collection methods are employed in the evaluations. They include:
   a Self-evaluation by a company rating team.
   b Rating by an independent assessor (third-party verification).
   c Questionnaire to the work force.
5 The company establishes goals for comparison with actual performance.

There is no scientific evidence that helps us in selecting a particular SHE management model for use in this application. We have already touched upon some of the models available on the market (Section 5. 7). They apply the principles of feedback control and use codified experiences from SHE practice as input. The models vary with respect to the number and types of elements and in the detailed definition of each element.

19.1.1 International Safety Rating System (ISRS)

The International Safety Rating System (ISRS) uses standardised questions during data collection and fixed criteria in evaluating the results. Specially trained auditors, who usually come from an outside organisation, do the rating. ISRS is based on the ILCI model of accident causation. It was primarily developed as a safety-performance measurement system but is now applied as an audit tool. Twenty topics are addressed, Table 19.1. For each topic, a number of questions and criteria for evaluation of the results have been developed. This makes it possible for the auditor to assess the company in a comprehensive and reproducible way. Dependent on the results, the company is given a score for each topic, ranging from 1 to 10, and the results are displayed in a way similar to Figure 19.1. In addition, the company is given an overall score or 'ISRS level' that also ranges from 1 to 10. The maximum achievable number of points is 12000 and the maximum number of points per topic is dependent on the importance of the topic in question (Table 19.1). To reach a specific level, the company has not only to achieve at least a certain number of points but also to achieve certain minimum results on each topic. It is thus not possible to compensate for a bad performance on one topic, say 'education of leaders' by a good performance on 'inspection'.

The ISRS levels are subject to goal-setting similar to the way the LTI-rate is employed. Typically, top management sets goals for improvements in the ISRS level for the following years. In order to meet the goals, middle management applies the ISRS manual as a comprehensive and detailed recipe for development of their internal safety-management system. This may be done at the cost of internal priorities and is an example of how a measurement system will affect the behaviour of the organisation being measured. Evaluations of ISRS suggest that it has a low effect on such SHE performance indicators as the LTI-rate (Eisner and Leger, 1988; Guastello, 1993).

Exercise: Evaluate ISRS in accordance with the requirements to SHE performance indicators in Chapter 11. Discuss the effects of introducing ISRS on management behaviour.

ISRS has stimulated work inside companies and branches of industry to develop their own rating system adapted to their needs.

Example: An oil company has developed a contractor's SHE qualification guideline, based on these principles (Statoil, 1997). Seven elements in the contractor's SHE management system are considered. They include (compare the E&P forum model in Figure 5.13): Leadership and commitment; Policy and objectives; Organisation, resources and documentation; Evaluation and risk

Table 19.1 Measurement of safety performance according to ISRS

Topic	Level 1	2	3	4	5	6	7	8	9	10	Points
1 Leadership and administration	×	×	×	×	×	×	×	×	×		1310
2 Education of leaders	×	×	×	×	×	×	×	×	×		700
3 Inspections	×	×	×	×	×	×	×	×	×		690
4 Job analysis and procedures		×	×	×	×	×	×	×	×	×	650
5 Investigation of accidents and near accidents	×	×	×	×	×	×	×	×			605
6 Observations of job execution								×	×	×	450
7 Emergency preparedness	×	×	×	×	×	×	×	×	×	×	700
8 Safety rules and work permits	×	×	×	×	×	×	×	×			615
9 Analysis of accidents and near accidents				×	×	×	×	×	×	×	550
10 Education of personnel	×	×	×	×	×	×	×	×	×	×	700
11 Personal protective equipment	×	×	×	×	×	×	×	×	×		380
12 Health services and industrial hygiene	×	×	×	×	×	×	×	×	×		700
13 Evaluation of safety work					×	×	×	×	×	×	700
14 Technical-change control			×	×	×	×	×	×	×		670
15 Personal communication	×	×	×	×	×	×	×	×	×	×	490
16 Group communication	×	×	×	×	×	×	×	×			450
17 Safety motivation			×	×	×	×	×	×	×	×	380
18 Hiring of personnel, job assignment	×	×	×	×	×	×	×	×	×	×	405
19 Control of goods and services	×	×	×	×	×	×	×	×	×		615
20 Leisure-time accidents							×	×	×	×	240

Source: Det Norske Veritas, 1994. Copyright Det Norske Veritas and reproduced by permission.

management; Planning and procedures; Implementation and performance monitoring; and Auditing and review. For each element, a number of questions are asked. The table below shows an example of a question and the scale applied in evaluating the results. Results are displayed in a profile, showing the mean rating on each element.

Whereas such a development may result in indicators with an improved 'face validity' and acceptance by the users, it is uncertain whether validity actually has improved. A study has shown that production personnel, safety personnel and managers tend to overestimate the effects of safety-management elements on the risk of accidents at the same time as they underestimate the effects of general and production-management elements (Tinmannsvik, 1991). When these personnel participate in the development of SHE rating systems, these biases will be transferred to the new system and result in poor criterion validity.

Table 19.2 Scale applied in evaluation of commitment to SHE through leadership

A – Unacceptable	B – Poor	C – Acceptable	D – Excellent
No commitment from senior management	SHE disciplines delegated to line management, no direct involvement by senior management	Evidence of active senior management involvement in SHE aspects	Evidence of a positive SHE culture in senior management and at all levels

19.1.2 Self-rating as a means of improving SHE management

The Safety-Element Method (SEM) is a rating system used internally in a company by a review team (Alteren, 1999). The method employs a matrix with six SHE management elements and five stages or levels of performance, Table 19.3.

For each element, the review team defines at what stage the company is operating. The results are plotted in the matrix. The review team then decides about ambitions for improvements.

A number of interesting questions arise when we compare rating systems based on third-party verification such as ISRS with the Safety Element Method. In each of the two cases, a different set of motivational factors at the management level applies:

- The introduction of ISRS requires a certain degree of commitment on the part of top management and the marshalling of adequate resources for the ISRS audits and for follow-up of results to meet management's goals. SEM does not involve commitment to the same extent in relation to resources and results.
- ISRS is prescriptive and gives detailed instructions on how to design a company's SHE management system. An advantage is that it is relatively easy to arrive at a solution by applying ISRS. SEM requires problem-solving and results in local solutions. We must expect that management will feel a greater ownership of the solutions that follow SEM than those that follow from ISRS.
- ISRS is based on independent evaluations and on detailed instruction on how to make the evaluation. It will be possible for management to compare the results of their own company with that of other companies applying ISRS. It is not possible to compare the results from SEM between companies in the same way. The application of ISRS will allow for benchmarking and will promote 'competition' between companies in a way that SEM does not.

It should be an interesting research topic to explore and analyse the differences in motivational factors between companies applying the two methods and how they affect the final results.

Table 19.3 Safety element method

Element	Stage 1	Stage 2	Stage 3	Stage 4	Stage 5
Goals, ambitions	Missing	Ambitions to satisfy regulations	Go beyond regulations	Go beyond regulations and match the best companies	Working to influence and improve regulations
Management	Modest obligations to safety work	Follow-up of accidents; Low accountability for weak SHE management; Mainly local treatment of risks	Actively engaged in safety work; Reactions to breaches of SHE instructions; Systematic safety work; Focus on technical and human failures	Safety is equally prioritised and followed up as production and quality; Comprehensive views and systematic approaches; Strong focus on organisation and management factors	Strong management commitment and obligations to improve safety culture; No self-satisfaction; Line management is good model
Feedback systems, learning	Causal transfer of experiences	Simple statistics; Many short-term corrective actions	Thorough statistics; Deviation control; Action plans and measures are worked out; Time schedules are kept	Proactive seeking for improvement; Continuous preventive measures; Thorough processes when working out action plans	Extensive and systematic exchange of experiences with other companies

Safety culture	Mastering risky challenges is the ideal	Little extra done to work safely, more essential to finish fast	Mainly seeking safe behaviour, sometimes chances are taken	Safe behaviour is a matter of course; The employees actively seek out each other's experiences	Always safe working methods, never breach of routines; All employees work actively to obtain a working environment without losses
Documentation	Small amount of formal routines	Satisfies minimum requirements	Comprehensive documentation; Improvements following audits	Plain and practical documentation; Procedures are accepted and followed by most employees	Documentation is well known, always up to date and followed
Result indicators	No result indicators on SHE	Absenteeism and accident statistics are the only indicators	Extensive use of SHE results as indicators	Co-ordinated and integrated goals; Relations between injury, damage and other losses are visualised	Has received international awards for the SHE and quality management

Source: Alteren, 1999

19.1.3 Tripod Delta

Tripod Delta is a SHE performance profile developed for Shell (Van der Want, 1997). Based on the Tripod accident model presented in Sections 5.2, eleven so-called general failure types were identified. Eight of these focus on conditions at the workplace, ordered in relation of their proximity to the victim:

1   Defences, i.e. failure to provide the necessary measures designed to mitigate the effects of human or component failures.
2   Error-enforcing conditions, e.g. a shift schedule that produces adaptation problems, poor physical working conditions or a remuneration system that promotes rule violations.
3   Training, i.e. deficiencies in the routines to provide the necessary knowledge, skills and awareness.
4   Procedures, i.e. unclear, unavailable or incorrect work instructions.
5   Design, i.e. a poor plant layout or a poor design of equipment and tools.
6   Hardware, i.e. material failure due to inadequate quality, equipment and tools that are not available, and failure due to ageing.
7   Maintenance management, i.e. failures in the system aimed at maintaining an adequate integrity of plant and equipment.
8   Housekeeping, i.e. failures in the routines to keep the plant tidy and clean and to dispose of waste.

The remaining three general failure types concern the general and SHE management system and include:

9   Communication, i.e. failure in the transmission of the necessary information for a safe and efficient operation between the different parts of the organisation.
10   Organisation, e.g. ill-defined safety responsibilities or other organisational deficiencies that allow warning signals to be overlooked.
11   Incompatible goals, i.e. failure to manage conflicting safety and production goals or incomparability between formal and informal rules.

The absence or presence of each of the eleven general failure types is determined through the employment of indicator questions, which are administered through a questionnaire to the workforce. There are about 2200 such questions. Each time a questionnaire is distributed at a plant, a random selection of 10 per cent of the questions covering all general failure types are selected and adapted to the conditions at the plant. The results are used to compute a so-called failure-state profile. Figure 19.1 shows an example.

Tripod Delta is intended for use in plants where the accident frequency rate is too low to give adequate feedback for further improvements. It provides feedback on potential causes of incidents.

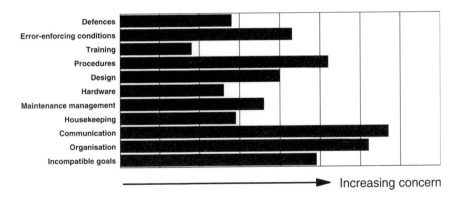

Figure 19.1 Example of a failure-state profile.
Source: Van der Want, 1997. Copyright 1997 by European Process Safety Centre and reproduced by permission.

The Tripod Delta performance-measurement system also includes a scale for measurement of top-management commitment. Its scale ranges from pathological to generative proactive, see Section 6.5. This indicator has some immediately attractive aspects. The emotions and defence mechanisms that will follow from the introduction of such an indicator for feedback control purposes must, however, be carefully considered.

19.2 Measurement of safety climate

In Chapter 5, we introduced the concept of SHE culture. It deals with shared values and beliefs of the members of an industrial organisation that determine their commitment to the organisation's SHE management systems and achievements. We will here use the term 'safety climate' to denote such aspects of an organisation that are possible to measure by use of a questionnaire-based survey and where the results meet statistical criteria for aggregation to the organisational level (Cox and Flin, 1998). Results of such attitude surveys are used as performance indicators at the organisational level.

There is no general agreement on the dimensions that the safety climate is made up of. Examples of dimensions that are mentioned in the research literature are (Zohar, 1980; Brown and Holmes, 1986; Dedobbeleer and Béland, 1991; Niskanen, 1994; Cox and Flin, 1998):

- Management attitudes and commitment, demonstration of concern for the safety of the employees.
- Involvement by the employees in SHE work.
- Communication on safety between different organisational levels and units.

- Risk perceptions within the organisation, attribution of causes of accidents.
- Perceived priorities of safety vs. production goals.
- Belief in effects of SHE work.
- Adherence to safety rules, acceptability of rule violations.
- Active search for new risks.

Although research into safety climate has been in progress for about two decades, the positive effects of measuring the safety climate for use in feedback to the organisation have yet to be demonstrated. Feedback and development of actions to affect the organisational climate should preferably take place in work groups, operating according to the principles of quality improvement projects.

Reason has analysed the effects of an inadequate safety culture on the risk of accidents from a barrier perspective (Reason, 1998). The three conditions that he focuses on are found in the list above on safety-climate elements. First, a poor safety culture will increase the frequency of human errors and rule violations and thus also increase active barrier transgressions. Second, it will result in complacency and in unwillingness to check and maintain passive barriers adequately. Third, the organisation that is characterised by an inadequate safety climate will be unwilling to report and follow up near accidents and identified deficiencies in the barriers.

A crucial question is, however, whether the establishment of safe behaviour is best achieved by focusing on attitudes as measured by safety-climate surveys. We must consider the other possibility, i.e. that it is more efficient to focus on behaviour directly, e.g. by use of the type of measures described in Section 18.2. When behaviour changes, attitude change will follow.

19.3 Measuring the degree of learning from incidents

This type of indicator focuses on one aspect of SHE management, i.e. the extent to which the organisation uses the experiences from unwanted events to prevent recurrence. In Section 10.6, we introduced Van Court Hare's hierarchy of the order of feedback. This has been applied in the development of a measure, based on an analysis of actions taken after accidents.

Figure 19.2 illustrates why this measure is interesting. It shows an example from an oil company. As indicated by the diagram, the company's organisation has behaved rationally by following up severe accidents more seriously. We want to use an indicator that catches the extent to which the organisation actually takes its severe accidents seriously in the way illustrated by the figure.

The indicator on the average order of feedback employed in accident investigations does this. It is based on an analysis of the actions documented in the accident reports. Each action is rated according to Van Court Hare's hierarchy from 0 to 4. If there is no action, a dummy action with 0 order is introduced. 'Actions' concerning general information (e.g. 'to take care') are given the same rating. The average order of feedback is then calculated for

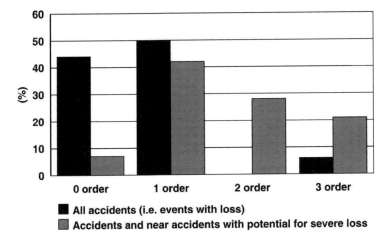

Figure 19.2 Distribution of actions taken after accidents by level of feedback.
Source: Blindheim and Lindtvedt, 1996.

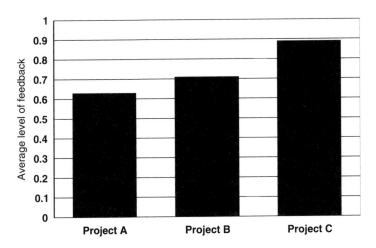

Figure 19.3 Weighted average level of feedback in accident reports from three offshore projects.
Source: Boe, 1996.

the total number of accidents within a defined severity category, e.g. lost-
time accidents or accidents with a potential for serious consequences.

Figure 19.3 shows an example. The project with the best performance in
this respect, project C, also experienced the lowest LTI-rate.

The aim of introducing the average level of feedback as an indicator
of SHE performance is to improve the organisation's ability to use the
experience from accidents to achieve continual improvements. We lack
experience on the possibility of manipulating this indicator.

20 Selecting key SHE performance indicators

20.1 Combinations of SHE performance indicators

We have seen that there are many different SHE performance indicators to choose between. Each of them has its advantages and weaknesses. It is recommended to use a combination of such indicators.

When making recommendations on the selection of SHE performance indicators, we have to consider not only conditions outside the workplace such as legislation and top-management priorities. We also have to consider how the application of such indicators may affect the behaviour of middle and lower management and the employees. We have earlier discussed how unilateral control by top management will have negative effects on the behaviour at these levels (Argyris, 1992). Top management may, for example, require compliance with unrealistic SHE goals that the lower levels do not feel any ownership of. The different departments may be able to show positive SHE performance results without really changing the basic conditions that control the risk of accidents. We also discussed how top management, by exercising more unilateral control as a reaction to such signs of superficial adaptation, would increase the maladaptive behaviour further at the lower levels.

Example: We will here look at a typical scenario, where the starting point is the application of the LTI-rate. Today, we find many companies with very ambitious SHE goals, e.g. an LTI-rate of zero. Each workplace then has to meet this goal and one single lost-time injury represents a deviation from the goal. The workplaces may respond by hiding the lost-time accidents through transfer of the injured persons to other types of work. Management's response is then to introduce ambitious goals concerning the TRI-rate, which also accounts for these types of accidents. Another approach is to introduce the ISRS system. This implies a significant increase in the degree of control of the workplaces by focusing on many different topics. Such indicators will, according to Argyris, when introduced as unilateral management directives, result in the organisation's superficial adaptation to the requirements without showing responsibility and commitment. Management may then respond by introducing measurements directed at these conditions, such as the measurement of safety climate.

The point is not here to warn against the implementation of ambitious SHE goals from the top down. Rather, the conditions for goal recognition and acceptance throughout the organisation have to be carefully considered. It is important that the goals are considered to be realistic and fair and that the associated measurement methods are regarded as meaningful. The ways the goals and indicators are introduced are of equal importance to their selection. It is also important that the causes of deviations from the goals are analysed at a sufficiently low level of the organisation. It is only here that adequate knowledge exists about the factual circumstances that explain the results.

Exercise: In Section 17.1, you were asked to draw a control chart for Ship Yard Inc. Based on these results:

1 *Discuss the problems connected with the application of the LTI rate and control charts in monitoring the safety performance of a company.*
2 *Propose other SHE performance indicators for use in addition to the LTI-rate. State the reasons for your recommendations.*

Example: A refinery has many years of experience in using the LTI-rate as its sole SHE performance indicator. The results now show a yearly LTI-rate fluctuation of between 0 and 2, dependent on whether there is a lost-time accident or not. The LTI-rate is determined to be inadequate in this situation, and management decides to introduce new SHE performance indicators along two new paths:

1 *To improve the follow-up of the risk of injury to personnel, the TRI-rate is introduced. The frequencies of lost-time injuries, restricted-work injuries and medical-treatment injuries are registered separately and in summary and displayed as shown in Figure 17.6. During the first year, the TRI-rate is 30, which makes it a new starting point for improvements.*
2 *Management also wants to follow-up on the developments in the risk of major accidents as well. A new high-potential incident rate is introduced. It is calculated as the number of such incidents per quarter year and is displayed in a control chart. Incidents involving gas leakage, overflow of tanks, fires and dropped objects inside process areas are considered as potentially serious. Figure 20.1 shows a display of the results for the first ten quarters.*

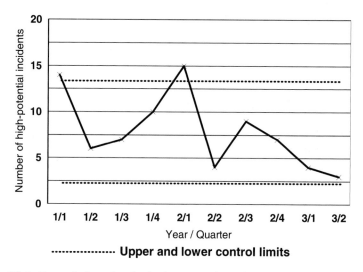

Figure 20.1 Control chart for the high-potential incident rate at a refinery.

20.2 Indicators of barrier availability

In Chapter 7, we introduced the concept of barriers against losses. Two examples were shown – one of barriers against exposure to hazardous chemicals and the other of the prevention of fires and explosions. We may use this concept to define a different set of performance indicators, i.e. that of barrier availability. This is defined as the probability that a barrier is available when it is needed and ranges from 0 to 1 (or from 0 to 100 per cent). These types of performance indicators have by their nature to be specific to the type of hazard that is to be prevented. Let us illustrate this principle by two examples.

Example 1: Figure 7.2 illustrated four different barriers against the exposure to hazardous chemicals. In a plant for the manufacturing of aluminium rims, different types of solvents are used in the paint shop. These solvents represent a hazard to the operators, who may inhale them or get exposed through the skin. Below are examples of indicators of barrier availability for this specific case:

- *Proportion of working time when operators are exposed to solvents with a hazardous class of less than 2.*
- *Proportion of working time when operators are exposed to a concentration of solvents in the working atmosphere of less than one-third of the threshold limit value.*
- *Proportion of working time when operators wear required personal protective equipment.*

All three indicators may be measured through sampling techniques of the type described in Section 18.2.

Example 2: Similarly, we may define performance indicators for the availability of barriers against fires and explosions in a refinery such as:

- Duration of significant gas leakage in percentage of total production time. (A more practical indicator is the frequency of significant gas leakage.)
- Availability of such safety systems as fire and gas detection and active fire protection.
- Share of evacuation drills where the personnel are able to show up at the meeting point within a stipulated time period.

Part V

Risk analysis

This Part reviews different risk-analysis methods. Risk analysis is a planned activity involving an identification and evaluation of accident risks in the workplace. We will start in Chapter 21 by establishing some common principles of the different risk-analysis methods. We will then in Chapters 22 to 25 go through four different methods that primarily are used in the analysis of the risk of occupational accidents: energy analysis, job-safety analysis and comparison analysis. One chapter is dedicated to the risk assessment of machinery. Finally, we will in Chapter 26 review the CRIOP method, where the interaction between the operators and the technical system are analysed when the system is in a disturbed state. The aim is here to prevent major accidents.

Risk analysis is a proactive approach in the sense that it makes us able to identify and assess accident risks before any losses have occurred. It has many similarities to the reactive approaches presented in Part III. It utilises all four sub-systems of the SHE information system, i.e. data collection (risk identification), processing (evaluation of risk), memory (e.g. reliance on historical accident data) and distribution of results for decision-making.

21 The risk-analysis process

21.1 What is risk analysis?

Accidents and near accidents are unwanted events occurring at random points in time. At the same time, they represent opportunities of learning about hazards and causal factors at the workplace and in the company. We should grab the opportunities when they occur and conduct adequate investigations. Workplace inspections and SHE audits are pre-planned activities that also represent opportunities of learning. They focus on deviations, contributing factors and root causes and not, in the first place, on hazards.

Risk analysis is complementary to these different activities. Risk analysis is a planned activity, where we want to identify and remedy hazards before accidents and near accidents happen. Risk analysis is thus anticipatory in nature and supports feed-forward processes rather than the feedback processes of, for example, accident investigations. By applying risk analysis, we will speed up the learning processes. Evaluations show that properly executed risk analyses of work systems will reveal hazards corresponding to many years of accident experience from the same types of systems (Suokas, 1985).

Risk analysis is a collective name for different methods involving the following activities (cf. CEN, 1996; ISO, 1999):

1 *Defining the analysis object* and determining its limits. The analysis object is a geographical area of a plant, a machine, a job, etc.
2 *Hazard identification* and evaluation of situations, where people (or other voluntary targets such as the environment or material assets) may get in contact with the hazard.
3 *Estimation of risk*, i.e. an assessment of the exposure to hazardous activities, the probability of accidents in the activities and the severity of losses.

A risk analysis is followed by an *evaluation of the risk* and the implementation of risk-elimination or -reduction measures where required. In this activity, we determine whether the risk is acceptable or not and whether there is a need for remedial actions to reduce the level of risk. We will here consider the whole process, including risk analysis and evaluation of risk.

There has been a trend among legislators to replace prescriptive require-ments to SHE at the workplaces with requirements of a goal-oriented type. Compliance with such requirements often has to be documented through a risk analysis. The European legislation on machinery safety is a typical example (European Council, 1989/98). It is the responsibility of the manu-facturer to document the safety of machinery through use of adequate risk-analysis methods. A European directive on protection against chemical hazards at work lays far-reaching obligations on employers to assess the risks of hazardous chemicals at work (European Council, 1998).

Risk analysis is applied as a mandatory tool in many countries in the prevention of major accidents, see Section 2.2. Legislation such as the Seveso Directives in Europe and the Clean Air Act amendment in the USA has supported this development (European Council, 1996; EPA, 1990).

21.2 Acceptance criteria for the risk of losses due to accidents

An *acceptance criterion* defines the highest accepted risk, as documented for example in a company's SHE policy. The risk is expressed as a combina-tion of the frequency of the occurrence of losses due to accidents per unit of exposure in a specified activity and the consequences of such losses.

Figure 21.1 shows the relation between the acceptance criterion and the SHE goal. Those combinations of frequency and consequence that were denoted 'high' represent a risk that falls above the acceptance criterion. A 'medium' risk of accidents falls within the ALARP region, i.e. between the acceptance criterion and the goal. ALARP expresses that the risk level has been reduced as far as reasonably practicable and that no further cost-effective measures can be identified. Only an accident risk below the SHE goal is denoted 'low'.

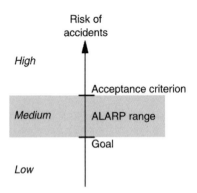

Figure 21.1 Relation between acceptance criterion and goal for the risk of accidents. In the range between the acceptance criterion and the goal, the risk should be as low as reasonably practicable (ALARP).

Risk-acceptance criteria are used as a reference in evaluating the results of the risk analysis and in determining the needs of remedial actions. They therefore have to be available prior to the start of the analysis. There are many different types of risk acceptance criteria, see NORSOK standard Z-013 for an overview (Norsk Standard, 1996/97/98/99). Here we will use acceptance criteria related to the risk of fatality or injury to personnel. They must meet the same standards as those of the SHE performance indicators discussed in Section 11.1. In other words, they must be:

- Quantifiable at least on an ordinal scale to be suitable for use in evaluating the results of the risk analysis and in decisions on remedial actions;
- Valid indicators of the risk of accidents;
- Sensitive to changes in the characteristics of the analysis object;
- Compatible with other goals; and
- Easy to understand and suitable for communication with management and the employees.

21.3 Methods of risk analysis

An important attribute of risk analysis is that it is systematic. Ideally, all possible hazards should be found and evaluated. The various risk-analysis methods apply different analytic tools and checklists to accomplish this goal. Often, they combine the use of analytic tools and checklists with brain-storming and group problem-solving techniques. There are many textbooks and standards describing various risk-analysis methods (see e.g. Suokas and Rouhiainen, 1993; Harms-Ringdahl, 1993; ISO, 1999). Table 21.1 below summarises some of the methods.

Figure 21.2 illustrates how the starting point, the directions and the scope of each method fit into the accident-analysis framework of Chapter 6. Two of the methods, Fault tree analysis and Comparison analysis are *deductive* in that they start with the unwanted event. They proceed by analysing the underlying incidents and deviations (Fault tree analysis) or contributing factors (Comparison analysis). Several of the methods are mainly *inductive* in that they start with a deviation and proceed by studying the effects of this deviation. This applies to HAZOP, Failure mode and effect analysis, Event tree analysis and CRIOP, although they also have a component of causal analysis. Coarse analysis and Job-safety analysis start with the hazard and use a combination of inductive and deductive analyses.

We will here focus on four different methods: Coarse or energy analysis, Job-safety analysis, Comparison analysis and CRIOP. They all involve ana-lyses of the effects of human performance on the risk of accidents. Use of experience of earlier accidents and incidents through discussions in problem-solving groups is another common denominator. We have earlier shown how group processes support in the identification and solving of SHE-related

Table 21.1 Overview of different risk-analysis methods

Method	Scope and aim	Analysis object	Basic characteristics	Measure of risk
Coarse/energy analysis (also called Preliminary hazard analysis)	Overview of hazards in an area by type, causes, consequences and risk	A geographical area of an enterprise	Checklist of hazards	Risk matrix (low/medium/ high)
Job-safety analysis	List of hazards connected to each step of a job by type, causes, consequences and risk	Individual jobs or jobs connected to particular equipment, etc.	Breakdown of job into steps; Checklist on hazards	Risk matrix (low/medium/ high)
HAZOP	Qualify process safety systems by identifying potential deviations and their causes and consequences	A process plant, as documented in e.g. piping and instrument diagrams (P&ID)	Breakdown of plant into process units (tanks, pipelines, reactors, etc.); Checklist of process parameter deviations	Consequences of deviations
Failure mode and effect analysis	Assessments of systems reliability	A technical system	Reliability model of the system; Checklists of failure modes; Failure data for components	System availability (%) etc.
Action error analysis	Overview of possible human errors and their consequences and causes	Tasks performed by an operator	Breakdown of a task into steps and identification of causes and consequences	Probability and consequences of human errors
Fault tree analysis	Estimate the probability of severe accidents with consequences to people, materials or the environment	A technical system	List of known top events; Logical model of causes (fault tree); Failure data for components	Frequency of top event with known consequence
Event tree analysis	Establish frequency distributions of potential accidents	A technical system	Checklist of critical initial events; Event trees; Event frequencies and probability distributions of different outcomes of each event	Combinations of frequencies and losses
Comparison analysis	Assess changes in accident frequencies for a new plant, as compared to an existing plant	A new or modified plant	Overview of accidents in existing plant by area, activity and type of event; Systematic assessments of effects of new design on exposure and probability	Accident-frequency rate

Table 21.1 cont'd

Method	Scope and aim	Analysis object	Basic characteristics	Measure of risk
CRIOP	Evaluate the contributions from human actions to the failures in barriers against major accidents	Process plant	Establish potential accident scenarios; Identify critical human actions; Human-factor analysis of the actions	Consequences of human errors

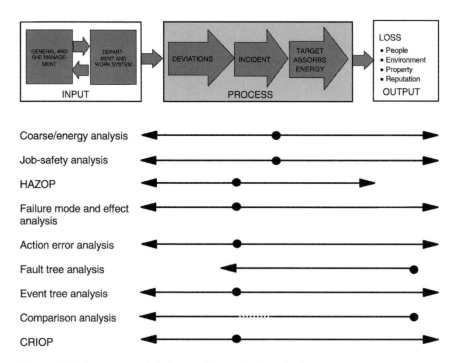

Figure 21.2 The scope of different risk-analysis methods.

problems in e.g. accident investigations. Here, the risk analysis sets the scene for a systematic review and documentation of the group members' collective experiences of accident risks at the workplace. Accident and incident data support the experience exchange but are not always necessary to accomplish the wanted results.

The first three methods analyse the risk of occupational accidents. They have a joint theoretical basis in the framework for accident analysis described in Chapter 6. The methods serve different purposes. Often, Coarse and Job-safety analysis are used in combination by starting with the Coarse analysis. A Job-safety analysis will follow when severe hazards have been identified and there is a need to go into details on how they may result in harm. We will also look into applications of Coarse and Job safety analysis in the prevention of machinery and chemical accidents. Both applications represent means of meeting new legislation in Europe.

We will also look into a fourth method, CRIOP. In this analysis, the man–machine interface is in focus. The aim is to reduce the risk of major accidents by identifying a need to improve the conditions for a safe handling of disturbances by the operators. In CRIOP, we will apply an information ergonomics model of the interactions between the human operators and the environment in a disturbed system.

Risk analyses are carried out at different phases of the analysis object's life cycle. When the analysis object is an existing workplace, we want personnel with knowledge and experience from this workplace to participate in the analysis. We also carry out risk analyses during the design phase of new machines and workplaces. In this case, personnel with experience from similar workplaces/machines need to be present.

All four methods follow a similar basic routine:

1 Planning of the analysis including establishment of an analysis team.
2 Preparation by delimiting the analysis object and reviewing the necessary documentation.
3 Group meetings where the analysis object is reviewed and the hazards are analysed and documented. In these meetings, the procedure of the selected method is followed.
4 Evaluation of the risk and development of remedial actions as needed. These activities are usually carried out by the group but may also be a separate activity.
5 Information to the concerned parties and follow-up and implementation of results.

22 Coarse or energy analysis

The aim of a Coarse analysis is to identify and evaluate the hazards of the different areas of a plant. The Energy model is its theoretical basis. Those hazards representing the potential of severe losses are focused upon. A Coarse analysis typically follows a stepwise procedure.

The steps of a Coarse analysis:

1   Planning
1.1 Selection of analysis object
1.2 Establishment of analysis team
1.3 Allocation of resources
- time (working hours)
- training, access to facilitator
- access to information
- administrative support

1.4 Scheduling
2   Preparation, incl. collection of documentation and familiarisation
3   *Execution of the analysis*
3.1 *Description of analysis object*
3.2 *Identification of hazards*
3.3 *Identification of causes*
3.4 *Risk estimation, evaluation of needs of further actions*
4   Development of safety measures
5   Documentation
6   Follow-up of results

This chapter gives a detailed description of each step of the Coarse analysis. We will apply the same basic steps in all the four risk-analysis methods presented here. We will not repeat the description of each basic step in such detail for the other risk-analysis methods.

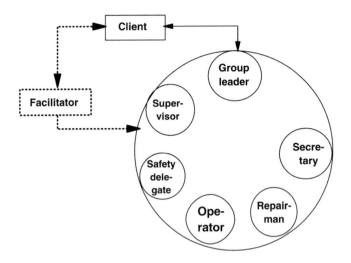

Figure 22.1 Typical analysis team.
Source: Adapted from Wig, 1996 and reproduced by permission from the author.

22.1 Planning

The *analysis object* is a geographical area of the plant. It is recommended to subdivide the plant into areas, following the established departments.

Example: In a Coarse analysis of an aluminium plant, the plant was broken down into the following 'analysis objects': carbon section for manufacturing of cathodes; electrolysis; foundry; and transportation. The latter is geographically spread over the whole plant.

A team is responsible for the analysis. Depending on the circumstances, the team will consist of the team leader, a secretary (may be combined with the team-leader position) and experts with specialised background, Figure 22.1.

The *group or team leader* has to be well acquainted with the Coarse analysis method and have management capabilities. The team leader is responsible for:

- assisting with the selection of other team members,
- preparation of the analysis and execution of it,
- representing the team with discussions with the client and the management responsible for the analysis object,
- submitting the analysis report.

The *secretary* of the team is responsible for documenting the team's analyses and for summarising these in the report. Often, the team leader takes on the responsibility of secretary as well.

The *participants* in the team have to ensure that all conditions of significance to safety are brought up in discussions and assessed. A Coarse analysis during the design phase aims at identifying and assessing hazards that may occur when the designed system has been put into operation. Participants will represent the engineering organisation (typically layout, mechanical and safety disciplines) and personnel with experience from similar systems in operation (production, maintenance). In an analysis during the operations phase, the team will include persons with direct experience of the job(s) to be analysed. It is recommended that the team size does not exceed seven persons.

There may be a need for a *facilitator* in case the team is not fully acquainted with the analysis method. Responsibilities of the facilitator include:

- guiding the team in the analysis method,
- assisting in collection of documentation,
- providing experience transfer from other teams.

It is the client's responsibility to ensure adequate allocation of resources. This includes time (working hours) for the analysis and training, access to facilitator (if necessary) and access to documentation and personnel for interviews.

A typical Coarse analysis of a an area (e.g. the foundry in an aluminium plant) will require a total of 40 to 60 man-hours, including planning (4–8 hours), team meeting (six persons for 6 hours), reporting (8 hours). Normally, one team meeting will be sufficient to execute a Coarse analysis. It will be possible to cover all activities including planning and reporting within one week.

Preparation consists of the collection and review of documents about the analysis object and familiarisation with it. A typical list of documents for a Coarse analysis includes:

- Layout drawings.
- Manning and organisation plans.
- Training schedule for new employees.
- General safety instructions for the department (if available).
- Accident and near-accident reports from the actual area (or from a similar area at another plant if the analysis is carried out during the design of new workplaces).
- Relevant regulations and standards.

22.2 Execution and documentation

In the description of the analysis object, it is recommended but not strictly necessary to establish a comprehensive list of all jobs within the area. Such a comprehensive list is especially valuable when Coarse and Job safety analyses are carried out in combination in order to secure a comprehensive review of

Table 22.1 Checklist of typical jobs

- Operation and supervision of machinery
- Normal, manual operations
- Inspection, sampling
- Handling of production disturbances
- Maintenance, incl. maintenance handling
- Housekeeping
- Transportation of materials (manually/by crane, truck etc.)
- Walking

Table 22.2 Checklist of hazards

- Vehicle(s) in movement*
- Contact with or caught in or between moving parts of machinery*
- Release of pressurised gas or fluid
- Contact with sharp object
- Impact from or squeezed between moving objects*
- Flying object, fragments
- Falling object*
- Fall on same level
- Fall to a lower level*
- Stumbling
- Hitting against
- Fire, explosion*
- Release of/contact with poisonous, corrosive chemicals
- Contact with electric conductor
- Contact with hot or cold surface
- Reduced oxygen content

---

* Hazards that dominate the statistics on fatalities (cf. Section 5.4).

the area. Also reasonably foreseeable jobs or activities that are not planned should be included (e.g. handling of disturbances). The checklist in Table 22.1 is used to support this task. It is recommended to visit the actual workplace, if possible, during this step.

22.2.1 Identification of hazards and causes

The team uses its experiences and imagination in the identification of hazards and possible accidental events within the area. The checklist in Table 22.2 supports this task.

In the Coarse analysis, we especially want to focus on hazards that may result in serious consequences. It is also recommended to focus on hazards that may result in many (but less severe) accidents. Consult the accident statistics for the actual plant or similar plants for accident concentrations. Results are documented on a record sheet according to Table 22.3.

Table 22.3 Record sheet for documentation of the results of a Coarse analysis. An example from a foundry at an aluminium plant is shown

AREA: *Foundry*

Hazard	Causes	Consequences	C*	F*	R*	Recommendations
Explosion	Cold metal with moisture in contact with melted metal in furnaces	Fatality	4	3	H	Pre-heating of cold metal
Burns from melted metal	Splashing during charging of furnaces	Serious burn injury	3	3	M	Remotely operated charging operation
Collision with vehicles	Insufficient line of sight in crossing	Crushing injury	4	2	M	Traffic regulation
Fall to lower level	Unsecured work at height on top of furnaces	Falling injury	3	3	M	Installation of work platform
Etc.						

Note: C = Consequence, F = Frequency, R = Risk.

Identify causes by again using the team's experience. Use the checklists of deviations and contributing factors at the workplace level in Chapter 6 as an extra control (see Tables 6.9 and 6.12). It may be necessary to call in expertise to the team's meeting in order to go deeply into special problems. Technical design is focused upon in analyses that are carried out during the design stage.

22.2.2 Risk estimation

Risk estimation serves as a basis for prioritising in relation to safety measures and the need for detailed Job safety analyses. Estimate the most serious *consequence* of the occurrence that from a realistic point of view may happen according to the scale shown in Table 22.4. The consequences are dependent on the amount and type of energy involved.

In establishing the risk estimation matrix according to Table 22.4, the following two anchorage points defining the acceptance limits have been applied:

- An LTI-rate of above 10 is considered to be too high and thus above the acceptance criterion. For a workplace with 100 employees working 2000 hours per year, an LTI-rate of 10 corresponds to two lost-time injuries per year. It follows that ten lost-time injuries per year is too high a number but one is below the acceptance limit.

Table 22.4 Matrix for risk estimation for a workplace with about 100 employees

Consequence		Frequency				
		1 per 1000 years (1)	1 per 100 years (2)	1 per 10 years (3)	1 per year (4)	10 per year (5)
First-aid injury	(1)	Low	Low	Low	Low	Medium
Lost-time injury	(2)	Low	Low	Low	Medium	High
Permanent disability	(3)	Low	Low	Medium	High	High
Fatality, one person	(4)	Low	Medium	High	High	High
More than one fatality	(5)	Medium	High	High	High	High

Table 22.5 Matrix for estimation of the individual risk

Consequence		Frequency					
		< 1 per 1000 years (0)	1 per 1000 years (1)	1 per 100 years (2)	1 per 10 years (3)	1 per year (4)	10 per year (5)
First-aid injury	(1)	Low	Low	Low	Medium	High	High
Lost-time injury	(2)	Low	Low	Medium	High	High	High
Permanent disability	(3)	Low	Medium	High	High	High	High
Fatality, one person	(4)	Medium	High	High	High	High	High

- A FAR of above 10 is unacceptable. With the same assumptions as above this defines the acceptance limit as lying between one fatality per ten years and one per 100 years.

If other acceptance criteria for the LTI-rate and FAR are applied, the matrix has to be re-scaled to suit this application. Similarly, the matrix has to be modified when applied to considerably smaller or larger workplaces. Table 22.5 shows the matrix for the individual risk of an operator working full-time. It applies the following frequency limits for intolerable risk:

> 10^{-3} per year for fatality
> 10^{-2} per year for permanent disability
> 10^{-1} per year for lost-time injury

Estimate the expected *frequency* of incidents following from each identified hazard with the assumed consequence by using the team's experience and accident statistics (if available). The *risk estimation* follows from the team's assessments of consequences and frequencies.

The team is responsible for evaluating the needs of remedial actions and further detailed Job-safety analyses. Tables 22.4 and 22.5 are usually interpreted in the following way:

High risk:	A risk above the acceptance criterion where action must be taken to reduce the risk.
Medium risk:	Reduce the risk as far as reasonably possible (ALARP principle).
Low risk:	No further risk reduction is required.

The need to implement safety measures is determined for each individual hazard and for the identified hazards of the department as a whole. If, for example, the summary frequency of identified hazards with risk of fatality exceeds one per ten years, the hazards taken together violate the acceptance criterion and risk-reducing measures have to be implemented.

22.2.3 Development of safety measures

The identification of causes serves as a basis for the development of safety measures. The following preferences should be made in the selection of measures:

1 Will the risk be reduced or eliminated by design?
2 Will safeguarding reduce the consequences?
3 Will the risk be reduced by access to personal protective equipment, safe work procedures and/or information and training?

22.2.4 Documentation and follow-up of results

All findings from the analysis are documented in the Coarse analysis record sheet according to Table 22.3. The following outline is proposed for the analysis report:

* Scope and purpose. A definition of the analysis object is given here.
* Conclusions and recommendations.

- Method: Participants in analysis team and reviewed documents are listed, assumptions are stated.
- Results. Reference is given to the full analysis form, which is given as an attachment.

The report should be subject to quality control before it is approved and distributed. As a minimum, this should include a review by the team members. In project work, quality control is accomplished through inter-disciplinary checks. For analyses of existing plants, the report should be sent to plant management for review. It is a management responsibility to follow up on results.

22.3 Establishing a database on potential accidents

A Coarse analysis results in information on hazards and potential accidents similar to that obtained though accident investigations. In many cases it may be practical to store the results in a computer for easy retrieval and summarisation. This approach offers several advantages:

- It will be easy to compare results from different plants. This will facilitate checking of the results in order to obtain common risk-estimation criteria.
- It will also be easier to learn from earlier findings and as a result to use accumulated experience on hazards and causal factors in new analyses.
- A joint analysis of the results from different departments with similar types of production will support in the identification of potential accident concentrations.

Example: An aluminium company conducted a Coarse analysis of its primary aluminium plants. An analysis team was established for each department that participated in the analysis. The teams were lead by SHE experts from the corporate staff. In addition the teams included the department supervisor and operations and maintenance personnel.

The teams identified in all about 800 different potential accidents. A database was established that contained the following information about each event:

- *Plant and department.*
- *Type of production and activity.*
- *Possible sequence of events.*
- *Type of event (compare Table 22.2).*
- *Estimated frequency, consequence and risk.*

The corporate staff analysed the events. Figure 22.2 shows results for the electrolysis departments. There were fewer than ten high-risk events requiring immediate attention and about 300 medium-risk events.

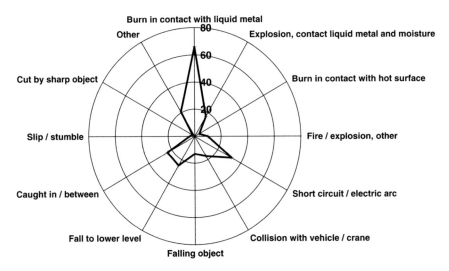

Figure 22.2 Distribution of all identified potential accidents in the electrolysis departments by type of energy involved.

The identified accident risks were grouped and the groups were ranked by accumulated risk level as a basis for prioritising further follow-up. For the electrolysis departments, the following types of events were considered to be of special concern:

1 Burn injury due to contact with melted aluminium.
2 Vehicle hitting an operator working adjacent to an electrolysis cell.
3 Burn injury due to electrical arc.
4 Fall to a lower level during maintenance of electrolysis cells.
5 Caught between heavy objects during crane operations.

Each department was responsible for follow-up of its own results. The results of general significance were later implemented in a company technical standard for use in the design of new plants and modifications.

23 Detailed job-safety analysis

The aim of a detailed job-safety analysis (JSA) is to identify and evaluate the hazards to which employees are exposed when performing work activities. It is similar to a Coarse analysis, but differs as to the level of detail by which the activities are broken down.

23.1 Analysis object

The analysis object is a job which is made up of a sequence of activities. In selecting jobs for the JSA, the following criteria are applied:

- Jobs where serious or frequent accidents or near accidents have been experienced before, or where there is a potential for severe consequences of accidents.
- Jobs into which many man hours are put.
- New or changed jobs, where the consequences to safety are uncertain.

23.2 Resource needs and scheduling

It is difficult to establish generally valid estimates of resource needs for a detailed JSA. The following examples can illustrate typical needs.

Example 1: JSA during the design phase of typical skid for an off-shore platform: One day for preparation and documentation by chairman/secretary and half a day for one team meeting, total about 40 man hours.

Example 2: JSA of operation and maintenance jobs of a more complex machine in operation: Four half-day meetings with 3–5 persons, total about 120–160 hours including time for planning and documentation.

Example 3: JSA during the kick-off meeting for a new job: A two-hour meeting with 2–4 persons, i.e. a total of 4–8 hours. A JSA record sheet is filled in during the meeting but no full documentation is provided.

A complicated JSA cannot be completed during one meeting and needs to be planned with fixed milestones. The following milestones are proposed:

Table 23.1 Record sheet for documentation of a job-safety analysis. A simplified analysis from an offshore installation is shown, involving crane transportation of drill pipe from the pipe deck to a ramp

Activity	Hazard	Causes	Consequences	C*	F*	R*	Recommendations
1 Fix the slings to the drill pipes	Fall from stack of pipes	3 m height of pipe stack; Icy conditions	Broken leg	2	3	H	Lower height of pipe stack
2 Lift the pipes to the pipe ramp	Hands squeezed by sling	Restricted view from crane	Hand injury	2	3	H	Replace slings by magnet lift
	Hit by falling pipe	Faulty slings	Crushing injury	3	1	M	Review routines for checking of slings
3 Release the slings	Hit by pipes	Restricted view	Crushing injury	2	2	M	See above
	Fall from pipe ramp	Lack of fall protection	Broken leg	2	2	M	Guard-rail

Note: C = Consequence, F = Frequency, R = Risk.

(1) Planning, collection of background materials; (2) Identification of activities; (3) Identification of hazards and causes; (4) Proposals for remedial actions; (5) Draft and final report. The total time-span should not exceed one week.

23.3 Description of the steps of the job

In this step we find the largest differences between the Coarse analysis and the detailed JSA. The JSA team reviews and describes the activities, one after the other in a sequence. It is thus necessary that personnel with operational experience participate in the team. Remember to include all activities, also those outside normal practices. What disturbances can occur and what will then be done? The checklist in Table 22.1 is used as an extra control to ensure that no activities have been forgotten.

The breakdown into activities should be performed at an adequate level of detail. Too coarse a breakdown makes the subsequent identification of hazards more arbitrary. If the breakdown is too detailed, the JSA will be time-consuming. Compare the activity description with the work instructions (if any). Do the instructions describe the actual performance of the work? Are 'dialects' present, i.e. does the way the work is done differ between different shifts. Preferably, the description of the job steps should be checked against the actual performance of the work.

23.4 Subsequent steps

The subsequent steps do not in principle differ from those of the Coarse analysis, see Section 22.2. In JSA, there are requirements to comprehensiveness and no hazards are too insignificant to be documented. All hazards will sooner or later result in injuries or near accidents, if they are not remedied.

In estimating the risk of accidents, working time has to be considered. Usually, the job is carried out by one or a few operators, and Table 22.5 is in this case most suitable. It represents the acceptance criteria for the total risk of an individual operator working about 2000 hours a year. The contribution from the analysed job has to be evaluated in relation to the total risk to which the operator is exposed.

Exercise: Establish a job-safety analysis team. Select a hazardous job that at least one of the team members is familiar with. Assign different roles in the group and perform a job-safety analysis. Document the results in a record sheet according to Table 23.1.

23.5 Accidental exposure to chemicals

According to the European legislation on chemicals at work, employers must assess the risks of chemical hazards to the safety and health of workers, considering effects due to both accidental and long-term average exposures (European Council, 1998). We will here focus on the assessment of accidental task exposure, and a variation of JSA may be used for this purpose:

1   Establish an inventory of all hazardous chemicals that are in use or are planned for use.
2   For each chemical, determine the potential health effects of being exposed, i.e. the toxicity of the chemical and severity of the consequences. Consult the Safety data sheet.
3   List all activities that are associated with each chemical and where there is a potential for significant exposure with detrimental health effects through inhalation and/or skin contact. Include activities during handling/transportation, storage, use, and waste disposal.
4   For each activity and chemical, estimate the degree of accidental exposure. The extent, frequency and duration of exposures should be taken into consideration. Use the record sheet according to Table 23.1 to document the results.
5   Assess the risk (probability and consequence) of developing an occupational injury or disease, considering the potential health effect of the chemical and the degree of exposure.
6   Evaluate the need for and prioritisation of remedial actions.

Table 23.2 Matrix for risk estimation for exposure to chemical agents

Consequence		Probability of developing an occupational disease				
		Very small *(1)*	*Small* *(2)*	*Moderate* *(3)*	*High* *(4)*	*Very high* *(5)*
Insignificant	(1)	Low	Low	Low	Low	Medium
Moderate reversible	(2)	Low	Low	Low	Medium	High
Serious reversible	(3)	Low	Low	Medium	High	High
Irreversible	(4)	Low	Medium	High	High	High
Serious irreversible, life-threatening	(5)	Medium	High	High	High	High

Steps 4 and 5 are not straightforward activities, especially when it comes to long-term effects, i.e. the development of occupational diseases. For accidental exposures to acute toxic/corrosive chemicals, the risk-estimation matrix of Table 22.5 can be used. A similar matrix for use with other chemical exposure assessments is shown in Table 23.2. The application of it involves a number of crucial decisions concerning:

1 Estimation of exposure. For exposure through inhalation, for example, the occupational exposure limit for each chemical component should be used as a reference value. The frequency and duration of exposure are also major determinants. For skin exposure, the affected area and the chemical's concentration are important factors.
2 Rating of the probability of developing an occupational disease and its potential severity. Here we distinguish between the chemical's potential health effects, i.e. how hazardous the chemical agents are to the workers' health:
  a Very toxic, carcinogenic, mutagenic, or reproductive toxicant.
  b Toxic, corrosive or allergenic.
  c Health hazardous.
  d Not classified in any health-hazardous category.

There are clear priorities in the selection of remedies in step 6 (European Council, 1998). Substitution of hazardous chemicals with less hazardous ones is the preferred solution. Engineering solutions to prevent exposure through containment at the source or through the provision of adequate ventilation (extraction/exhaust ventilation preferred to dilution of contaminants) come next in priority. Only if these solutions are not feasible or do not provide adequate protection, does the use of personal protective equipment come into question.

23.6 Systematic mapping of hazards within an organisation

Detailed JSA is applied in the analysis of individual work activities that for some reason have been focused upon. It is also applicable in a systematic and detailed mapping of the hazards of a department. Below is an example of how this may be accomplished:

1   All activities in the department are identified and listed. This is done, for example, by reviewing the responsibilities of the different positions in the department.
2   All activities are reviewed by application of Coarse analysis. The results are used as a basis for decisions on which activities to focus upon in detailed JSA.
3   The critical activities are analysed.
4   Measures are developed and implemented for each activity.
5   After the measures have been implemented, the activity is subject to an update of the analysis by observing the actual performance of the activity. The aim is to check that the analysis describes the actual situation adequately and that the measures are adequate.

The analyses seen together will represent a 'total map' of the hazards in the different activities in the department. This 'map' will be a powerful tool in the work to reduce the risk of accidents in the department. It is important that it is updated with changes in the activities and with new accident experience. If this is done systematically, the 'map' will represent an important carrier of the department's collective experiences on accident risks (see Section 15.3).

24 Risk assessments of machinery

24.1 Requirements as to risk assessments

The EU Machinery directives define goal-oriented safety requirements to machinery (European Council, 1989/98). Machinery manufacturers have to carry out risk assessments to document that they meet these requirements. Documentation on risk assessments has to be available before the manufacturer can issue a Declaration of Conformity and mark the machinery with the CE sign. The harmonised standards EN 292 and EN 1050 describe the basic requirements for the risk assessments (CEN, 1991; CEN, 1996).

The legislation accounts for the fact that machines interact and form production systems of varying complexity. *Machinery* is in the regulations defined as an assembly of linked parts or components, at least one of which moves, with the appropriate actuators, control and power circuits joined together for a specific application. The legislation also applies to *assemblies of machines*. These consist of machines arranged and controlled so that they can function as an integral whole for a specific application. A *sub-assembly* is a mechanised assembly that cannot function independently, and that is intended for incorporation into a machine.

In Section 7.5 we went through the basic barrier philosophy applied in the European machinery-safety legislation. We will here repeat the stepwise procedure to be followed by the manufacturer to ensure a machine is safe for use:

EN 292 defines the duties of the designer in identifying the necessary safety measures. A stepwise procedure must be followed, including:

1 Determination of the intended uses of the machinery, lifespan, space requirements etc.;
2 Identification of the hazards and assessment of the risk;
3 Removal or limitation of the hazard;
4 Design guards and/or safety devices against any remaining hazards;

5   Information and warning of the user; and
6   Consideration of any necessary additional precaution.

In determining the intended uses (Step 1), the manufacturer shall account for reasonably foreseeable misuse. This does include normal carelessness but not deliberate misuse.

Exercise: A tool manufacturer develops and markets a hand-held drilling machine. This has to be CE marked, which makes it necessary for the manufacturer to document the evaluations that have served as a basis for the selection of safety measures. Questions:

1   *Execute a job-safety analysis of the drilling of a hole in a concrete wall by using this drilling machine.*
2   *Propose measures to reduce or eliminate the hazards that have been identified through the analysis and by applying the strategy for selecting safety measures according to Table 7.2.*
3   *Discuss the relation between the method applied to answer questions 1 and 2 and the stepwise procedure according to EN 292 that the designer must follow to identify the necessary measures. Are there other means of following this procedure than the one proposed above?*

24.2 Method for risk assessment

The harmonised standard EN 1050 details the requirements for the manufacturer's risk assessment. The aim is to ensure that it is consistent and systematic. EN 1050 also proposes different methods for use in analyses of hazards and risk estimations.

In Norway, the offshore industry has developed guidelines for follow-up of the documentation of machines in project work (NORSOK S-005, see Norsk Standard, 1999). The guidelines are also applicable to land-based industry. We will here present an application of the method and illustrate it with an example from the purchase of a new wire-rolling mill at an aluminium plant. There are also other guidelines on applications of risk assessments to machinery, for example, those developed by the United Kingdom Health and Safety Executive (HSE, 1997b).

The principal steps of the method are shown in Figure 24.1. It reflects the progress of a typical process for the design of new plants and modifications of existing plants. Step 1 is carried out before the placement of purchase orders and Step 2 before layout of the machinery has been finally decided.

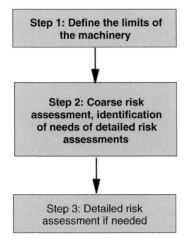

Figure 24.1 The main steps of the total risk assessment.

Step 3 is carried out in a later phase and is based on detailed engineering drawings.

24.2.1 Determination of the limits of the machinery (Step 1)

In Step 1, the organisation that has the overall responsibility for the design of the plant including procurement of equipment (i.e. the project contractor or the main contractor) has to define the limits of individual machines and assemblies of machines in all areas of the plant. The aim of Step 1 is to clarify the responsibilities for issuing the Declaration of Conformity. Within this responsibility lies the responsibility to perform risk assessments.

Industrial plants are often very complex and consist of many different machines and assemblies of machines. The limits of each machine or assembly of machines have to be clearly defined at an early stage in order to define the responsibility for the Declaration of Conformity clearly in the purchase order. It usually lies with the machine manufacturer. This principle also applies when the machine is connected to the plant's process-control system. It happens, however, that the organisation responsible for overall design and procurement (the project or the main contractor) assembles different machines into an integral whole. In this case, the project or main contractor is responsible.

The definition of the limits of machines and assemblies of machines in Step 1 is critical. There are conflicting interests between the supplier and user that have to be handled. Seen from the user's point of view, it is important that all interfaces between individual machines are analysed in case they may give rise to new hazards. A supplier of industrial systems, on the other hand, is interested in a breakdown of responsibilities to the next level, i.e. to the suppliers of the individual machines.

Example: A main contractor was responsible for a turnkey delivery of a new wire rolling mill to an aluminium plant. This consisted of the following parts:

- *A casting wheel.*
- *Two sets of rolling mills.*
- *Reels for the collection of wires.*
- *Material-handling equipment.*

The main contractor referred to the sub-suppliers of these parts as being responsible for documentation of machinery safety and issuing of Declarations of Conformity. The client did not accept this answer and argued that the different parts interacted to serve a common purpose and had overlapping danger zones. In negotiations, it was decided to consider the whole rolling mill to be an assembly of machines. Material-handling equipment was defined as machines and the supplier was responsible for the risk assessments. The other parts were sub-assemblies and were documented by suppliers' declarations. The main contractor was responsible for the risk assessment of the rolling mill as an integral whole.

Below are some guidelines in determining the limits of machines and assemblies of machines:

1 Different machines are considered as being part of an assembly of machines when
   a They interact to serve a common purpose
   b They are connected through a common control system
   c The danger zones overlap or the activity of one machine will automatically initiate hazardous movements in the interconnected machine.
2 A machine is considered to be an individual unit and not part of an assembly of machines when
   a It is controlled manually by an operator
   b It has a separate control system
   c The danger zone of the machine is separated from the danger zone of other machines.

The machinery manufacturer must document the following items prior to the start of the risk assessments:

- Description of the machine (model, type, number, etc.) and layout drawings and other relevant technical documentation (P&IDs, isometrics, block diagrams, etc.).
- Description of use of the machine including maintenance, transportation, etc.

- Assumptions about the user, qualifications, etc.
- Life span, mean time between change of components, etc.
- Space requirements as defined by the footprint and need for access during operation, inspection and maintenance.
- Ambient conditions at the place of installation (indoor/outdoor), lighting, etc.
- Interfaces with the rest of the plant concerning physical arrangements, control system and energy supply. In some cases, the machinery manufacturer will have to secure the necessary interface information from the main contractor.
- Accident and health-impairment statistics of similar machines, if available.

For machines that have been documented before by risk assessments, these need not be retaken. It is important, however, to consider all changes in design and assumptions about uses before deciding whether a new risk assessment is needed or not.

24.2.2. Coarse risk assessment (Step 2)

In the Coarse risk assessment, the machinery manufacturer or the party responsible for the assembly of machinery identifies the main hazards and necessary risk-reducing measures. The purpose is also to identify the need for detailed risk assessments. This activity has to be performed at an early design stage before layout freeze and placement of major purchase orders of sub-deliveries. A team similar to the Coarse analysis team is responsible for the analysis, see Chapter 22. The procedure shown in Figure 24.2 is applied and the results are documented in a record sheet according to Table 24.1.

All team members need to have a basic understanding of the machinery. In addition, the individual team members should represent the following knowledge and competence:

- Design engineers with detailed technical knowledge of mechanical design and design of control and power supply systems.
- Personnel with experience of how similar machines are operated and maintained.
- Safety-engineer knowledge and experience about the risk-analysis method to be used, safety regulations and accidents likely to occur in practice.

The team identifies work to be performed in the danger zone of the machinery for each type of hazard listed in column one and documents the results in column two of a record sheet according to Table 24.1. It is here necessary to review the different activities in which personnel may come in contact with the hazard. It is important to consider the risks caused by sudden start of remotely controlled or automatic machinery when personnel are in the danger zone. The team then evaluates the types of hazards listed in the first

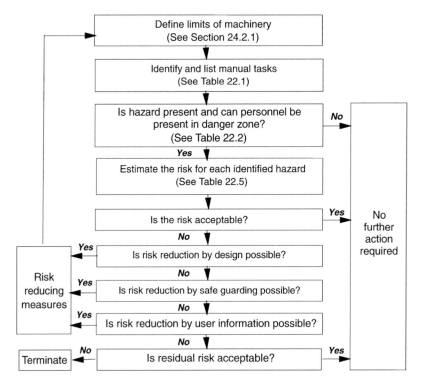

Figure 24.2 Coarse and detailed risk assessments (Steps 2 and 3).

column and documents identified hazards and hazardous situations in the third column. Results of risk estimations are documented in column four. It is important to consider experiences with similar types of machinery in assessing the frequencies of the accidents. The matrix on individual risk usually applies in this case, see Table 22.5. The comments to the application of this matrix in detailed job-safety analyses in Section 23.4 also apply here.

At this stage in design, the team concentrates on evaluations of layout and the selection of sub-assemblies, components and safeguards. When high and medium risks are identified, measures have to be evaluated. These are documented in the fifth column. The strategy for selection of safety measures according to EN 292 referred to above must be considered. The team also has to consider the effects of the measures on the usability of the machinery and the costs of the measures.

It is not necessary to repeat earlier risk assessments, provided that they are of an adequate quality. These are referenced in the last column of Table 24.1. Decisions on the need for new detailed risk assessments are also documented here. The decisions have to be implemented in purchase orders to sub-suppliers and in internal action plans.

Table 24.1 Record sheet for documentation of the Coarse risk assessment. Only the mechanical hazards are detailed. The table shows a simplified example of the hazard identification of the wire rolling mill

Hazards and hazardous situations (Ref. EN 292 and EN 1050)			Risk			Safety measures in design, safe guarding	Need for detailed RA?
Checklist	Activities in danger zone (work tasks, misuse, etc.)	Description of hazards	C*	F*	R*		
Mechanical • Crushing • Shearing/cutting/stabbing • Entanglement/drawing in • Trapping • Impact • Friction/abrasion • Fluid ejection	Guiding of bars during start up Removal of bars that have got stuck Cleaning	Drawn in between the rollers	3	3	H	Automatic start-up sequence Selection of highly reliable components	Yes
	Maintenance	Getting squeezed while lifting rolls	2	2	M	Access to overhead crane	Yes
Electrical		Not identified					
Thermal	Same as for mechanical	Bar breaks and hits operator	2	3	H	Automatic start-up sequence	Yes
Noise/vibration	All activities	High noise level from rolling mill (> 90 dBA)	3	2	H	Replacement of gear boxes with low noise emission types	No
Radiation		Not identified					
Materials and substances	Cleaning of rolls	Contact with solvents	1	3	M	Arrange for steam cleaning	No
Ergonomics	Maintenance	Heavy lifting of rolls	2	2	M	Access to overhead crane	No

Note: C = Consequence, F = Frequency, R = Risk.

Table 24.2 Record sheet for documentation of the detailed risk assessment. The example shows simplified results of the detailed risk assessment of one of the rolling mills

No.	Activity	Hazard and cause	Risk			Design measures/ safe guarding	Assumed precautions by user	Remaining Risk			Status
			C*	F*	R*			C	F	R	
1	Start up/shut down										
1.1	Manual guiding of bar from feeding wheel to first pair of rollers when automatic system is out of order	Hand squeezed between rolls	3	1	M	Not required, since this is a low frequency event	Use of personal protection	2	1	L	
2	Normal operation and disturbance handling										
2.1	Removal of bars that have got stuck	Caught between rolls that start to move in wrong direction. Wrong operation of pendant control	3	2	H	Interlock to prevent wrong operation	User instructions for removal of stuck bars	3	0	L	
3	Cleaning										
3.1	Cleaning of rolls	Caught between rolls moving in suction direction	3	3	H	Interlock preventing use of suction mode during cleaning		2	0	L	
4	Repair and maintenance										
4.1	Shifting of rolls	Strain injury due to heavy workload	2	2	M		Lifting instructions	2	2	M	Overhead crane provided

It is advisable to break down the machinery into manageable functional parts prior to the analysis when analysing complex assemblies of machinery. The interfaces must also be carefully considered in the identification of hazards.

24.2.3 Detailed risk assessment of the machinery (Step 3)

The method that is selected for the detailed risk assessments may vary, depending on the types of safety problem involved. If, for example, the reliability of safety systems is a concern, fault tree analysis or failure mode and effect analysis may be the appropriate method. Here, a detailed risk assessment similar to the job-safety analysis of Chapter 23 is presented. It follows the general outline of the coarse analysis of Figure 24.2 but is performed at a more detailed level. The aim is to study manual activities in the danger zones of the machinery in detail in order to take the necessary precautions. A team will be responsible for this analysis as well. Results are documented in a record sheet according to Table 24.2.

As compared to the JSA, there are the following differences:

- A checklist of hazards according to Table 5.3 is used in the hazard identification.
- A first risk estimate is made.
- When the risk is 'medium' or 'high', measures in design are evaluated by considering the strategy for selection of safety measures according to EN 292, see Chapter 7. The current state-of-the-art in the design of similar types of machinery should be consulted.
- If 'medium' and 'high' risks are still remaining, the necessary operational precautions to reach an acceptable 'low' remaining risk are identified. These may include adherence to certain job procedures, use of personal protective equipment, permit-to-work system, etc. Please note that 'high' risks should only be remedied by operational procedures on an exceptional basis.
- Design measures, assumed operational precautions and the remaining risk are documented.

25 Comparison risk analysis

Comparison risk analysis fills a gap among existing methods concerning the analysis of the risk of occupational accidents (Kjellén, 1995). It is used during the design of new plants (and modifications) in order to predict the occupational accident-frequency rate for the plant during operation. Results are expressed as relative changes in the accident-frequency rate in relation to the experienced rate of a 'reference plant' that has been in operation for some years. The method was originally developed to meet the Norwegian risk-analysis regulations for the offshore industry. They require oil companies to develop acceptance criteria for the risk of losses due to accidents and to prove that the criteria have been met through the use of risk analysis (Norwegian Petroleum Directorate, 1990).

25.1 Acceptance criteria for the risk of occupational accidents

Historical data on accident-frequency rates for a particular plant often show a downward trend. A trend of this type can be explained by operational conditions, such as improved experience and SHE management. Risk-acceptance criteria for use during design should not be sensitive to such trends. Rather then defining the risk-acceptance criterion as a fixed accident-frequency rate, the following criterion is suggested:

'The plant should be designed and the operation of it planned in such a way that operation's goals concerning the accident-frequency rate is not more difficult to meet than for existing plants.'

This means that there must not be a deterioration of the 'technical' safety level for new *plants*, as compared to existing *plants*.

Example: This criterion was implemented in the SHE programme of an offshore project for the design and construction of a new floating production and drilling installation. The selected concept involved a number of challenges concerning the risk of occupational accidents:

- *Floater movements*
- *Large open areas, where the personnel are exposed to wind and rainfall*
- *Increased handling of heavy drilling equipment*
- *Reduced crew*

All these factors were expected to increase the accident-frequency rate. The acceptance criterion implied that compensatory actions had to be taken in design to meet the criterion. These included actions to reduce the exposure of the personnel to hazards and actions to reduce the probability of accidents.

25.2 Risk-assessment model

Comparison analysis is based on a model for the assessment of the accident risk in an organisation. This is determined by (1) the frequency of hazardous activities that are performed by the organisation, (2) the probability of accidents when a hazardous activity is performed, and (3) the consequences of the accidents (Kjellén, 1995). The activities are defined by a discrete set of types of activities, $\{A_i\}$. Associated with each type of activity is the possibility of the occurrence of injuries. We will analyse the risk of reportable injuries, i.e. either lost-time injury (LTI) or total recordable injury (TRI). The injuries belong to the discrete set of types of injuries, $\{I_{i\alpha}\}$.

There is a probability ($p_{i\alpha}$) of the occurrence of an injury ($I_{i\alpha}$) during the performance of the activity A_i. The probability (p_i) of the occurrence of an injury during the performance of the activity A_i is thus:

$$p_i = \sum_\alpha p_{i\alpha}$$

During a given time period (e.g. one year), the activity A_i is on the average carried out N_i times. The expected number of injuries of type $I_{i\alpha}$ in the activity A_i during this period equals:

$$E(Z_{i\alpha}) = p_{i\alpha} \times N_i$$

Similarly,

$$E(Z_i) = p_i \times N_i$$

where $Z_i = \sum_\alpha Z_{i\alpha}$

Let e_i denote the mean number of working hours in the activity A_i during the time period in question. The injury frequency rate (λ) during this period equals:

$$\lambda = \frac{\sum_i E(Z_i)}{\sum_i e_i} \times 10^6$$

We assume that there are two workplaces, the 'new workplace' and the 'reference workplace'. The types of activities of the two workplaces belong to the same set $\{A_i\}$. The injury frequency rate and the distribution of accidents by activity and type are known for the 'reference workplace'. We introduce the following relationships:

$$(E(Z_{i\alpha}))_{new} = k_{i\alpha} \times (E(Z_{i\alpha}))_{ref}$$

where $\quad k_{i\alpha} = \dfrac{(p_{i\alpha})_{new}}{(p_{i\alpha})_{ref}} \times \dfrac{(N_i)_{new}}{(N_i)_{ref}}$

The 'correction factor', $k_{i\alpha}$, is estimated for each type of activity and injury. This estimate is made up of two parts. First, the team evaluates effects of changes between the 'new workplace' and the 'reference workplace' on the number of times an activity is carried out (exposure). After that, the team evaluates the effects of these changes on the probability of the occurrence of the injury. The expected number of injuries in activity A_i during the time period in question for the 'new workplace' is calculated from the formula:

$$(E(Z_i))_{new} = \sum_{\alpha} k_{i\alpha} \times (E(Z_{i\alpha}))_{ref}$$

It is assumed that the total number of working hours during the time period in question ($\sum e_i$) is known for both the 'new' and the 'reference workplaces'. The expected injury frequency rate for the 'new workplace' is given by the equation:

$$\frac{\hat{\lambda}_{new}}{\hat{\lambda}_{ref}} = \frac{\left(\sum_i (Z_i)\right)_{new}}{\left(\sum_i (Z_i)\right)_{ref}} \times \frac{\left(\sum_i e_i\right)_{ref}}{\left(\sum_i e_i\right)_{new}}$$

25.2.1 Assumptions

An expert panel estimates the effect of the differences between the 'new workplace' and the 'reference workplace' on the risk of injuries. The panel employs a number of assumptions in these judgements when they evaluate the effects of design and operational conditions on the risk of injury. Of the different operational factors that affect this risk, only changes in manning level, in activity plans and in specific work procedures are considered. No credit is given to other planned operational factors such as changes in the SHE management system or in safety training or information.

Table 25.1 Decision rules in risk estimation

Relative change in probability	Evaluation
0	The risk of injury has been completely eliminated.
0.25	Major reduction in the probability of injury.
0.5	Large reduction in the probability of injury.
0.75	Moderate reduction in the probability of injury.
0.9	Small reduction in the probability of injury.
1	No change in the probability of injury.
1.1	Small increase in the probability of injury.
1.25	Moderate increase in the probability of injury.
1.5	Large increase in the probability of injury.
2	Major increase in the probability of injury.

Factors in design or operation of the 'new workplace' that have not been determined at the time of the analysis are assumed to be equal to those of the 'reference workplace'.

The probability of the occurrence of an injury when an activity is carried out is assumed to be independent of the number of times that the activity is carried out. Effects of experience are thus disregarded. It is further assumed that the level of manning does not affect the expected probability of injuries in an activity. Manning is only considered when calculating the LTI-rate.

The expert panel's judgements are of two types. In assessing changes in exposure, $(N_i)_{new}/(N_i)_{ref}$, the activity plans for the 'reference workplace' and the 'new workplace' are used as input. Mechanisation that results in a removal of the operator from the danger zone is also considered. These assessments are usually rather uncomplicated. If an activity has been completely eliminated at the 'new workplace', for example through mechanisation, this ratio takes the value 0.

Assessments of changes in the probability of injuries, $((p_{ia})_{new}/(p_{ia})_{ref})$, are more complex, since they have to consider a number of different factors related to design and operation. To simplify these assessments, the team uses the decision rules according to Table 25.1.

25.3 The steps of the analysis

The analysis involves the establishment of a database of injuries of the 'reference workplace'. This is manipulated in order to 'simulate' a database of expected injuries in the 'new workplace'. It is performed in four steps, Figure 25.1.

In step 1, the database for the 'reference workplace' is established. A database on injuries is used in the analysis. The injuries in this database have to be categorised by type of system, type of activity and type of accidental event. From this database, a subset of data for the 'reference workplace' is

INPUT

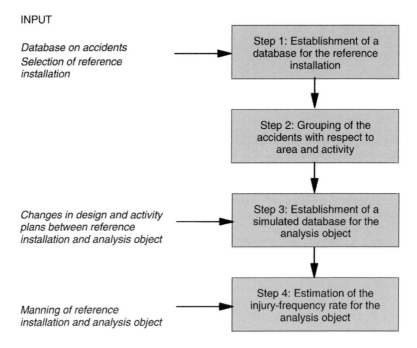

Database on accidents
Selection of reference
installation

Step 1: Establishment of a
database for the reference
installation

Step 2: Grouping of the
accidents with respect to
area and activity

Changes in design and activity
plans between reference
installation and analysis object

Step 3: Establishment of a
simulated database for the
analysis object

Manning of reference
installation and analysis object

Step 4: Estimation of the
injury-frequency rate for the
analysis object

Figure 25.1 The steps of a Comparison risk analysis.

established. Two different sets of criteria are used in defining this database. First, the 'reference workplace' must be reasonably similar to the 'new workplace' concerning technology and activity plans. Second, the 'reference workplace' database must be of an adequate size. The size determines the degree of resolution in the subsequent analysis.

Example (continued): An analysis team carried out a comparison analysis of the floating production and drilling installation mentioned in Section 25.1. For this purpose, a database of about 100 injuries from two existing platforms of a similar type was set up. The database included so-called total recordable injuries, i.e. those resulting in medical treatment and transfer to another job in addition to the lost-time injuries.

The purpose of step 2 is to establish the distributions of injuries of the 'reference workplace' by type of activity and injury, $(E(Z_{i\alpha}))_{ref}$. Injuries of the 'reference database' are grouped into activities. A hierarchical breakdown of the activities is used in this analysis. For each activity, the injuries are grouped by type of accidental event. The analysis is truncated when the breakdown has reached a meaningful level of detail for each type of activity and event.

Step 3 involves the establishment of the simulated 'new workplace' database. First, the 'correction factor', $k_{i\alpha}$, is assessed for each combination of activity and event type. This is done by an expert panel that consists of

Table 25.2 Checklist of factors that affect the risk of accidents

Factors affecting job frequency		Factors affecting probability of injury	
Design	*Organisation*	*Design*	*Organisation*
Walking distances	Manning	Layout, gangways	Job experience
Production regularity	Activity planning	Lifting equipment	
		Access, ergonomics	
		Physical working	
		environment	
		(lighting, noise)	
		Outdoor operations	
		Safety measures	
		Guarding	

personnel with operational experience from the 'reference workplace' (subject-matter experts), designers from the project and experts on risk analysis. The panel's task is to assess the effects of differences in design and operational plans between the 'new workplace' and the 'reference workplace' on:

1 Changes in the probability of injury each time an activity is carried out $((p_{i\alpha})_{new}/(p_{i\alpha})_{ref})$; and
2 Changes in job frequency $((N_i)_{new}/(N_i)_{ref})$, i.e. the number of times each activity (A_i) is carried out per year.

The checklist in Table 25.2 of different factors that affect the risk of accidents is used in this evaluation together with the decision rules according to Table 25.1.

The mean number of injuries by type of activity and event for the 'new workplace' $((E(Z_{i\alpha}))_{new})$ is then established on the basis of these assessments. For new types of activities in the 'new workplace', the expected number of injuries is determined by comparing with injury data from similar activities at the reference workplace. The same applies to activities that have been redesigned such that new risks are introduced. Detailed job safety analyses may be required to identify and assess new risks.

The team documents its judgements in a record sheet according to Table 25.3. It is important that all judgements are carefully recorded. Since expert judgements play such an important role, the analysis has to be transparent and allow for later review and verification.

Example (continued): The analysis of the floating platform covered all areas of the platform. Table 25.3 shows an example of the results for the kitchen. This part of the analysis was carried out by a team consisting of a supervisor and a catering worker from one of the reference platforms, an architect responsible for design of the living quarters of the new platform and a working-environment specialist.

Table 25.3 Extracts of results from the comparison analysis of kitchen work

Activity	Historical data for reference installations		Predictions for new installation			Comments
	N	Type of injury	Change in job freq.	Change in probability	N	
Cutting of meat, vegetables	6	Cut injury	0.7	0.9	3.8	Reduced cutting work Better layout and lighting Floater movements
Handling of hot pots, pans, etc.	2	Burns	1	1.25	2.5	Floater movements

Table 25.4 Results of a comparison analysis

Activity	Expected number of accidents per year for the new installation	Historical data on the number of accidents per year for reference installations
Catering	4.2	4.5
Operation	1.8	1.5
Drilling	4.2	3.8
Maintenance	3.8	4.7
Walking	1.7	1.7
Others, unknown	0.4	0.3
TOTAL	16.1	16.5

In step 4, the injury-frequency rate is estimated for the 'new workplace'. The number of injuries is summarised in order to establish the overall mean number of injuries for the 'new workplace' and the 'reference workplace' $((\Sigma E(Z_i))_{new}/(\Sigma E(Z_i))_{ref})$. Next, the known manning level at the 'reference workplace' and the planned manning level at the 'new workplace' are used to establish $(\Sigma e_i)_{ref}$ and $(\Sigma e_i)_{new}$. λ_{new} is then calculated.

Example (continued): The total analysis predicted the occurrence of on average 16 injuries per year at the new installation, Table 25.4. There was a 2 per cent reduction in relation to the historical data from the reference installations. Maintenance accounted for the largest reduction and this was explained by reduced maintenance due to improved planning and improved access to equipment during inspections and material handling. Increases were expected in drilling due to more extensive work and more handling of heavy materials.

Calculations of the total recordable injury frequency rates for the new and the reference platforms showed a small decrease for the new platform. It was concluded that the acceptance criterion was met. The analysis, however, revealed a number of activities where an increase in the injury frequency of more than 20 per cent was expected. The ALARP principle called for actions to reduce the risk of accidents in these activities in particular. They included:

- *Handling of hot objects in the kitchen (floater movements),*
- *Cleaning of equipment in utility areas (more hydraulic equipment),*
- *Crane operations (reduced view from crane cabin, floater movements), and*
- *Pipe handling in drilling (increased job frequency, floater movements).*

26 CRIOP

CRIOP (Crisis Intervention in Offshore Production) focuses on the actions of process plant operators and the role they play in maintaining barriers against major accidents. The method was developed in the late 1980s for use during the design of new offshore installations and was influenced by experiences from the Piper Alpha disaster in 1988 (Ingstad and Bodsberg, 1989). It has been further developed to include later technical developments and user experiences.

A CRIOP analysis consists of two parts. Part I is a standard design review of the plant's control centre. It employs checklists, covering applicable codes, standards and regulations. The following areas are checked:

- Arrangements
- Man–machine interfaces
- Communication systems
- Physical working environment
- Control- and safety-systems
- Work organisation
- Procedures
- Training programme

CRIOP Part II helps us in analysing new accidents that may happen in the future in detail rather than at the summary level of the traditional technical risk analyses. The possibilities of transgressing barriers and the operators' contributions are analysed successively, compare Figure 7.3 for the case of risks of fires and explosions. It starts by looking into the operators' interventions in the process to control disturbances that, if not properly handled, may cause major accidents. In cases where there is a reasonable probability that the situation may deteriorate, the analysis proceeds by looking into the operators' handling of the emergency situation.

CRIOP is used in verifications of design, manning, organisation and emergency preparedness to ensure that these are adequate from a human decision-making point of view. When applied in the detail-engineering phase,

CRIOP will focus on the design of arrangements, man–machine interfaces and control and communication systems. CRIOP is also used as a preparation for start-up. Here the focus is on organisation, manning, procedures, operator training and emergency preparedness.

We will here focus on Part II. A typical *CRIOP scenario* starts with a disturbance that calls for a manual intervention. It goes through a phase of lack of control and is followed by a loss of control of energies with the potential for major losses such as fires, explosions, collisions, dropped objects, etc. Next, there is a phase where the emergency organisation tries to limit the damage. The scenario ends when the energies have been brought under control. For a scenario to develop through all these phases, a number of barriers must fail. A crucial point is thus the extent to which the operators are able to identify potential barrier failures and to take the necessary actions to re-establish barrier integrity or to mitigate the effects of the failure.

The first thing to do is to select the scenarios for the analysis, Figure 26.1. This is a very critical part of the analysis that determines the validity and credibility of the results. The selected scenarios should represent significant areas of concern to the involved parties. It is important that the selection of scenarios is made jointly by the client and project. Accident statistics from similar plants and results of the technical risk analysis of the new or modified plant are used as input.

The CRIOP scenarios are selected on the basis of the following criteria:

- They should be realistic. The involved parties should feel that they might occur.
- They should have the potential of major losses. The scenarios should represent hazards that contribute significantly to the total risk of major accidents as indicated by the technical risk analysis of the plant.
- Human actions should be critical to the outcome of the scenario. It should make a difference whether the operators carry out the correct actions or make mistakes.
- The involved parties should be uncomfortable with their knowledge about the existing solutions. In practice, this means a focus on new solutions.

Example: The design of a new type of floating offshore production, drilling and living quarter installation was taking place. The project and the operations organisation were concerned about the layout and the communication systems of the installation. More specifically, they wondered whether the different crews on the installation (process, drilling and marine crew) would be able to co-ordinate their actions in a crisis situation. An analysis team selected the following two scenarios for the CRIOP analysis, representing the need for co-ordination between the different crews:

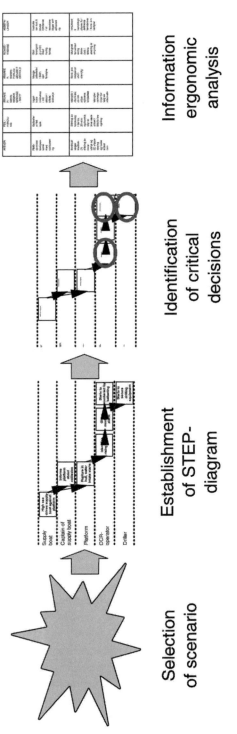

Selection of scenario → Establishment of STEP-diagram → Identification of critical decisions → Information ergonomic analysis

Figure 26.1 The steps of the CRIOP scenario analysis.

- *Blow-out in connection with drilling*
- *Collision with supply vessel*

The analysis team needs access to various types of documentation about plant design during the analysis. Typically, this includes:

- General layout drawings and layout of control cabins including controls and displays.
- Process and instrument diagrams, shut-down logic (cause and effect diagrams).
- VDU pictures.

The detailed analysis of each scenario starts with the establishment of a STEP-diagram, see Section 13.4. The analysis team draws experience from previous accidents and risk analyses in this work as well as from the day-to-day practices in similar plants in operation. It is important that personnel with detailed knowledge about the practices in handling disturbance participate in the analysis. Several barriers often have to fail for a major accident to occur in a well-defended system. The analysis team has to assume that certain malpractice takes place, and it is important that the operations personnel in the analysis team confirm that it is realistic. The STEP-analysis may be terminated under certain conditions:

- The participants conclude that the scenario is unrealistic.
- The analysis team comes up with reliable remedial actions that will stop the scenario from developing into a critical phase.

Example (continued): Figure 26.2 shows the first part of the scenario involving collision with supply boat. It takes place in bad weather, but the situation allows drilling and unloading from a supply boat to take place simultaneously. Heavy sea drives the supply boat against the platform and the collision that follows results in the hull being holed. The central control room (CCR) initiates process shut-down and actions to prevent heeling. The drilling crew starts to secure the drilling equipment. Due to extensive water intake, the platform heels and this results in uncontrolled movement of equipment. A mobile chemical injection skid hits the first stage separator and damages it. Gas escapes from the separator containment and is ignited. The crew initiates the emergency procedures. After 45 minutes, the situation is under control.

A problem encountered in CRIOP analyses is that the operators participating in the analysis lack the necessary imagination to envisage rare events. They base their experiences on events occurring with a frequency of up to 0.1 per year. Failures of technical barriers (e.g. a deluge skid on an offshore platform) or severe operator errors are usually much less frequent.

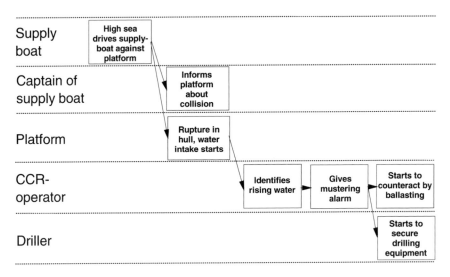

Figure 26.2 First parts of a STEP diagram of the scenario involving collision between a supply boat and a floating offshore installation.

It is the combination of such rare events that lead to major accidents. It has proved important to base the scenarios on actual occurrences to make them credible.

The scenario and the conclusions have also to be documented in cases where the analysis has not been fully completed. An initial STEP-analysis together with the conclusions why the analysis was terminated may prove as valuable as a full analysis.

Next, critical actions by operators are identified and analysed. These are actions where more than one action alternative exists and where human errors may cause severe barrier failure. A record sheet according to Table 26.1 is used for the documentation of this part of the analysis. The critical actions, the identified human errors and consequences are documented in columns one, two and five respectively of the form.

Checklists in Tables 26.2 and 26.3 support the human-factor evaluation. First, the analysis team evaluates whether the human error is related to wrong detection or diagnosis of the disturbed situation or to erroneous execution of action. Thereafter, the team looks into causes. Here a variation of Swain's checklist in Table 6.11 on human performance-shaping factors is used. The checklist includes items related to working conditions, physiological and psychological stresses and expected individual characteristics of the operators. The results of the human-factor evaluation are documented in columns three and four. Finally, the team makes an overall evaluation and comes up with recommendations on remedial actions.

Table 26.1 Record sheet for documentation of the CRIOP scenario analysis. Extracts of the results from the supply-boat scenario of Figure 26.2 are shown

Critical action (See STEP diagram)	Human errors	Causes in the man–machine interface (Table 26.2)	Performance-shaping factors (Table 26.3)	Consequences	Recommendations
Marine operator identifies rising water in a ballast tank.	Delayed diagnosis of location and rate of leakage.	No cause identified, man–machine interface is adequate.	Information overload due to low manning (only one CCR operator).		A second operator to be available within short notice.
Marine operator counteracts heeling through ballasting.	Ballasting of wrong tank.	No clear instructions for action in case of leakage.	Many activities in CCR, many disturbances.	Considerable heel (> 10°), extensive material damage.	Instructions to use trim tanks only. This will limit the consequences.
Driller orders securing of drilling equipment.	Decides to secure the equipment through normal shut-down rather than through emergency procedure (dropping the casing and shutting the shear ram).	The driller lacks sufficient information from the CCR to evaluate how time-critical the situation is.	No direct communication between CCR and driller's control room (DCR). Drilling procedure is incomplete.	Possibility of escalation of the accident into a blow-out in case of serious heel.	Review communication equipment between CCR and DCR. Define action in drilling procedures.

Table 26.2 Checklist of phases in human-information processing

---

1   Detection
1.1   Has the operator the necessary information available (via instrument, communication, alarms, etc.) to detect unwanted events/changes in time?
1.2   Does this information enable the operator to evaluate the criticality and qualities of the situation adequately?

2   Diagnosis
2.1   Is it easy for the operator to retrieve the necessary additional information in order to diagnose the situation correctly and in time?
2.2   Is the situation free from conflicts or other difficult evaluations?
2.3   Is the operator's workload acceptable?
2.4   Is it possible to consult colleagues for advice?

3   Action
3.1   Is it possible for the operator to carry out the necessary manual actions in due time and without errors?
3.2   Is it possible for the operator to communicate the necessary information to other parts of the organisation without misunderstanding and in due time?
3.3   Is information promptly available about the action and its effects and is it possible to recover wrong action?

Table 26.3 Checklist of performance-shaping factors

Workplace design	**Physiological stresses**
Physical working environment	Fatigue
Arrangements	Platform movements
Man–machine interface	Noise
Communication equipment	
Organisation of the workplace	
Operating procedures	
Manning	
Working hours	
Supervision	
Authority	
Psychological stresses	**Individual characteristics**
Workload	Experience, knowledge
Risk of large unwanted consequences	Skills
Monotony	Motivation
Alertness requirements	Physical conditions
Conflicting goals	Team cohesiveness

In CRIOP, the team does not make any probability assessments. It is thus not possible to evaluate the effectiveness of remedial actions in terms of risk reduction. CRIOP may be combined with other risk analysis methods such as event tree analysis, which is suited for calculations of probabilities. A probability assessment requires access to human reliability data.

Part VI

Putting the pieces together

This Part illustrates different applications of SHE information systems. Chapter 27 is built around a case, the Ymer offshore platform with Norskoil as operator. The case is fictive, but it draws experiences from a number of real field developments on the Norwegian continental shelf. We will look into how experience feedback is accomplished in different phases of a platform's life cycle, i.e. design, construction and operation. The purpose is to demonstrate the use of different methods presented in previous sections of this book as integral parts of a SHE management system. We will illustrate both the prevention of occupational accidents and the prevention of major accidents.

In Chapter 28, we will move to the field of traffic safety. We will illustrate the application of different principles and methods of experience feedback in this new setting. This is done from three perspectives, that of the trucking company, the truck manufacturer and the roads administration.

27 The oil and gas industry

27.1 Accidents in offshore oil and gas production

There are major accident risks involved in the development and operation of offshore oil and gas fields. The Bravo blow-out in 1977, the capsizing of the Alexander L. Kielland in 1980 and the fire and explosion that destroyed Piper Alpha in 1988 remind us about this fact. A blow-out may result in extensive losses of human lives and damage to the environment. The Piper Alpha catastrophe demonstrates that fires and explosions may cause a high number of fatalities and extensive monetary losses.

Ordinary occupational accidents are also a concern in this industry. Norwegian accident statistics from the period 1989–98 show that a majority of the fatalities in offshore activities were so-called ordinary occupational accidents, when we exclude helicopter accidents (Vinnem, 1999). There were in total 18 fatalities in 16 different accidents. For permanent installations, the level of FAR was 2.2 and the LTI-rate was about six. The risk level on mobile drilling units was considerable higher with a FAR level of 21.

Each part of an offshore platform has its own characteristics when it comes to the types of accident that occur. In drilling, for example, contact with objects or machinery in motion is a predominant accident type. Handling accidents predominate in catering.

SHE is a central issue in all activities in connection with the development and operation of offshore fields. We will here demonstrate how the different tools described in Parts III to V may be put into a system for use in SHE practice during the different phases of an offshore platform's life cycle. We will concentrate on design, construction and operation. The selected case, the Ymer field with Norskoil as operator, is fictive. It draws experience from real installations in operation and project work.

27.2 The Ymer Platform

27.2.1 Design

Ymer is a semi-submersible production, drilling and living quarters (PDQ) platform, Figure 27.1. The platform is located directly above the subsea

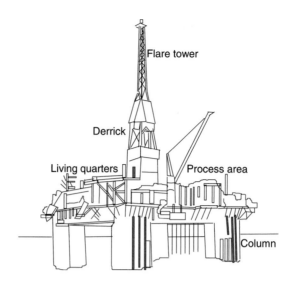

Figure 27.1 The Ymer platform.

completed wells. They are connected to the platform via flexible risers. The oil is exported by pipeline.

The platform deck is built as an integrated deck structure. This is located on the top of a substructure, consisting of four columns and a rectangular pontoon. The platform is fixed to the seabed by sixteen anchor chains.

The drillfloor and the derrick (drilling tower), with pipe handling systems and the flare tower on the top, are located in the centre of the platform. In this area, we find the remaining systems, for drilling and mud treatment. Subsea equipment such as valve assemblies (X-trees) and blow-out preventer (BOP) are stored and handled in the moon-pool area below the drill floor.

The aft of the integrated deck is dedicated to the process systems. There is an oil- and gas-processing train with three-stage separation for production of 15000 Sm^3 stabilised crude per day. All gas is compressed and re-injected during the initial production phase.

The living quarters are built as a separate module at the front end of the platform. Here we find the accommodation areas and the offices. Most of the utility systems for electricity, water, air, etc. are located in the integrated deck below the living quarters. There are two main generators adjacent to the derrick.

27.2.2 Organisation and manning

The manning level of the Ymer platform is lower than on other comparable platforms. This is obtained by simple and functional design and by letting machinery suppliers take more responsibility for maintenance of their own delivered equipment. During the drilling period, Ymer is manned by 86 persons, of whom 56 work dayshift and 30 at night.

Work similar to that carried out on Ymer is found in land-based industry as well. The operation and maintenance crew (23 persons) has similar responsibilities to those of e.g. an oil refinery. Drilling involves the handling of heavy tools and equipment and is to some extent comparable to construction work. The catering crew's responsibilities are very similar to those of the employees of an ordinary hotel.

The main difference to land-based industry is the location of the workplaces about 100 km out in the North Sea, and the shift schedule following from that. All employees work offshore for two weeks in 12-hour shifts and after that they have three or four weeks off duty.

27.3 Prevention of accidents in design

According to basic safety-engineering principles, safety by design is the first choice in the prevention of accidents. A safe design allows platform management to reach SHE goals without undue efforts. An unsafe design, on the other hand, has to be compensated for through operational safety measures that may have a negative impact on production and on the quality of the working life of the employees.

27.3.1 The phase model for offshore field exploration and development

In order to understand how SHE is managed in an offshore project, we first have to look at how the total project period is split into separate phases. Figure 27.2 shows the phase model applied by Norskoil in the Ymer project. In order to continue from one phase to another, the project has to pass so-called decision gates. This involves meeting certain criteria, which are different for each decision gate.

A project division within Norskoil carries out project work. At the first decision gate (DG1), the market organisation of Norskoil, i.e. the client, commissions the project division to start up project work. Beforehand, the client has made the necessary initial evaluations of the potential resource basis of oil and gas and of the market. The scope of work of the *feasibility study* also has to be defined. This phase aims at demonstrating whether it is technically and economically feasible to develop the field. The project team establishes detailed reservoir models in order to assess the oil and gas reserves and to establish production rates over the field's expected lifespan (production profiles). In parallel, the project evaluates different field-development concepts such as fixed platforms, floating platforms and subsea developments for tie-in to existing platforms in the area. This evaluation is based on information about existing infrastructure, water depth, reservoir models, etc. The activities are documented in a Report of Commerciality.

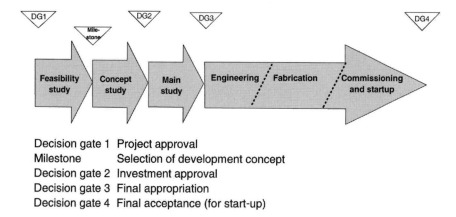

Decision gate 1 Project approval
Milestone   Selection of development concept
Decision gate 2 Investment approval
Decision gate 3 Final appropriation
Decision gate 4 Final acceptance (for start-up)

Figure 27.2 Phase model for offshore field exploration and development.

Provided that field development is judged to be feasible, one development alternative is selected and further matured in the *Concept study phase*. The aim is to provide an adequate decision basis for investment approval to be made by the board of directors. The aims of the SHE activities in the first two phases are to ensure that the selected concept is feasible with respect to SHE and that the authorities and Norskoil's SHE goals and requirements are met in a cost-efficient way. It is not necessary to choose the best concept from a SHE point of view, but the rationale for making the selection has to be documented.

Norskoil decides to select a semi-submersible platform concept for further development of the Ymer field. At the second decision gate (DG2), Norskoil's board of directors makes the project 'go' decision. To pass this gate, the investment plan has to show adequate profitability. This implies that the capital and operational expenditures as well as the revenues from production of oil and gas are known with adequate certainty. It is here necessary that the project's scope has been adequately defined. The platform concept and the subsea and pipeline installations must be defined at an adequate level of detail.

The basic aim of the *main study phase* is to develop the necessary basis for government approval and for the main contracts. The project matures the concept further and establishes a contractual strategy, technical specifications and other tender documents. A Plan for the development and operation of the installation and an Environmental Impact statement are developed and transmitted to the authorities. They are based on input from different SHE studies, including a coarse Risk and Emergency Preparedness Analysis (REPA). A primary aim is to provide input to the documentation required by the authorities. The different SHE analyses and studies are also important tools in concept optimisation and serve as input to technical specifications.

The procurement department establishes a list of qualified tenderers for engineering, procurement and construction (EPC). They must be able to offer adequate standard semi-submersible platform concepts and must be able to mobilise a project organisation with adequate qualifications and experience. Tender documents are developed and an invitation to tender is sent out. Bids are received and evaluated. An EPC contractor is selected among the tenderers.

At decision gate three (DG3), Norskoil's board of directors makes the final appropriation of the terms of the main project contracts. The *Project execution phase* consists of engineering (including procurement), fabrication, commissioning and start-up of oil and gas production. The contract governs the relations between Norskoil and the Contractor. The technical requirements for platform design are specified in the contract. They include a design basis, technical specifications and drawings, etc., and serve as input to *engineering*. The Contractor is responsible for the design of the platform and procurement of equipment. Norskoil's project team follows up on the quality of the work and on budget and schedule. This team also handles design changes. An important task in the follow-up is to ensure that regulatory and company safety requirements are complied with in design. The engineering phase results in detailed drawings, purchase orders for equipment and procedures for mechanical completion checks and commissioning of the different systems and parts of the installation. Mechanical completion (MC) and commissioning are activities that aim to ensure that the delivered modules and systems of the installation meet the specified requirements. By and large, MC refers to visual inspections of the platform by use of checklists, and commissioning refers to dynamic testing of equipment and systems during operational conditions.

Construction of the different modules and structures takes place at different yards. Equipment is installed and the different modules are assembled together with the substructure. The platform is mechanically completed and commissioned.

The platform is *installed* together with the subsea systems at its location in the North Sea. Norskoil as the operator has the main responsibility for safety during all the offshore activities.

Decision gate four (DG4) involves the go-ahead for start-up of production of oil and gas. Norskoil's operations organisation is responsible for this decision. All essential processes, drilling, utility and safety systems must be ready, as well as the organisation and the procedures for the operation of the platform. *Production* also involves maintenance and modifications of the installation. Again, Norskoil as the operator has the main responsibility for safety during all these activities.

27.3.2 SHE management principles

Norskoil has the overall SHE responsibility in all phases of the platform's lifecycle. The company's overall SHE policy states that Norskoil must comply with all regulatory requirements. Beyond that, Norskoil must be one of the foremost operators on the Norwegian continental shelf in relation to safety, health and environment (SHE) work and proven results of such work. Management states that a high SHE standard is a prerequisite for Norskoil to maintain a competitive business operation.

Norskoil has broken down its policy into SHE goals and action plans for the different project phases. They relate to:

- *Technical safety*, i.e. protection and emergency preparedness against major accidents due to fire, explosion, blow-out, falling objects, ship collisions, etc.;
- *Working environment*, i.e. work-related factors that affect the health and well-being of the personnel, such as accident risks, noise, chemical substances, ergonomics, and psycho-social conditions;
- *Environmental care*, i.e. prevention of toxic releases to the air, sea and ground; and
- *Reliability*, i.e. factors affecting oil and gas production regularity.

Norskoil applies common SHE management principles throughout the project phases in order to control and manage the different SHE aspects. The control-loop model of Figure 27.3 illustrates these principles.

The SHE requirements as stated in the Regulations and in the Contract are focused upon. Norskoil has to translate the company's SHE policy and goals in relation to design into binding contractual requirements. There are two types of SHE requirements, goal-oriented and prescriptive. The *goal-oriented* aim at defining satisfactory protection against hazards, without being specific about the detailed technical solution. Regulatory requirements, especially for the offshore industry, are often of this type. It is up to the contractor to choose any solution that meets the goal-oriented requirements, but the contractor has to document that the selected solution is in compliance. This is done e.g. through the application of risk analysis. The results from these analyses serve as input to design.

The *prescriptive requirements* specify the detailed design solutions. Design reviews are applied to verify that these requirements are met.

The most important 'SHE requirements' as to design are:

- Regulatory requirements.
- Norskoil's overall SHE policy and goals.
- Acceptance criteria for the risk of losses due to accidents.
- NORSOK and other European and international standards (Norsk Standard, 1996/97/98/99).
- Norskoil's technical specifications.

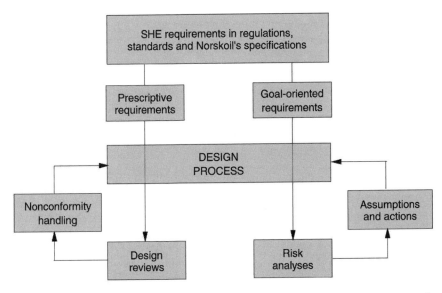

Figure 27.3 The role of different control and verification activities in meeting goal-oriented and detailed requirements to design.

27.3.2.1. Acceptance criteria for the risk of losses due to accidents

According to Norwegian risk-analysis regulations, oil companies have to develop acceptance criteria for the risk of losses due to accidents. Risk analyses should be used to verify that new design solutions meet these criteria. Norskoil applies the following acceptance criteria for the Ymer platform:

1 *Frequency of fatalities.* The average fatality rate (FAR) must not exceed 10 fatalities per 10^8 work hours.
2 *Loss of barriers*
   a *Explosion barriers.* The annual frequency of impairment of explosion barriers between areas must be less than 1×10^{-4}.
   b *Loss of evacuation possibilities.* There must be an upper limit on the frequency of loss of means of evacuation. This applies to evacuation from areas outside the area immediately affected by the accident. The average number of personnel in the area determines this limit, Figure 27.4.
3 *Frequency of occupational accidents.* The frequency of occupational accidents must not be higher than for comparable platforms in operation. Norskoil also uses a risk matrix similar to the one shown in Table 22.4 in determining the acceptability of occupational accidents.
4 *Environmental risk.* The frequency of accidents that impose serious damage on the ecological system must not exceed 10^{-4} per year.

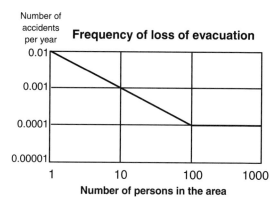

Figure 27.4 Acceptance criteria for loss of evacuation possibilities due to major accidents.

The different acceptance criteria apply to losses to personnel and the environment. There are no acceptance criteria concerning losses of material assets. Here, the safety measures are implemented on the basis of cost–benefit considerations. Two of Norskoil's acceptance criteria for risks to personnel use losses of barriers as the undesired outcome rather than injuries or fatalities. The aim here is to guide the risk analyses towards the identification and evaluation of design aspects that have a known relation to risk.

27.3.2.2 *Experience transfer between operation and design*

When designing new production systems, a careful balance has to be made between SHE measures in design and operation. In practice, we are rarely able to eliminate or guard against all hazards by design. Instead, we must rely on barriers that require constant vigilance by operations management and operators to be effective (compare Ashby's law of requisite variety in Section 10.5). The task of controling barriers though procedural measures and measures directed at the individual workers must not be too complex and demanding.

Experience transfer between the design and operations communities must help find an adequate balance between measure in design and operation. It must also help in:

1　Reduction of the complexity of the operational control tasks through:
　　a　Identification and elimination or reduction of hazards;
　　b　Design of barriers between the hazard source and the system's users that require an acceptable effort to be established and maintained;
　　c　Design of feedback mechanisms to ensure that the barrier efficiency is maintained.

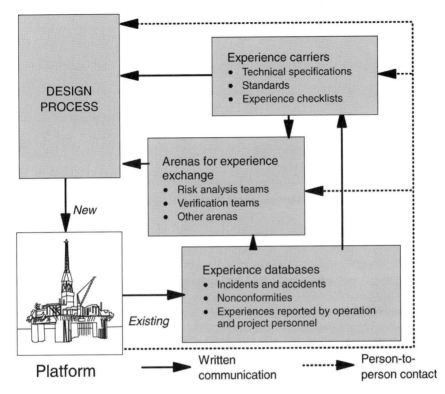

Figure 27.5 Experience flow from operations to project to support the design
process.
Source: Adapted from Wendel, 1998.

2 Information to operations on how to establish and maintain barriers for
which they are responsible.

Experience exchange between operations and project has to subordinate
itself to the basic SHE management principles of Figure 27.3. There are
different experience transfer mechanisms between operations and project,
Figure 27.5. These are both of a formal type, involving documentation on
paper or in databases, and of an informal type, involving person-to-person
communication.

In the formal transfer of experience, the *experience carriers* play an important role. They include Norskoil's technical specifications and experience
checklists. The latter are less binding to the contractor. The experience
carriers express operations experience in a language and format that is
comprehensible to the design community. The process by which specific
and detailed experience data are extracted, summarised and evaluated for

implementation in the experience carriers is crucial. This does not come by itself. Rather, the organisation has to use a structured approach by establishing procedures for regular reviews of the experience databases.

The *experience exchange arenas* represent another means of experience transfer. The individual experiences are challenged in group discussions and combined in ways that promote the development of a joint understanding and qualified experience.

We find an incident and accident database among the *experience databases*. Although the project personnel have access to this database, there are few spontaneous queries of it. The experiences are considered too diluted and incomprehensible and there are concerns about the data quality. The database has proven valuable when used in black-spot analyses.

The informal communication, through integration of operations personnel in the project organisation, is also shown. A relevant question is whether the experiences that are transferred through these person-to-person contacts represent the operation organisation's experience rather than personal preferences of the integrated personnel.

27.3.2.3 Control and verification activities

In the internal control regulations for the Norwegian offshore industry, it is stated that the oil companies are responsible for ensuring that the regulatory requirements as to health, safety and the environment are met. Other regulations, such as the risk-analysis regulations and the regulations concerning systematic working-environment activities detail the internal control requirements (Norwegian Petroleum Directorate, 1990; Norwegian Petroleum Directorate, 1995). Norskoil has an interest in controlling and verifying that the company's SHE policy and goals are met.

In design, the EPC contractor must develop SHE management systems to control and verify compliance with all contractual requirements. They shall be based on the principles and requirements in the quality management standard family ISO 9000. Design reviews, risk and reliability analyses, inspections and tests are examples of control and verification activities. Norskoil's project team participates in some of the activities and carries out independent control and verification activities as well.

The NORSOK standard S-002 on Working environment reflects these basic SHE management principles. Chapter 4 of the standard lists the various types of control and verification activities that must be performed by the contractor to cover all essential working environment factors. We here find requirements as to coarse and detailed job-safety analysis and as to quantitative risk analysis. In Chapter 5 we find the working environment requirements as to design.

The project's and Contractor's SHE management systems are documented in the respective organisation's *SHE programme*. This is a document that describes:

- SHE objective and goals and acceptance criteria for the risk of losses due to accidents.
- List of references to SHE requirements and to applicable procedures for control and verification.
- Organisation and responsibilities for implementation of SHE requirements in design and for control and verification activities.
- Plans for control and verification activities, deliverables included.

The project's SHE manager is responsible for the SHE programme and has to update it for each new phase. It must reflect the criteria that have to be met at each decision gate. The initial SHE programme, which is established at the start of the feasibility study phase, must address the following questions (DG1):

- Has the SHE scope been adequately understood? Are there specific SHE challenges as to sensitive environmental resources, climatic conditions, water depth, etc. that have to be addressed at an early stage in the project's activity plan?
- Have the applicable SHE requirements and other steering documents (regulations, policy, goals, acceptance criteria, technical specifications, etc.) been identified and are they of adequate quality?
- Have the necessary authority contacts been established?
- Is the project organisation adequate and is it manned with SHE personnel with adequate qualifications?

At the next decision gate (DG2), the project must be able to show that there are satisfactory answers to the following questions:

- Have all significant SHE problems been adequately identified and addressed?
- Has it been documented that the selected concept is feasible with respect to SHE and that all significant SHE problems have been resolved with adequate certainty with respect to cost and schedule?
- Have the plans for authority contacts been adequately defined and initiated?

The SHE programme for the main study phase will address the need to mature the development concept further with respect to SHE and to develop the necessary tender and contract documents. At the end of this phase, the project must be able to give adequate answers to the following questions in order to pass decision gate three (DG3):

- Has the selected concept been adequately documented with respect to SHE to meet Norskoil and regulatory requirements? Does the concept comply with the risk-acceptance criteria? Has the environmental impact been adequately documented?
- Have the necessary authorities' approvals been received before entering into binding contracts?
- Have the SHE requirements as to design and control and verification activities been adequately defined in the contracts?
- Are the Contractors adequately qualified with respect to SHE in order to perform the work?
- Has Norskoil established an adequate project organisation for follow-up of the contracts? Is adequate experience transfer from Operations ensured?

The project's SHE programme for project execution will include the necessary activities for follow-up and verification of contract work. At the end of this phase, the project must be able to hand over adequate documentation to the operations organisation to meet the criteria for decision gate four (DG4). In all, operations has to be able to give positive answers to the following questions before the platform is ready for start-up and introduction of oil and gas into the process systems:

- Are the technical systems, the operational procedures and the organisation and personnel ready for start up with respect to SHE?
- Have the necessary authority approvals been received?

27.3.3 Prevention of major accidents

27.3.3.1 Concept selection and definition

In the concept-selection process, different pros and cons of the alternative concepts are evaluated against each other by use primarily of economic criteria. It is not necessarily the best concept from a safety point of view that is selected, but the concept that from an overall evaluation has the greater merit. The basic safety requirements and risk acceptance criteria have to be met by the selected concept.

Major accidents involve the accidental release of considerable energy and the potential for multiple fatalities and major material damage. In offshore oil and gas production, there are a few types of hazards or energies associated with the risk of major accidents. For a floating PDQ platform, they include fires and explosions due to rupture of process containment, blow-out or maritime accidents (hull puncture due to ship collisions, dropped objects, extreme weather or ballasting errors). We will here focus on the prevention of fires and explosions, since they dominate the risk picture. Figure 7.3 illustrates how fires and explosions are prevented through the implementation of independent barriers.

The project screens different development concepts. Risks that may make a concept unfeasible or result in high extra costs are focused on in special studies. An independent consultant carries out coarse **Risk and emergency-preparedness analysis (REPA)** for the two development alternatives that are considered (NORSOK Standard Z-013, see Norsk Standard, 1998). They are a standard production ship and a standard semi-submersible PDQ platform. The aim is to get a total overview of the risks involved for concept selection and to check compliance with the acceptance criteria. It is important to remember that the REPA and the acceptance criteria evaluate the integrity of passive and embedded barriers in particular. These are to a large extent determined in the early project phase.

A REPA consists of two parts. The risk analysis involves the following activities (Norsk Standard, 1998; Suokas and Rouhiainen, 1993):

1 System description, including description of technical systems and relevant operational activities, part of life cycle to which the analysis applies, and personnel and environmental assets exposed to risk.
2 Identification of hazards and listing of initial events such as gas leakage. This work is based on generic information about the type of concept, and operational experience plays a minor role.
3 Accident modelling, consequence evaluation and assessments of probabilities. Here platform design and operational modes serve as important input. Probability assessments are based on generic accident statistics, averaging over different types of platform design, barrier types and safety management systems.
4 Evaluation of risk and comparison with the risk-acceptance criteria.

The risk analysis gives input to the emergency-preparedness analysis. It identifies the so-called dimensioning accidental events. They are major accidents (such as fires, explosions and ship collisions) which generate the most severe accidental loads that the safety barriers must be able to withstand. Dimensioning accidental events are used as the basis for design and for the planning of emergency response. The aim of the emergency-preparedness analysis is to verify that Norskoil's emergency-preparedness requirements are satisfied for all dimensioning accidental events.

Table 27.1 shows the results of the REPA for the two development alternatives. The REPA gives the ship alternative the best rating. In this alternative, an independent rig carries out the hazardous drilling and well completion and maintenance activities far away from the rig. Risks on the rigs are excluded from the analysis, which gives the ship alternative an advantage. A production ship also has a safety advantage through its intrinsic design. There is good separation between safe and hazardous areas, where

Table 27.1 Examples of results from the coarse REPA for the two development alternatives for Ymer

Risk aspect	Norskoil's acceptance criterion	Risk level assessed for ship	Risk level assessed for semi with drilling
Loss of main safety function	1×10^{-4}	3.3×10^{-4}	6.4×10^{-4}
Frequency of escalation	1×10^{-4}	5.7×10^{-5}	6.5×10^{-5}
Fatal accident rate	10	6.1	6.9
Catastrophic environmental damage	1×10^{-5}	$< 1 \times 10^{-6}$	$< 1 \times 10^{-6}$

the hazardous areas are always located downwind. Main safety concerns in the ship design involve:

- The turret, through which oil and gas from the subsea installations are brought up to the ship. There are high risks of leakage of oil and gas in the swivel inside the turret. Ignited gas leaks may cut off escape ways from the rear of the ship to the living quarters.
- Risers, which in some positions of the ship are located beneath the living quarters. Ignited leaks will expose the muster area and lifeboats.
- Ship collisions in general, and especially the shuttle-tanker activities close to the stern of the ship.

There are existing ship concepts that are considered feasible, provided that a collision warning system is introduced. It is also considered realistic, that further design development of the turret will result in a feasible solution.

Major safety concerns with the semi are:

- The location of the subsea wells underneath the platform, which means that oil and gas will reach the sea surface under the platform in case of blow-out.
- Burning riser leaks, which will expose the platform substructure and may cause it to collapse.
- Separation of safe and hazardous areas. Especially, the barrier between the drilling areas and the safe haven (living quarters) is critical.

According to the results of the REPA, the anchor-release system of the semi has to be designed such that the platform can withdraw to a safe location outside a burning pool fire on the sea surface in case of blow-out. The risers and substructure have to be protected. The derrick must be able to withstand a burning blow-out during the time it takes to relocate the platform to a safe position outside the pool fire on the sea. With these measures, the semi is considered feasible.

The coarse REPA for the semi alternative is handed over to the authorities as an attachment to the PDO. When a contractor has been selected, the contractor's standard semi concept is subject to a new REPA.

Although operations are involved in evaluating the results of the REPA, their experience has a limited effect on the results. The method employed in the REPA does not invite to the use of operational experience, since this is too restricted to have any significant impact on the end results.

During Concept definition, the platform concept and subsea systems are further matured and optimised. A major safety challenge is to contribute to cost-efficient solutions within critical safety areas.

The contractor updates the REPA to include the results of concept optimisation. Norskoil's project SHE manager supervises this work. It is shown that the concept meets all applicable acceptance criteria, provided that a number of assumptions are met. These define the accidental loads from fire, explosion, dropped objects and ship collisions that the semi must be able to withstand. Another set of assumptions defines the safety and communication systems that have to be operable during an accident. There are also assumptions regarding the collision warning system and the withdrawal of the platform from the subsea wells in case of blow-out. The REPA also results in a number of recommendations based on the ALARP principle. The assumptions and recommendations serve as input to design development and to specifications and need to be followed up in later project phases.

The REPA is used as a formal tool in defining the EPC contractor's obligations. Consequently, there is a need to clarify its limitations. It has been recognised that the REPA does not cover aspects in the interaction between design and operations, simply because its resolution is too low.

27.3.3.2 *Project execution*

The contractor is responsible for engineering. The Ymer project has a technical team in their offices for follow-up of the work. A main task is to ensure that the contractor implements the regulatory and Norskoil's safety requirements and risk-acceptance criteria in design in a satisfactory way. It follows that the assumptions from the REPA must also be implemented. It cannot be expected that all safety issues and questions have been resolved during the concept-definition phase. The project thus has to handle engineering-modification requests (EMRs) from the contractor and deviations from regulatory and company requirements (nonconformities). In resolving the EMRs and nonconformities, the main safety philosophy for the concept must not be violated.

Norskoil's technical team also issues variation orders (VOs), i.e. changes in relation to what was agreed in the contract. The cost consequences of these are negotiated with the contractor. It is the technical team's duty to represent Norskoil's understanding of regulatory and Norskoil's own safety requirements and acceptance criteria, and to be able to interpret these in discussions with the contractor.

Norskoil's project SHE manager monitors and follows up on all assumptions, recommendations and observations from the REPA and other related studies and evaluations. A computer-supported follow-up system is used for this purpose. Nonconformities in design are evaluated and documented by the responsible discipline. They are checked and approved by the SHE manager. It is here important to ensure that there are no violations against important safety philosophies and assumptions. A layout change, for example, in the process area involving re-location of equipment must be checked against assumptions regarding explosion venting.

The **REPA** is updated by the contractor to take design development into account and to evaluate the effects of changes. The results are compared to the risk-acceptance criteria. It is a concern that the resolution of the REPA is often not sufficient to be able to discriminate between good and inadequate solutions to active safety barriers such as gas and fire detection. This has to do with the fact that the REPA employs statistics on failure rates that are averaged over a large number of installations with different characteristics. It is important that the SHE management together with the technical team arrive at a consistent understanding of how to implement the general results of the analysis.

Example: Cable trays of aluminium are used in the design of process areas. During a fire, these trays will collapse earlier than steel trays. It cannot be derived directly from the analysis whether this is acceptable or not.

Other important issues that are brought up as a result of the updated REPA are:

- Requirements as to fire protection of the derrick to avoid collapse. A minimum of 20 minutes exposure of the derrick to fire without collapse is specified.
- It must be possible to withdraw the semi outside a pool fire on the sea in case of blow-out. It is concluded that the anchor system must be designed for a withdrawal of at least 135 m from the location of any well.
- Fire protection of risers to avoid escalation of a jet fire from one riser to neighbouring risers.

As a result of this update of the analysis, the specifications concerning accidental loads and safety requirements to systems during accidents are revised. Engineering has to assess the cost-consequences of the revised requirements and the modifications of design that follow, before they are implemented.

Other risk analyses are carried out that cover operational aspects. **Hazop** is an important tool in this respect (Suokas and Rouhiainen, 1993). It focuses on the process arrangements and the process control and shut-down systems.

The **CRIOP scenario analysis** studies interactions between design and operation by analysing the involvement of the operators in potential accident scenarios, Chapter 26. The method is used to verify design, manning, organisation and emergency preparedness. The CRIOP analysis of the Ymer platform looks into scenarios involving blow-out and collision with a supply boat. It identified concerns relating to:

- Location of important emergency-preparedness functions for drilling separate from the similar functions for the operation and marine systems.
- The availability of the communication link between the driller's control room and the central control room in a crisis situation.
- The possibility to get feedback on whether the marine riser has been released or not in a crisis situation.
- The location of displays of importance in a crisis situation outside the control room operators' line of sight.

Barrier availability is a concern due to insufficient feedback on barrier failures (Rasmussen, 1993). The availability is affected by such factors in design as equipment configuration and quality. Different operational factors such as inspections and testing procedures also affect it. Norskoil has defined a maximum safety unavailability for different safety systems:

- Process-control and process-safety systems
- Fire and gas-detection systems
- Emergency shut-down and gas relief systems
- Fire and evacuation alarm, emergency communication systems and emergency lighting
- Emergency power
- Ballast systems

The safety unavailability is defined as the probability that a safety system does not operate as intended. To verify compliance, **reliability calculations** are made. They are based on input from the equipment vendors and from operations. Requirements as to equipment reliability to be met by vendors as well as requirements as to inspection and testing intervals come out of the calculations.

27.3.4 Prevention of occupational accidents

27.3.4.1 Concept definition

Occupational accidents are not a major concern in concept selection. The risk of this type of accidents is mainly decided by the detailed design solutions developed in later phases. Early on, the project focuses on some basic characteristics of the different development alternatives that have implications for the possibility of developing acceptable solutions in later phases. They include:

- Use of open solutions, where the personnel are exposed to wind and rainfall when operating process and utility systems;
- Movements of the floater (ship or semi-submersible platform) due to heave; and
- General arrangements and their effects on walking and transportation.

The project carries out coarse working environment evaluations including a **Coarse analysis**. These serve as input to concept selection. Norskoil's operations department participates in these evaluations.

The weather vaning characteristics of the ship are an advantage. They improve the climatic conditions (wind, rainfall) in naturally ventilated process areas and thus the risk of accidents. Access for inspection and maintenance in the turret is a concern, due to very congested conditions.

Accident risks in drilling activities on the semi are a concern. A special challenge is to secure safe handling of heavy subsea equipment in the moon pool area in the centre of the platform. Advantages with the semi are the shorter distances to the different workplaces and less movement due to heave.

Figure 27.6 shows how the different analyses and evaluations of the risk of occupational accidents are co-ordinated with the design process (Kjellén, 1990; Kjellén, 1998). They are part of a total evaluation of all SHE aspects. Each activity has been defined and scheduled in Norskoil's and the contractor's SHE programmes for the Ymer project. During conceptual design, the experience checklist is established and the coarse analysis is carried out.

Norskoil develops experience checklists for hand-over to the EPC contractor. They are based on **accident-concentration analyses**, where accident data from existing installations have been used. An accident database is employed in the analysis. Norskoil's project team reviews the findings together with operations personnel from these two platforms. Jointly, they come up with recommendations on how to prevent the identified accident concentrations by measures in design.

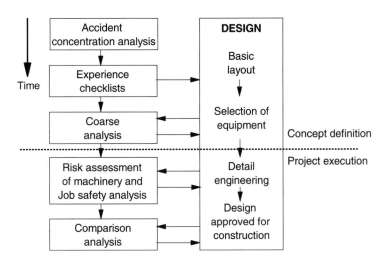

Figure 27.6 Use of different risk analysis methods in design. Arrows pointing to the right symbolise the flow of recommendations to design and arrows pointing to the left symbolise information input to the analyses.

Example: There were 49 drillfloor and derrick injuries in the database from two existing platforms. Nine of these involved moving of heavy tools between the store and the drillfloor. The experience checklist identifies hazards in connection with this activity and recommends that a continuous monorail is installed from the store to as close to the centre of the drillfloor as possible. By using this monorail, manual handling of heavy tools will be made safer.

At the end of the concept-definition phase, layout drawings that describe the platform in sufficient detail are made available. The EPC contractor performs **Coarse analyses** of all areas of the platform, i.e. process, utility, drilling and the living quarters. They invite Norskoil's operations personnel to participate in the analyses. The analysis teams concentrate on expected problem areas related to layout, such as manual materials handling, work in open areas, access to equipment, location of noisy equipment, and solitary work. A number of layout changes are introduced as a result of this evaluation.

Example: A coarse-analysis team reviews the handling of heavy tools between the store and the drillfloor. They look especially into the layout of the store and the location of the monorail. They also review the line of sight from winch control panels into the danger zone for material handling. The analysis team decides about the exact position of the monorail and the location of winches.

27.3.4.2 Project execution

The vendors are responsible for the **Risk assessments of machinery**. The reports from these assessments are part of the required documentation to issue a declaration of conformity with the machinery regulations. The contractor defines the documentation on risk assessments to be delivered by the vendors for review and verification. This includes the risk assessments for machinery and assemblies of machinery that are critical from a safety point of view (cf. NORSOK S-005 in Norsk Standard, 1999).

Example: The review by the contractor reveals inadequacies in the vendor's risk assessments for drilling equipment. There is a risk of getting caught in a rotating shaft of the remotely controlled manipulator arms for pipe handling during inspections. Machinery guarding has to be improved. The assessment also reveals insufficient provisions for isolation of hydraulic equipment. There is a risk of accidental start of the equipment during maintenance.

The EPC-contractor is responsible for the **Detailed job-safety analyses** (JSA). They cover the parts of the platform design that are the responsibility of the contractor including interfaces with equipment. These analyses focus on tasks with a high risk of accidents (i.e. a high probability of accidents and/or severe consequences).

Example: Manual tasks in the derrick are a concern, especially those carried out outside work platforms. Here a riding belt connected to a special winch has to be used. This operation is dangerous due to the risk of getting caught in or between pieces of equipment in the derrick. The JSA team lists all maintenance tasks in the derrick and reviews each task with a focus on the risk of getting caught by pipe-handling equipment. A primary safety measure is to eliminate the use of the riding belt. As a result of the analysis, the team decides to move lubrication points so that they are within reach from work platforms. Additional work platforms are also installed to improve access.

In a **Comparison analysis**, the number of accidents per year of operation is estimated for the Ymer platform. Accident statistics for the two reference installations serve as a basis for these estimates. Three different teams representing the different areas of the platform make the estimates. All teams consist of project and operations personnel with thorough knowledge of the area. Results show that the total number of accidents per year is expected to decrease somewhat compared to the reference platforms. Reductions in the number of accidents in maintenance in particular contribute to this decrease. This has to do with improved equipment handling during maintenance, improved equipment reliability and improved maintenance plans.

Example: The Comparison analysis indicates that there will be a small increase in the frequency of drilling accidents from on average 5.8 accidents per platform year at the reference platforms to 5.9 at Ymer. This increase is mainly explained by increased drilling activities, whereas the probability of accidents per activity has been reduced. Due to expected improvements in other areas, the acceptance criteria are met.

27.4 Construction-site safety

There are two main SHE concerns in this phase. One is to ensure that the platform itself has a satisfactory quality with respect to safety, health and environment. A second concern is to ensure that the construction work is carried out in a safe way. We will here focus on this latter aspect. It is in Norskoil's interests to avoid accidents during construction that:

- Have cost or schedule impact. A severe accident may delay start-up by up to one year, which will cause losses in the order of billions of NOK for Norskoil and the other owners.
- Reduce the quality of the product, i.e. the platform.
- Result in injury to the contractor's own personnel at the site.
- Result in loss of good will to Norskoil.

Also, bad safety statistics are regarded as a sign of sloppy management control at the site on the part of the construction contractor. Sloppy management of this type will also manifest itself in bad quality and delays (Grimaldi, 1970; Levitt and Samelson, 1993; Kjellén *et al.*, 1997).

27.4.1 SHE management principles

It is Norskoil's SHE policy to work with contractors in order to achieve satisfactory SHE results. The aim is to avoid accidents resulting in serious harm to personnel or the environment or in extensive material damage or delay. Norskoil's SHE goal for construction is a LTI-rate of five or less. The company uses its SHE management system for contractors to accomplish this goal, Figure 27.7. Regulatory requirements and the NORSOK standard S-CR-002 together with Norskoil's contractor SHE policy define the requirements to the contractors (Norsk Standard, 1996/97/98/99).

Contractors are obliged to set up and manage SHE programmes for construction to ensure that the regulatory and contractual requirements are complied with. According to the NORSOK standard, the following should be documented in this programme:

- Policy and goals;
- Organisation and assignment of responsibilities for work for the project;
- Activities for follow-up on safety, including workplace inspections, meetings, audits, introduction of new employees;
- Applicable safety rules for specified activities such as scaffolding, housekeeping, fire protection, etc.; and
- Emergency preparedness organisation and routines.

NORSOK states detailed requirements concerning the reporting to the oil company of incidents and accidents, and concerning safety information and training of company personnel at the yard.

27.4.2 Step 1: Pre-qualification

The basis for a successful follow-up on SHE in construction is laid during pre-qualification. The contractors who will be allowed to tender for the fabrication contracts are selected. The project team may select only highly qualified contractors with excellent SHE-results from before. If this strategy is chosen, the prospects for a safe execution of contract work will be good. It will, however, reduce the number of bidders considerably and thus reduce competition and increase costs. An alternative strategy is to accept contractors with low SHE qualifications and performance. In this case, Norskoil has to put more effort into follow-up of the contractors during the execution of the work in order to ensure an acceptable SHE performance. Again, this strategy will have a cost impact through a more costly follow-up and also because of more accidents.

Norskoil has decided to put rather stringent SHE requirements to the bidders to be qualified. A pre-qualification questionnaire has been developed that is used during **audits** of the potential bidders. This contains questions concerning the potential bidder's SHE management system, resources

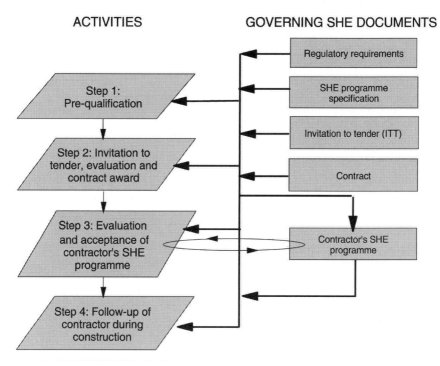

Figure 27.7 Norskoil's principles for follow-up of contractor's construction site
       safety.

and organisation, and accident statistics in previous work (see example in
Section 19.1). Bidders with unacceptable results on any of the main items
considered relevant are not allowed to bid.

*Example: The Ymer contract is a EPC contract, where one contractor
will be responsible for the design and fabrication of the complete platform.
This contractor will in turn sub-contract parts of the work to other contractors. One of the potential bidders for the Ymer contract had allied
itself with other yards to be able to tender for the complete contract. One
of the yards in this alliance did not perform well on SHE management and
performance in the pre-qualification audit. The Ymer project management
decided to accept this bidder alliance in order to achieve adequate competition for the contract.*

27.4.3 Step 2: Tender evaluation and clarification, contract award

In the next step, Norskoil's project team evaluates incoming bids. In addition
to the economic evaluation of the bids, the project team makes a technical
evaluation of the tenders. Here, the proposed organisation, the competence
of its members and plans for the work are focused upon. The bidders must

submit a preliminary SHE programme for the work and the names of the
SHE manager and other key SHE personnel with the tender.

Questions concerning the bids are discussed with the different bidders
in so-called bid-clarification meetings (BCM). Any concerns about the SHE
performance or SHE resources are brought up here.

*Example (continued): Norskoil addressed the poor SHE programme and
performance of one bidder's sub-contractor with the bidder during the BCM.
It was concluded that the poor SHE performance (an LTI-rate above 100)
could partly be explained by the workmen's compensation system in the
country of the sub-contractor. The workers actually received higher pay
during sick-leave due to accidents than they received if they were sick for
other reasons. Deficiencies in the yard's SHE management system were also
brought up. The bidder offered to provide additional personnel to follow up
this particular sub-contractor on SHE. This offer was included in the tender.*

27.4.4 Step 3: Evaluation of the SHE programmes

Prior to the start of construction work, Norskoil's SHE manager for the Ymer
project evaluates the EPC contractor's and sub-contractors' SHE programmes
for construction work. The EPC contractor has the overall responsibility
for SHE and the sub-contractors' SHE programmes are subordinated to this
programme. Norskoil's SHE manager will comment on items where the
programme does not meet the project's (i.e. NORSOK) requirements.

*Example: Norskoil's SHE manager identifies unclear line responsibilities in
the contractor's SHE programme. It is stated that contractor's safety engineer
is responsible for SHE. The contractor is requested to clearly define SHE as
a line responsibility according to the requirements in NORSOK S-CR-002.
After the contractor has implemented the comments, the programme is
accepted by the project.*

27.4.5 Step 4: Follow-up during construction

The EPC contractor establishes an organisation for follow-up of SHE in
construction. A site safety co-ordinator reporting to the contractor's fabrication manager does most of the practical work. Norskoil's SHE manager
monitors this work and intervenes if the SHE performance does not meet
the project goals.

At the start of construction, Norskoil arranges seminars on construction
safety with the contractor's management and site safety personnel. Norskoil's
policy and SHE goals are focused upon, and the necessary SHE motivation
to reach the goals is discussed. The seminar results in mutual agreements on
information and follow-up activities.

Norskoil's project team meets regularly with the EPC contractor's management. Here, SHE performance is brought up as a fixed item and areas
of concern are highlighted. The EPC contractor has similar management

meetings with sub-contractors, where SHE matters are also discussed. These regular management meetings are the main occasions for decisions on remedial actions where the SHE performance is inadequate.

There are regular **workplace inspections** at the yards. The frequency and scope of these inspections are defined in the different SHE programmes. The contractor's site safety co-ordinator participates occasionally in these inspections. Also, Norskoil is invited to participate but does so only exceptionally, since the company only has permanent site representation at the EPC contractor's own yard.

Example: The contractor's site safety co-ordinator identifies sloppy housekeeping conditions at one of the yards during the regular workplace inspection. He decides to monitor this yard more closely and participates in subsequent inspections. When the situation does not improve, this concern is brought up at the monthly management meeting. It is decided in the meeting to stop work for clean-up. The yard has to pay for it.

It is the responsibility of Norskoil's SHE manager to **follow up on the accident statistics** from the yards. He monitors the SHE performance in this way and compares the results with the project's goals. NORSOK S-CR-002 defines the requirements as to reporting of accident statistics from the EPC contractor to Norskoil. Based on this information, Norskoil follows up on the following SHE performance measures: LTI-rate, LTI'-rate (i.e. the frequency of lost-time injuries with potential for severe harm, see Table 17.2) and the TRI-rate.

Example: Figure 27.8 shows the development of these statistics for the Ymer project as a whole. The SHE manager recognises that the LTI-rate does not develop in a satisfactory direction and brings this up in contacts with the EPC contractor's site safety co-ordinator. Jointly, they identify the yard responsible for the poor results and decide to execute a joint audit of this yard.

SHE audits are the main tools of Norskoil's SHE manager for follow-up on safety in construction. The contractor's site safety co-ordinator also performs audits of this type. The audit team checks actual performance against regulatory and contractual requirements and the yards' SHE programmes. The audits usually consist of two parts, i.e. a system part and a technical part. In the system audit, the SHE manager checks SHE documentation and practices in relation to the commitments in the SHE programme. The technical audit is carried out as workplace inspection. The audits follow a schedule and each audit covers a separate theme. Below are examples of audit themes:

- Reporting of accidents and incidents
- Emergency preparedness
- Cranes and lifting safety
- Safe work practices, compliance with safety rules
- Housekeeping
- Fire precautions
- Handling of hazardous chemicals

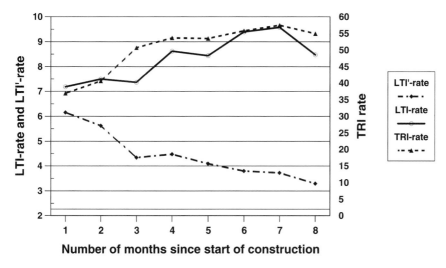

Figure 27.8 The lost-time injury frequency rate from start of construction.

Example: The SHE audit of a yard identifies three nonconformities and seven observations. Poor housekeeping and unsecured work at height represent typical findings. They are brought up with the yard for corrective action. The yard proposes to improve housekeeping by introducing new containers and regular housekeeping activities. Norskoil accepts the corrective actions.

Norskoil also carries out ad-hoc audits in cases of serious incidents or other negative experiences.

Example: A fire occurs in the air intake filter of one of the main generators. A welding spark ignites the filter. Norskoil's SHE manager performs a SMORT investigation of the fire by following the established audit procedures. This reveals serious misunderstanding and disagreements within the contractor's organisation on safety responsibilities and on emergency-preparedness procedures.

27.5 Safety during plant operation

Ymer's platform management faces a number of SHE challenges, especially during the first period after start-up. The organisation is new and unfamiliar with the platform and the management system. Construction work on the platform has not been fully completed at start-up and continues in parallel with production and drilling. It is of major importance to avoid construction work on systems that have not been de-energised. Here, the permit-to-work system plays an important role.

After the construction team has been demobilised, the platform goes into a period of routine operation. It is here important to maintain a high alertness to SHE challenges and to avoid complacency, although the risk is not as high as during the early phase after start-up.

27.5.1 SHE management principles

We here choose to describe and analyse the SHE management principles applied at Ymer during the operation phase in relation to the well-known quality control loop, Figure 27.9. We view safety against occupational accidents as a quality aspect that platform management has to satisfy.

SHE is a line responsibility. It starts from the top of the organisation, i.e. the platform manager. Norskoil as principal enterprise also has overall responsibility for SHE in work on the platform that is carried out by the drilling and construction contractors.

Norskoil has documented its SHE management system for the operation of the Ymer platform. It is not collected in a handbook but is integrated into the total platform-management system that is available via Intranet. It covers the following elements:

- Top-management commitment and leadership;
- Policy, goals and acceptance criteria;
- Organisation and responsibilities;
- A list of references to applicable regulatory requirements;
- Requirements to safety systems/barriers;

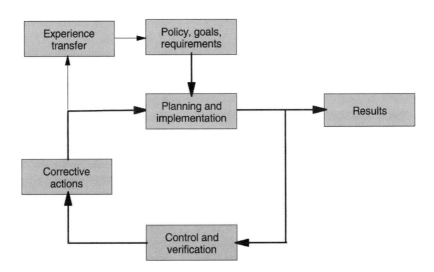

Figure 27.9 The SHE control loop.

- Procedures and work instructions for the identification and evaluation of SHE risks and for handling of changes and emergency response;
- Procedures and work instructions concerning SHE performance monitoring and follow-up of SHE activities;
- Periodic action plans that detail the implementation of the overall SHE policy and goals; and
- Auditing and review activities.

Let us relate these different items to the SHE control loop of Figure 27.9. The overall SHE requirements are defined in the section on policy and goals. The list of applicable regulatory requirements also belongs to this part of the control loop together with the more detailed requirements at lower levels, defined in technical requirements, procedures, work instructions, etc. The goals are more operationally specified. Norskoil updates these on a regular basis. A long-term goal is to reduce the lost-time injury (LTI) frequency rate to zero.

Implementation is ensured through the Ymer platform organisation as defined by its responsibilities and qualifications. There are yearly SHE action plans that break down the policy and goals into tangible actions.

Control and verification is accomplished through hazard identification and assessment, SHE performance monitoring and auditing. Here Ymer applies some of the tools described in Parts III–V of this book.

Ymer's emergency preparedness organisation and routines are documented as a separate part of the SHE management system. This plan describes the action plans for different accident scenarios that have been defined in advance.

The drilling and construction contractors have developed their own SHE management systems that are subjected to Ymer's system. These cover the control of hazards in special activities under their responsibility.

27.5.2 Policy and goals

Norskoil's overall SHE policy has been implemented in Ymer's SHE management system. In addition, Ymer's management has defined goals for the SHE performance of the platform.

Example: Ymer has defined the following SHE performance goals related to accidents for year six after start-up:

- *LTI-rate < 3*
- *TRI-rate < 10*
- *The number of reported near accidents per reported injury > 30*
- *Less than five events with potential Grade 4 or 5 (compare Table 6.4)*

27.5.3 Implementation

The SHE action plans represent a cornerstone in the implementation of SHE policy and goals. Both Ymer's own organisations and contractors develop such plans. This is done jointly by management and the employees. Management is responsible for the allocation of the necessary resources. The actions are implemented and followed up through clear responsibilities and deadlines. Ymer's management applies the principles of the Deming circle in this work.

Example: The drilling contractor experiences problems with a new unit for injection of drilling mud. It consists of screw conveyors, a mill, tanks and pumps. Personnel operating this unit are exposed to a high noise level and levels of oil mist above the threshold limit value when remedying operational disturbances. There is also a risk of getting caught in the screw conveyor. The drilling employees have submitted near-accident reports on this hazard. Jointly, management and the employees decide to look into a redesign of the unit and put this up as an item on the action plan. A problem-solving group is established for this purpose and it receives a budget for its own work and for remedial actions. The driller is responsible for the group.

Norskoil has realised that the attitudes and behaviour of its management personnel are of basic importance to the successful implementation of the SHE policy. Ymer management is trained in practising a set of rules for SHE management behaviour. They involve:

- Motivating the employees to take responsibility for SHE. This is to be accomplished through education and training and through empowerment of the employees by their participation in SHE-related activities and decisions.
- Marshalling correct behaviour through management's own vigour and example.
- Meetings and communication on SHE policy, goals and results.
- Visibility and trustworthiness in matters relating to SHE by participation in workplace inspections and investigations of serious accidents and by requesting accountability for SHE-related decisions and results.

Example: The production manager observes a repairman working on a pipe rack 3 m above the deck without the required fall protection. On speaking to the repairman, he is told that this is a high-priority job and that the scaffolders have not been available in due time. The production manager stops the job and oversees the erection of scaffolds so that the job can be completed in a safe way.

Regular SHE meetings with all personnel are important occasions for communication between management and the employees. Management uses these meetings for information on SHE matters (new procedures, instructions, etc.) and for discussions on experiences and new goals and action plans. The employees bring up SHE concerns and experiences in the meetings.

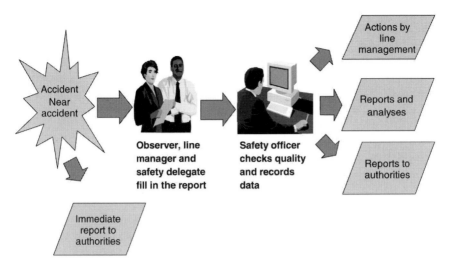

Figure 27.10 Registration and follow-up of accidents and near accidents.

27.5.4. Control and verification

27.5.4.1 Reporting of accidents and near accidents

Ymer uses a computerised database for storing and retrieval of reports on accidents and near accidents, Figure 27.10. Two procedures describe the routines, one the routine reporting and follow-up of events and the second the investigation routines in cases of serious incidents. The following types of events and conditions are reported and stored in the database:

- Accidents resulting in injury to personnel, material damage, oil/chemical spills, production stops.
- Near accidents.
- Gas leaks.
- Unsatisfactory conditions.
- SHE nonconformities.

Management emphasises the importance of reporting all potential problems. The reporting is promoted through simple reporting routines and an efficient follow-up. All reports are assigned actions for follow-up with a responsible person and deadline. The database is used for the follow-up of actions.

One of the safety officer's tasks is to assign a risk score to each reported event. The supervisors formerly carried out this scoring, but this routine was abandoned due to a reluctance to use high-risk scores. High-risk events are followed up closely by top management. Results and experiences of

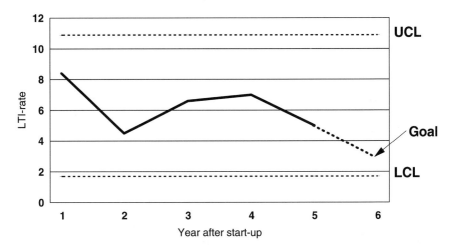

Figure 27.11 Control chart for the Ymer platform, showing the upper control limit (UCL) and the lower control limit (LCL).

a general value are fed into an experience database for use both in new projects and in operations on other platforms.

Problem-solving groups are employed in the development of remedial actions. The aim is to use the knowledge of the personnel directly involved to solve safety problems.

Information is distributed through periodic summaries of accident statistics, status reports on safety measures, special reports, etc. There are different uses of this information such as input to decisions on remedial actions, evaluation of the effects of these actions, and monitoring of SHE performance. The database also provides mandatory reports on accidents to insurance companies and the authorities.

The *periodic summaries* on accident statistics are distributed quarterly. They provide feedback on SHE performance to line management and the SHE organisation. Ymer's quarterly report contains control charts on the accumulated LTI-rate and the TRI-rate since the start of the year, distribution of accidents by accident type, injured part of body, etc. An edited summary of each accident case is also presented. The database helps in the production of standard reports and makes it possible to tailor these to the needs of each department.

Example: Figure 27.11 shows the control chart for the LTI-rate at the end of year five after start-up. After a rate of about eight during the first year, it hovers around six. The goal for year six is three, i.e. a reduction by about 50 per cent in relation to the average value for the last four years. Still, this value falls within the control limits.

The database also gives feedback to decision-makers on the responsibilities, target dates and *status of remedial actions*. This is a means of ensuring timely implementation.

Line managers, SHE personnel (including safety representatives) and decision-makers responsible for design, procurement, training and education, etc. have access to the database for *queries*. This possibility is utilised infrequently. When used, the majority of the questions are relatively uncomplicated. They mainly concern information on individual accident cases relating to a specific type of machine or job, type of hazard, etc. The database is well suited for the retrieval and summarisation of relevant accident cases. Data on accidents are also used in special studies such as risk analyses and accident concentration analyses.

27.5.4.2 Workplace inspections

There are different types of scheduled workplace-inspection activities. The department manager and the safety representative make bi-weekly inspections of their area. They follow a joint plan for the whole platform and cover different items each time. Housekeeping is a mandatory item on the agenda. Most of the identified problems are remedied on the spot and not recorded. Actions that are not executed immediately are documented for follow-up. Action status is reviewed at SHE meetings.

Ymer practices one-on-one rounds at the different levels of the platform organisation. Approximately twice a year, Norskoil's assistant director for production makes a one-on-one round with the platform manager. They are not notified in advance and the focus is on safe behaviour and housekeeping. The platform manager and the department managers make such rounds on a monthly basis.

The working-environment committee for the Ymer field also makes a workplace inspection once a year. The focus here is on ergonomics and industrial hygiene, and experts covering these areas participate in the inspections.

27.5.4.3 Control of barrier availability

Norskoil has established a comprehensive programme for periodic inspection and testing of barriers. Table 27.2 shows an overview of the different safety systems and components that are subject to inspection and testing.

The inspection and testing activities are subject to detailed planning. Procedures define the criticality of the technical components and systems and the inspection and testing intervals. Based on these procedures, yearly inspection and testing programmes are established. The results are followed up with immediate corrective actions where required and with more long-term measures. The status is reported to the authorities on a regular basis.

It is important that the tests are realistic and that they cover all subsystems to check the availability of the system as a whole. The fire and gas detection system, for example, is tested all the way from the detector, which is activated by gas, light or smoke, to the display in the central control room.

Table 27.2 Overview of different barriers and method for control of barrier availability

Type of barrier (cf. Figure 7.3)	Safety system/component	Control method
Process control	Process control system	Periodic testing through partial shut-down
High quality containment	Pipes and vessels	Periodic inspection, non-destructive testing and measurement
Fire and gas detection	Fire and gas detection system	Periodic testing and simulation
Emergency shut-down and pressure relief	Emergency shut-down system Process safety system	Periodic testing Leak testing of safety and emergency shut-down valves
Isolation of ignition sources	EX-barriers on electrical equipment Emergency shut-down of equipment	Periodic inspection and testing of shut-down of non-classified equipment in exposed areas
Ventilation	Mechanical ventilation system Fire doors	Periodic testing and inspection
Fire and blast walls	Fire and blast walls	Periodic inspection and testing of tightness
Integrity of load-bearing structure	Load-bearing structure Passive fire protection	Periodic inspection
Active fire protection	Active fire-protection system	Periodic inspection of valves and fire-fighting equipment Periodic testing of fire pumps, sprinkler systems and foam skids
Escape, evacuation	Emergency lighting Emergency communication system Life boats and rafts	Inspections Periodic testing Periodic evacuation drills
Common	Emergency power	Periodic testing

27.5.4.4 Risk analyses

Ymer has implemented coarse and detailed job-safety analyses as a systematic measure to control the hazards at the workplaces. All employees are involved in these activities. They participate in the listing of all activities on the platform. Each activity is assessed and documented with respect to:

- How often it is performed and number of personnel involved.
- Possibility of injury/loss.
- Risk estimation based on expected frequency and consequences.
- Decisions regarding remedial actions and detailed job safety analysis.

The documentation is reviewed yearly to identify the need for updating. Job-safety analyses are always performed before the start of new jobs that may be critical from a safety point of view. The supervisor reviews the need for such analyses as a routine procedure in the permit-to-work system. The following types of jobs have to be assessed in a job-safety analysis before they can start:

- Entering of tanks, where there is a risk of poisoning or suffocation.
- Heavy lifting involving risks of dropped objects.
- Non-routine maintenance of systems containing hydrocarbons.

The Risk and Emergency-Preparedness Analysis (REPA) is also subject to updating with new incidents and in connection with modifications. The aim in the latter case is to check that the modifications do not violate any of the acceptance criteria for the risk of major accidents.

27.5.4.5 Audits

SHE auditing is used as a tool for independent examinations of Ymer's SHE management system. Ymer uses an adapted version of ISRS, and a third party carries out regular audits to establish a rating of Ymer's SHE management system.

Norskoil's audit department carries out independent audits of all platforms where the company is responsible as operator. They follow a schedule and cover different items. The audit team is led by an authorised auditor and includes experts in various fields.

Example: Ymer has to go through major modification work involving upgrading of one of the compressors. A contractor will be responsible for the work. Since this is a non-routine operation, the platform manager initiates an audit to make himself familiar with the suitability of the SHE management system in this new situation. The audit team consists of a lead auditor, a platform manager from a platform that recently has undergone a similar activity and a project SHE manager. They audit both Ymer's and the contractor's SHE management systems by conducting interviews and reviewing documents.

The conclusions are rather serious. Ymer's management has not adequately foreseen the SHE challenges involved in carrying out modifications in parallel with ongoing production. There is a need to upgrade the permit-to-work system so that it can handle a much higher activity level. Systematic risk assessments must be performed on all modification work and procedures need to be established on the basis of the results.

28 The trucking industry

28.1 Accidents in road transportation

In road transportation, the driver controls large amounts of energy. From elementary physics we know that there is a linear increase in the kinetic energy with the weight of the vehicle and a quadratic increase with its velocity. These basic facts show up in the accident statistics. In Norway, traffic accidents account for about a third of the fatalities at work (see Section 5.4). Traffic accidents in general constitute a significant public-health problem. Swedish estimates show that out of 100 inhabitants born in 1956, about fifty will be injured in a traffic accident and about one will die from it (Englund *et al.*, 1998). Traffic accidents reduce the expected lifespan by half a year.

The risk of sustaining a traffic fatality varies considerably between different countries. In the UK, Norway and Sweden, there are about 0.7 fatalities in traffic accidents per 10 000 inhabitants per year. The corresponding figure for Germany is 1.2 and for the USA it is 1.6. These figures represent significant improvements in traffic safety during the last decades. In the USA, for example, the figure of 1.6 represents an improvement by almost 50 per cent since the late 1960s.

We will in this chapter focus on the prevention of accidents in commercial truck transportation. In the USA, there were according to the accident statistics from 1990 about 700 fatalities among drivers and passengers of medium and heavy trucks (Loeb *et al.*, 1994). This made up about 2 per cent of the total number of fatalities in traffic accidents. The crash-involvement rate for medium and heavy trucks (i.e. the number of vehicles involved in crashes per 100 million vehicle km) was about half that of passenger cars. The fatal crash-involvement rate was, on the other hand, about 45 per cent higher than that for passenger cars, indicating the significance of the large energies involved in truck accidents. Only 13 per cent of the fatalities were truck occupants as compared to 77 per cent being an occupant of another vehicle involved in the crash.

Table 28.1 Some common measures of the risk of traffic accidents

Measure	Definition
Health risk, general	Number of fatalities (or injuries) per inhabitant and year
Health risk, traffic	Number of fatalities (or injuries) per million hours in traffic
Traffic-accident risk	Number of traffic accidents per million vehicle km
Traffic-injury risk	Number of injuries (incl. fatalities) per vehicle km
Crash-involvement rate	Number of vehicles involved in crashes per 100 million vehicle km
Fatal crash-involvement rate	Number of vehicles involved in fatal crashes per 100 million vehicle km

28.1.1 Measures of the risk of traffic accidents

In the previous Section, we used different measures of the risk of traffic accidents. Let us rehearse the definition of the risk of accidents from Section 17.2 and look closer into how it is applied in the field of traffic safety. We defined the risk of accidents as a combination of measures of the probability or frequency of accidents involving losses per unit of exposure to a specified activity, and the extent of the losses (consequences). In occupational safety, we use the number of man-hours as a measure of exposure. In calculating the risk of traffic accidents, we usually use other exposure measures. When we want to assess the effects of transportation accidents on the *health of the general public*, exposure is measured as the number of inhabitants at risk during a specified period. We are also concerned with the risk of accidents in relation to the *transportation work* carried out. Here we use, for example, the number of vehicle km as an exposure measure. Table 28.1 above summarises some common traffic-risk measures.

It follows from the definition of the risk of traffic accidents that there are three different means of reducing the losses due to traffic accidents, seen from society's point of view. These include:

- Reducing exposure, i.e. the extent of road traffic;
- Reducing the probability of accidents per unit of exposure (e.g. vehicle km); and
- Reducing the consequences of the traffic accidents.

We will come back to these different means in the next Section, when we discuss how the different 'components' of the traffic system affect the risk of accidents.

28.2 The man–vehicle–road–environment model

We touched upon Haddon's phase model of traffic accidents in Section 5.3. It divides an accident sequence into three distinct phases: pre-crash phase; crash phase; and post-crash phase. During the pre-crash phase, the road users interact in ways that normally follow well-controlled patterns but occasionally result in traffic conflicts. Injury and damage occur during the crash phase. This is when there is an uncontrolled energy exchange between the involved vehicles and between vehicles and other road-users and/or obstructions in the environment. Losses may be limited through actions in the post-crash phase including first-aid, fire-fighting, medical treatment, etc.

During the pre-crash phase, the actions of the road users have significant effects on whether an accident will occur or not and also on its consequences. We use the driver–vehicle–environment model according to Figure 28.1 to analyse the situation. Here, the driver is regarded as one of the components of the traffic system, and the model focuses on the information processing of the driver. The driver receives and processes information mainly from the traffic environment. The driver acts on the information and these actions affect the movements of the vehicle, which in turn affect the traffic environment. The driver registers changes in the traffic environment and acts on these in an ongoing process.

Section 8.1 presented research into human information processing. Much of this research originates from the traffic-safety field and is of special relevance here, see Rothengatter and Vaya (1997) for an overview. In the 1960s, research focused on the driver's ability to process information and to act accordingly. Results showed that there were 'filters' in the various stages of information processing. These resulted in limitations in the driver's ability to handle complex and rapidly changing driving tasks. A number of measures

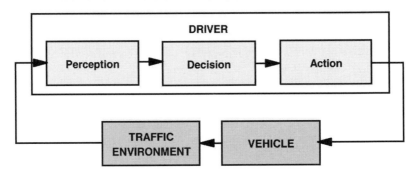

DRIVER–VEHICLE–ENVIRONMENT SYSTEM

Figure 28.1 A model of the driver–vehicle–environment system.
Source: Englund *et al.*, 1998. Copyright 1998 by Studentlitterature and reproduced by permission.

were taken to improve safety by reducing the complexity of the driving task. These included, for example, an improved driver–vehicle interface and measures in the traffic environment to make it less demanding.

It turned out that these measures did not give the expected effects. Drivers utilised the reduced demands in other ways than to increase safety, for example, by increasing speed. More elaborate models had to replace the simple information-flow model to take into account the fact that the driver, at least partly, was able to determine the level of risk himself. These models took into account that the driver's actions are not only affected by incoming information but also by the driver's motives and previous experience and knowledge. Of the underlying *motives*, meeting the objectives of the driving task (i.e. movement of people or goods from point A to point B), avoidance of accidents, economising on expenditure of time and money, pleasure-seeking, etc. are of relevance. The models also took into account how driving behaviour was affected by the driver's *knowledge* about the conditions in the environment (i.e. vehicle, road standard between points A and B, traffic intensity at the time of day, etc.) and about the effects of his own actions on it.

It follows from these revised models that changing behaviour is not merely a question of information and campaigns. The design of the traffic environment is at least as important. When improvement in safety is an overriding societal objective, investments in the road network should not invite the driver to utilise the improved standard to reduce travelling time at the cost of safety.

Another aspect that came into focus in traffic research in the 1980s is the way the driver develops ingrained or automatic behaviour patterns based on his experiences from interactions with the traffic environment. Section 8.1 presented Rasmussen's skill–rule–knowledge framework of human cognitive control of the environment. When applied to a traffic situation, it illustrates the development of driver skills. In a novel situation, the driving task involves problem-solving, and existing rules or behaviour patterns are not applicable. At the next level of performance, the driver initiates different ingrained rules of behaviour patterns dependent on his judgement of the situation. At the skill-based level, the driver's behaviour is automated and consists of well-ingrained behaviour applied in well-known situations.

This development is rational and reduces the mental load of the driving tasks. From the point of view of traffic safety, it is important to direct the development of skills in a direction that promotes safety. It is also important to design the traffic environment in a way that takes the needs of the novice driver as well as the experienced driver into account. The novice driver should be given a realistic understanding of the challenges that the traffic environment represents. Simultaneously, the attention of the experienced driver should be directed at recent changes in the traffic environment.

Let us now turn to each of the three 'components' of the driver–vehicle–environment model. Research has explored the effects of different characteristics of each component on the risk of traffic accidents. Some of the research results are summarised below with special focus on transportation safety.

28.2.1 The driver

The driver affects the accident risk in many ways. He has a crucial influence on road use and thus determines exposure. In professional driving, the decision about road use is usually made by the employer. Driving behaviour has a decisive influence on the pre-crash phase and hence on the probability of traffic accidents. Although the crash phase is to some extent controlled by the driver, previous actions to control speed and direction will affect the outcome.

Research early focused on the possibility to identify accident-prone drivers, i.e. drivers who due to certain personal characteristics were more likely to be involved in traffic accidents than others. The aim was to improve traffic safety by identifying such individuals, e.g. by use of psychological tests, and excluding them from the traffic environment. Accident statistics showed that some drivers experienced more accidents than others. Although this so-called *accident-proneness theory* is often accepted as credible by the layman, it has been criticised by the research community for several reasons and is today of little practical significance (McKennan, 1983). The critique may be summarised as follows:

- Differences between drivers as to accident experience are to a large extent statistical artefacts that may be explained by pure chance.
- A small part of the differences between drivers in accident experience may be explained by personal characteristics. It has turned out to be difficult, however, to identify the specific characteristics of accident-prone drivers. This makes it difficult to develop valid and reliable tests.
- The use of psychological tests to exclude accident-prone drivers would involve many instances of wrong diagnoses due to an inadequate correlation between test results and accident risk. It is politically unacceptable to deny people the right to drive (including the right to work as professional drivers) on such a fragile basis.

Age and driving experience are temporary personal characteristics that according to the accident statistics are significant determinants of accident risk. Statistics show that the risk of accidents per km driving follows a so-called U-curve, with a minimum risk around the age of 50 (Loeb *et al.*, 1994). Both young drivers (age 18 to 24) and old drivers (age above 75) experience an accident risk that is many times higher that of middle-aged drivers. Increased experience has a greater impact on accident risk than increased age (Maycock, 1997). Drivers who first experience traffic at a young age will start at a very high accident risk but will show rapid improvements during the first year of driving. Older inexperienced drivers will start at a lower risk level but will improve at a slower rate.

Fatigue is also a significant determinant of accident risk. The driver's experience of fatigue is related to the time of day, driving schedule, eating habits, physical activities, etc. The commercial driver has only limited control over these different factors and has been the focus of research into the effects of fatigue on the risk of accidents. Increases in the risk of accidents during certain periods of the 24-hour cycle are to a large extent explained by the circadian rhythm (i.e. 'body clock'). Studies of truck drivers show that accident risk doubles during the period between midnight and 6 o'clock in the morning, as compared to other parts of the day and night, see Folkard (1997) for an overview. The risk of single accidents is higher during the night and the consequences are more severe due to a higher speed. Also drivers are more susceptible to accidents after driving between 12 and 15 hours. The likelihood of accidents increases with driving time. Different studies show that accident risk increases by a factor of two or more for drivers working more than a 'normal' working day of eight to ten hours.

Fatigue as a causal factor has been studied in accident investigations. A distinction is made between falling asleep as a direct cause of the accident and a reduction in the driver's general performance as a result of fatigue. Different studies of truck accidents estimate that the driver has fallen asleep in between 1 per cent and 10 per cent of the accidents (Englund *et al.*, 1998). A study of fatal truck crashes concluded that fatigue was a causal or contributing factor in a third of the cases (Loeb *et al.*, 1994).

There is clear evidence of the negative effects of *alcohol* on traffic safety, and the risk of accidents increases with the alcohol contents in the driver's blood. A Swedish study showed that the accident risk increased by a factor of four at an alcohol blood concentration of 0.04 per cent and a factor of 40 at a concentration of 0.15 per cent (Englund *et al.*, 1998). Although the portion of drivers that are affected by alcohol is relatively low in Sweden (less then 0.2 per cent), alcohol is a contributing factor in about a third of fatal traffic accidents. The significance of alcohol as a causal factor varies between different countries, depending on alcohol consumption habits.

Alcohol and other types of drugs are also a concern in commercial driving. A US study showed that about a third of fatally injured drivers had positive results in toxicological tests, where 13 per cent of the drivers tested positive for alcohol, 13 per cent for marijuana and 9 per cent for cocaine (Loeb *et al.*, 1994).

28.2.2 *The vehicle*

Vehicle design affects both the probability and consequences of accidents. A distinction is made between the active safety features of the vehicle and its crashworthiness. The *active safety features* aim at making it possible for the driver to control the vehicle in a safe way and to avoid traffic conflicts (Englund *et al.*, 1998). They include:

- Viewing conditions from the driver's position. These are determined by seating position and design of windows, windscreen wipers, rear mirror, headlights, etc.
- Design of controls, primarily steering wheel and brake pedal, clutch, and accelerator pedals. These must be designed so that the driver has rapid and error-free control of the speed and direction of the vehicle.
- Vehicle stability against skidding and overturning.
- Vehicle controllability, i.e. the way the vehicle follows the driver's intended course, acceleration or retardation.
- Design of displays to keep the driver informed about important driving parameters such as speed and possible technical failure.
- Ergonomic conditions, i.e. seating position and conditions, climate control, noise, vibrations and other factors that, if adequately designed, make the driver's conditions comfortable and reduce fatigue.

A recent trend is to introduce different in-vehicle driver support systems to reduce the mental load of the driving task. These include, for example, automatic speed control, anti-collision radar, anti-lock braking system (ABS) and even automatic driving.

The different active safety measures listed above will typically reduce the complexity of the driving task. As discussed above, changes in driver behaviour may offset the positive safety effects of these measures. The driver may, for example, utilise such features as a comfortable seating position, good acceleration and low noise and vibration levels to prolong the driving time and to increase speed. Reduced feedback on driving speed due to low noise and vibration levels and good acceleration is here a special concern.

The introduction of the concept of *crashworthiness* was determined by the needs of minimising injury to the driver and passengers in case of a collision. This is accomplished by allowing the colliding vehicles to absorb as much of the kinetic energy as possible and to reduce the energy transfer to the driver and passengers to a minimum. To accomplish this, vehicles are equipped with crash deformation zones in the front and rear and a stiff cabin, offering protection to the belted driver and passengers. During the first phase of a collision, the deformation zone is compressed by about half a metre. This phase lasts for about 100msec for vehicles with a speed of say 50 km/hour. During the second phase, the body comes into contact with the protection system (seat belts, air bags) or the interior of the cabin. It is here important to minimise injury by moderating the rate and concentration of the energy absorption.

Vehicle maintenance is also a contributing factor to the risk of accidents. Although research is not fully conclusive, some studies suggest that mechanical defects and especially brake failures are relatively common in traffic accidents and contribute to up to a third of all truck accidents in the USA (Loeb *et al.*, 1994). Other 'technical' factors such as load characteristics and centre of gravity of the loaded truck also affect the risk of accidents.

28.2.3 *The traffic environment*

The traffic environment affects the exposure to traffic-accident risks as well as the probability and consequences of traffic accidents. An important aspect of the traffic environment is the technical design. It includes road standard (width, curvature, surface), design of crossings, signal regulation, speed limits, etc. (Englund *et al.*, 1998).

The road network is divided into *road standard* categories to meet the needs of different users in relation to travelling distance and speed. It is thus common to distinguish between national, regional and local roads. The following design recommendations apply, for example, to national roads in Sweden to support safe and efficient travelling over long distances:

- Full or partial separation of pedestrians and cyclists from motor-vehicle traffic.
- Speed limit of 90 or 110 km/h.
- Specified minimum viewing distance and road curvature, requirements for a clear view for overtaking.
- A limited number of intersections per km.
- Crossings in one plane or with overpass or roundabout, signal regulation not allowed.

The *speed limit* is one of the most basic traffic technical measures to improve safety. Although the significance of the vehicle speed as a determinant of traffic accident risk is indisputable, a trade-off has to be made between safety and accessibility. Speed directly affects the probability of an accident through its influence on the stopping distance and other factors affecting the driver's ability to avoid a crash. In case of a crash, the energy involved and thus the extent of injury and damage is to a large extent determined by the speed. Evidence from police-investigated fatal accidents in the USA suggest that 'driving too fast or in excess of posted speed limits' was a contributing factor in as much as 22 per cent of the accidents (Loeb *et al.*, 1994). Statistics show that a reduction of the average speed by 10 km/h on public highways reduces the fatal accident rate by almost 40 per cent (Englund *et al.*, 1998). Not only high speed in an individual vehicle but lack of co-ordination of speed between different vehicles (speed variance) is a determinant of accident risk (Loeb *et al.*, 1994). The reason is that vehicles moving at similar speeds are less likely to collide.

The aim of speed limits is to achieve an adequate speed adaptation on the part of the drivers. Research shows that for speed limits to be effective, they have to be regarded as reasonable by the drivers, and there must be a fair chance of getting caught in case of speed violations (Englund *et al.*, 1998). This issue points up the general concern about how traffic technical safety measures affect behaviour. Recent research aims at identifying some general principles for the design of technical safety measures that promote safe driving behaviour. Besides good speed adaptation, they include such issues as:

- Obvious rules for yielding in intersections.
- Adequate viewing distances.
- Simple and clear rules for interactions between road users.
- Integration of unprotected road users where complete separation is not possible.

Also *snow, rainfall and lighting conditions* affect the risk of accidents. Compared to driving during ideal conditions (dry road surface, daylight), the accident frequency (per vehicle km) increases by a factor of two during night driving and by a factor of two to five during rain or snowfall (Englund *et al.*, 1998). The combination of night driving and slippery roads results in an increase in the injury frequency rate (per vehicle km) by up to ten times.

28.3 Sources of information on traffic-accident risks

Figure 28.2 shows an adapted version of the classical iceberg theory according to Heinrich (Section 13.3). Traffic accidents resulting in losses (fatality, injury or property damage) are usually visible in the statistics from police investigations, insurance claims, health-care statistics, etc. Events below the waterline occur more frequently but are usually not, apart from traffic violations, registered and available as a basis for experience feedback to improve safety. Before we analyse the different feedback mechanisms connected to the three actors in the transportation safety system, we will look into some important sources of information on traffic-accident risks. The presentation applies to Sweden but should be valid in many other countries as well (Englund *et al.*, 1998).

28.3.1.1 Standard reporting of traffic accidents

Police investigations are the primary source of data for the planning of traffic technical measures. The primary aim of these investigations is, however, to clarify the question of culpability. As soon as the police are notified about an accident, notes on the occurrence are made and registered. A special form is used for this purpose. Dependent on the severity of the accident and the number of vehicles involved, the accident is subject to an extensive investigation. This often includes interviews with involved drivers and witnesses, and documentation of the site of the accident by use of photos and sketches, statements from experts, etc. Copies of the police-investigation reports are sent to the road-administration authorities for analysis and use in statistical compilation. The accident data are linked to other databases on driving licences, vehicles and surveys of travelling habits, etc., and the results are used in periodic reporting as well as in research.

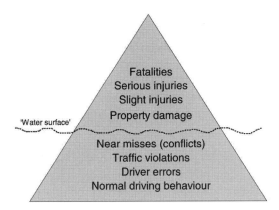

Figure 28.2 The driver-error iceberg.
Source: Maycock, 1997, page 66. Copyright 1997 by Elsevier Science and reproduced by permission.

Two types of variables of particular relevance for the planning of technical measures are the place of occurrence and the accident type. In order to identify the location as exactly as possible, a reference system has been developed for all public roads and streets. Each intersection has been assigned a unique identification, and the distance from the closest intersection defines the location of the accident site. The accident-type classification addresses the traffic situation and planned manoeuvres immediately before the crash.

Medical records provide the input to *registers on medical treatments* by the health-care system. The registers include demographic data on the patient (such as sex, age and place of residence) and data on diagnosis and treatment. The external cause is also registered for injuries by using the World Health Organisation's so-called ICD-10 codes (International classification of diseases version 10). It is here possible to identify road-traffic causalities.

The *insurance companies* maintain statistics on traffic accidents involving insurance claims. A primary purpose is to provide a basis for the establishment of insurance payments. The registers include data on the policyholders, the vehicle, injury and damage, liability, etc.

When a commercial driver is injured in a traffic accident, the *employer* is responsible for documentation of the accident for internal use and for reporting to the insurer and the authorities (cf. Sections 2.2 and Chapter 13).

28.3.1.2 Special research techniques

Multidisciplinary in-depth accident investigations are used to provide detailed knowledge about traffic accidents. An investigation team consists of experts in different areas. They apply a documented investigation method in order to ensure that all relevant factors are covered and that the results of the investigation are reproducible (i.e. other experts using the same method will come up with the same results). In-depth traffic-accident investigations have many similarities with the in-depth investigations of workplace accidents discussed in Chapter 13. A difference is that the traffic-accident investigations usually do not assess the overall SHE management system in the investigation. Haddon's phase model provides an important basis for the investigation team. It studies such factors as:

1 Pre-crash
   a The interactions between the road users, vehicles and traffic environment
2 Crash
   a Injury and damage mechanisms
   b Environmental factors
   c Vehicle design
   d Effects of safety systems
3 Post-crash
   a First-aid
   b Ambulance services
   c Medical diagnosis, treatment and rehabilitation

In using in-depth accident investigation techniques, the investigators have to address methodological issues such as:

• Should the investigation be carried out 'on the scene' as soon as possible after the occurrence, or 'after the fact', i.e. when data collection takes place some time after the occurrence? We have already discussed this theme in Section 13.6 in connection with the investigation into workplace accidents.
• How should the accident cases be selected for in-depth investigations? It is common to distinguish between statistical and clinical or case-study methods. In the first case, the accidents that are investigated should provide a representative sample of the total accident population. In the second case, on the other hand, the accidents that are investigated should provide in-depth knowledge about a certain type of accident of particular interest.
• How should the scope of the investigation be defined? Here, the client's aim of the investigation is decisive. Investigation carried out on behalf of vehicle manufacturers, for example, will focus on vehicle design.

There is abundant experience from in-depth investigations. A review of such experience has identified a number of advantages in this technique (FERSI, 1994):

- Identification of 'new' traffic-safety problems and the generation of new hypotheses for research.
- More detailed information about accident mechanisms and causes than those revealed through statistical analyses.
- Identification of causal factors such as inadequate design of vehicles or the traffic environment.
- Information on how 'normal' driving behaviour may result in an accident.
- Input to the design of remedial actions.

Traffic-conflict techniques are used to improve the statistical basis for decisions on safety measures, i.e. the aim is similar to that of near-accident reporting (cf. Chapter 13). There is one important difference, however. Whereas near accidents are usually reported by the involved personnel, traffic-conflict techniques utilise specially trained observers located directly in the traffic. There are differences between the traffic-conflict techniques developed in different countries. A comparison between eight different European techniques showed that they all evaluated the severity of the conflict in a similar way although there were differences in the scales applied (Englund *et al.*, 1998). The Swedish traffic-conflict technique, which is a representative example, applies two central concepts:

- The pre-crash time, which is the timespan that remains from the start of the avoiding manoeuvre to the occurrence of the crash, provided that the road user (driver, etc.) had continued with unchanged speed and direction.
- Conflict speed, which is the speed of the road user at the start of the avoiding manoeuvre.

Figure 28.3 shows the definition of serious traffic conflicts by combining the two variables. Studies of serious traffic-conflict rates and fatal-accident rates in different cities and countries show good correlation, indicating that this definition has an adequate validity.

This technique has been applied in evaluations of different traffic technical measures such as the design of pedestrian crossings. Although the technique is rather expensive, it has proven to be the only feasible method in instances where rapid quantitative feedback data is required. Another advantage is that it may be combined with studies of the behaviour of road users prior to the conflict.

Finally, we shall look briefly at an example of a method for *self-reporting of driving habits* (Hatakka *et al.*, 1997). This is a questionnaire-based method, where the respondents are asked concrete questions about personal driving habits, feelings, motivation and self-evaluations of personal risks and skills. The method has been applied in different studies and the results have been correlated with accident data. Results show that problematic drivers and driver groups may be identified and that it is possible to predict drivers'

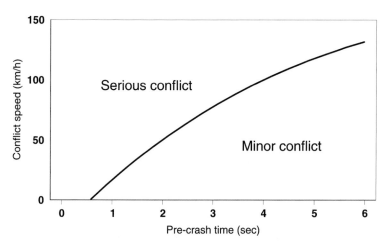

Figure 28.3 Borderline between serious and minor conflicts according to the Swedish traffic-conflicts technique.
Source: Englund *et al.*, 1998. Copyright 1998 by Studentlitterature and reproduced by permission.

traffic violations and accidents. The best predictors are closely correlated with motivational and strategic aspects rather than with self-evaluations of skills and risks.

28.4 Feedback mechanisms

In this Section, we will look into the feedback mechanisms that promote organisational learning and the implementation of safety measures within the trucking industry, Figure 28.4. The situation is not so simple as in the previous example from the oil and gas industry, where the operator of the installation had overall responsibility for safety.

As indicated in Figure 28.4, different organisations share the responsibility for safety in transportation on the public road network. At first glance, the split of responsibility mainly follows the division into the different 'components' of the driver–vehicle–environment model. Whereas the trucking company is mainly responsible for the driver, the truck manufacturer is responsible for vehicle design and the public-road administration for design, maintenance and clearing of the road network. A closer look shows that the situation is not quite so simple. The trucking company is responsible for the purchase of trucks and for maintenance of the vehicles. Selections of route and scheduling of transportation activities are other examples of decisions at the discretion of the trucking company. The legislators have overall responsibility for traffic safety from society's point of view, and determine the limits within which each organisation may act. Seen from a feedback-

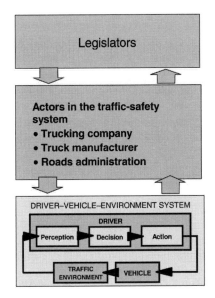

Figure 28.4 Different feedback mechanisms in safety in road transportation.

control perspective, the situation is rather complicated. Each organisation defines its own policy and goals as to transportation safety. It has limited possibilities of improving safety above a certain level, determined by the ambitions of the other organisations. This is because each organisation only exercises control over a certain part of the totality of factors that affect the safety level (compare Ashby's law of requisite variety in Section 10.5). The decisions made by the other organisations are merely to be regarded as parts of the outer context (see Section 2.2).

28.4.1 The trucking company

Typical decisions at the discretion of the trucking company that affect the risk of transportation accidents are selection of transportation route and scheduling of transportation tasks, driver selection, training and supervision, and truck specification and maintenance. Studies on the significance of the trucking company for the risk of trucking accidents have been summarised by Loeb *et al.* (1994). They include in-depth accident investigations and statistical analyses of accident data in the USA. Results of the in-depth investigations identified deficiencies in the trucking company's follow-up of the driver or the vehicle in about two-thirds of the investigated cases of driver-fatal truck accidents. The results and interpretations of the review of the statistical studies are summarised below. They show that the following firm characteristics are associated with a high accident-frequency rate:

- Non-compliance with safety regulations relating to driver qualifications, driving of motor vehicles, accident reporting and vehicle inspection and repair.
- High spending on insurance and safety, made necessary by high accident rates.
- High spending on maintenance due to an old truck fleet.
- Small firms and owners that at the same time were operators of the trucks.
- A low return on investments.
- Low driver experience and high turnover rates.
- Lack of or inadequate SHE management organisation and practices, including inadequate routines for checks on new drivers, accident reporting and disciplining of drivers for violations.

Let us look at a concrete example of how a large Swedish logistic company follows up on traffic safety. The company has about 6000 employees involved in the development, selling and production of transportation and logistic services. It runs its own storehouses and terminals but buys transportation services from separate trucking companies. The company has developed an environment and safety management system, mainly based on the requirements in the international standard series ISO 14000 on environmental management. It is the company's policy to prevent incidents and accidents that have a negative impact on health, environment and material assets. The number of incidents is followed up and compared to the company's goal.

The company applies a performance measure to traffic safety, where exposure is measured by the paid transportation work in tons and kilometres. There are 7–10 traffic crashes per year involving trucks in the company's service, giving a truck involvement rate of 1.5 to 2.2 per million ton km. There is no systematic follow-up of traffic accidents in general, since this responsibility is taken care of by the police, insurance company and truck manufacturer. Instead, the logistic company has developed a system for follow-up of the individual trucking companies through the pre-qualification, bid and contract evaluation procedures. The individual trucking companies have to develop their own quality and environmental management handbooks. They have to implement the logistic company's policies regarding drug use, and their programme for technical inspections and testing of vehicles. It is, for example, required that all drivers are tested for drug use when first taken into employment, and that 25 per cent of the drivers are tested yearly on a random basis.

Each trucking company is subject to a yearly performance review. It includes different aspects regarding quality, customer satisfaction, traffic safety, goods handling, personnel, environment and drug use. In the area of traffic safety, results of drug tests, inspection programmes and reports on incidents are reviewed. The results are compared with earlier reviews and with reviews of other trucking companies. An incident-reporting system has been implemented, where so-called 'improvement reports' are registered and used in a follow-up of the trucking companies. Damage to goods is carefully registered and more than 99.95 per cent of all deliveries must be error-free. Serious incidents are subject to in-depth investigations by a so-called crisis group.

The approach selected by the logistic company to control the safety performance of the trucking companies is well in line with research findings. It puts emphasis on the trucking companies' SHE management systems. Follow-up is carried out through audits and performance reviews similar to the methods presented in Chapter 14. The logistic company also puts emphasis on two aspects relating to the driver and the vehicle respectively, i.e. drug use and maintenance (especially the standard of brakes). The methods used in performance measurement, i.e. percentage of positive findings, are similar to the behavioural sampling techniques presented in Chapter 18.

28.4.2 The truck manufacturer

Let us next look at a major European truck manufacturer. It has 20 000 employees and delivers 80 000 heavy trucks a year with a weight above 16 tonnes. The company has the ambition to be a world leader in terms of active and passive safety in vehicles. The company management's motives are commercial as well as ethical. Safety is a cornerstone in the company's marketing strategy. It is claimed that the safety features of the company's trucks reduce accident risk and increase driver efficiency.

To ensure experience transfer from traffic accidents to design, the company has for some 30 years operated a dedicated SHE information system (Avedal, 1998). The company's Accident Investigation Team plays a key role in this system. The team has five members, including technical personnel and a medical doctor, and is located at the company's main office. It is on standby 24 hours a day. There is co-operation with the local police and rescue services, and the team is notified about all traffic accidents involving the company's trucks at a distance corresponding to a maximum of 1.5 hours driving from the head office. These accidents are investigated on the scene. To increase the size and geographical diversity of the data, similar arrangements have been made for the north-western part of Europe and for an area in the USA. Here, the accidents are investigated a couple of days after the accident occurred.

The team investigates all relevant factors concerning the vehicle, the traffic environment and the driver and other road-users. Results are stored in a database and are used in the design of new trucks. About 1500 accidents are stored in the database as a result of almost 30 years of investigations. Data are compared with the official statistics from different countries to check how representative they are. Below are some examples of the use of accident data in design.

Official statistics from Sweden show that about two-thirds of the fatally injured victims of accidents involving trucks are car occupants. The truck manufacturers' own data support this finding and give detailed information about collision mechanisms. The company has introduced a low bumper and front under-run protection system to reduce injuries for car occupants.

Out of the 700–800 truck drivers killed each year in Europe, single-vehicle accidents are the most frequent accident type and account for about half of the injuries and fatalities. These are typically accidents where the vehicle drives off the road and rolls over, or rolls over due to high speed on a curve. An important design criterion has been to maintain a survival space in the cabin around the occupant after roll-over. The cab strength is thus important. This is tested by use of a pendulum test.

The company's own investigations show that seatbelt use is the single most important safety measure and reduces the injuries of truck drivers by about 60 per cent. Use of air bag in combination with seat belt will have an additional injury-reducing affect of about 21 per cent (Svenson and Vidén, 1994). In addition to safety belts, one safety feature that reduces the consequences of roll-over accidents is an impact-friendly and solid cabin interior.

Truck-to-truck accidents are the second most common accident type involving injuries or fatalities to truck occupants, and account for about a third of the accidents. Investigation results show that cab strength is very important also in this case, since the cab has to absorb large amounts of impact energy. Also interior design is essential. Trucks are equipped with pedals and knee bolsters, a steering wheel that allows deforming in a collision, seat belt and air bags.

Experiences show that in-depth accident investigations provide important experience data from traffic accidents for use in design of vehicles. There are several examples of safety features that originate from these investigations. Many of these features have been taken over by the legislators and made mandatory for all truck manufacturers.

28.4.3 The roads administration

We will here use the Swedish National Road Administration (SNRA) as an example. This authority has overall responsibility for traffic safety in Sweden and has direct responsibility for government roads. In 1997, the Swedish parliament decided that the so-called 'zero-mindset' or 'vision zero' philosophy regarding traffic fatalities and severe injuries should guide long-term traffic-safety work (Swedish National Road Administration, 1999). SNRA has the main responsibility for the traffic safety results, although it does not have all the necessary traffic-safety measures at its command.

The 'vision zero' philosophy has been translated into an operational goal. This is to accomplish continuous reductions in the frequency of severe traffic accidents. More specifically, the goal is to reduce the number of traffic fatalities by 50 per cent in a ten-year period, starting from 1996.

SNRA has a Traffic-Accident Data System at its disposal. It is used to store all police-reported traffic crashes on the government and municipal road and street network. The following types of information are stored about each accident:

- Date and time of day.
- Detailed definition of place of the accident.
- A sketch of the accident site, including road and vehicle positions.
- Narrative description of the sequence of events.
- Type of road and traffic regulations.
- Weather, road surface, traffic environment and lighting conditions.
- Types of vehicles and other traffic elements (pedestrians, wild animals).
- Involved persons, injuries sustained, suspected alcohol involvement.

The data system is used for follow-up of traffic-safety results and to provide support in making decisions that develop and implement traffic-safety measures. This work is supported by extensive search, analysis and presentation capabilities in the data-system software.

The traffic-safety work has been organised into a number of result areas in order to make it more focused. SNRA has defined performance indicators for each area to follow up on progress. There is a known relationship between the changes within each area, as measured by the performance indicator and the risk of traffic accidents. Table 28.2 shows an overview of some of the performance measures and the goal for each for year 2000.

Results show that there has been a downward trend in the number of traffic fatalities and severe injuries in the period 1990 to 1998. Since 1996

Table 28.2 Performance indicators for different traffic-safety areas. Detailed performance measures have not been defined for the traffic environment

Area	Measure	Target for year 2000 (Basis year is 1994)
Loss-based performance indicator:		
Fatalities	Number of fatalities per year	−33%
Process-based indicators:		
Drink-driving	Number of drunk drivers recorded by police checks	−27%
Reduction in speeding offences	Average occurrence of speeding in total vehicle mileage	−35%
Reduction in other driving offences	Proportion of drivers that violate other traffic regulations	−50%
Use of seat belts	Proportion of drivers and passengers that use seat belts	95%
Causal-factors-based performance indicators:		
Safe road environment in urban and rural areas	Proportion of road and streets kilometres that do not meet specified safety requirements	Reduction
Safe cars	Increase in crash-worthiness index for all cars	+12%

Source: Swedish National Road Administration, 1999

there have been no improvements, however, indicating that the efforts to implement the 'vision zero' philosophy have not given immediate results. The lack of positive results coincides with a similar lack of improvements for other performance indicators as well. In 1998, only the frequency of reported cases of insobriety showed a reduction. No statistically significant changes have taken place in the other areas. This is an illustration of the difficulties involved in meeting safety goals without having adequate control of the various variables affecting safety performance.

The approach selected by SNRA is interesting, since it illustrates how a combination of performance indicators may be applied for a systematic follow-up of the different aspects that affect the risk of injuries. This approach is similar to the follow-up of barrier availability by use of performance indicators discussed in Section 20.2.

Part VII

Improving the corporate SHE information system

This Part brings us back to the starting point, i.e. that of the design and development of a corporate SHE information system. Chapter 29 presents a model for the development and improvement of such a system. It becomes a focal point for the integration and application of the various principles and tools presented in previous Chapters on corporate SHE practice. Development work is based on the principles of the Deming circle. It is output-driven in the sense that the needs that the system serves define the starting point. Important aspects in the design of the system are reviewed in Chapter 30. Some final remarks are made in Chapter 31.

29 The improvement process

Traditionally, the design and development of corporate SHE information systems has been a concern for the SHE department in the company only. The SHE experts of the company have been the owners of the system. They have felt that they need better tools to store and retrieve accident reports and to generate periodic accident summaries. Usually, the introduction of computer support has been a key issue.

We will here discuss a more holistic approach, where the development of a SHE information system is a concern for the whole company. This approach is based on the assumption that an improvement in SHE management, where the SHE information system is a vital part, is a strategic issue for the company. It has the potential of giving substantial positive contributions to the economic results of the company. This is accomplished through a better 'control climate' resulting in reduced expenditures for personnel, energy, raw materials, maintenance, etc. Improved goodwill in relation to stakeholders (the employees, unions, authorities, customers, general public, etc.) is another long-term effect.

We here assume that the company is of a size and complexity to use modern information-technology network solutions for the general handling of information. This includes a common database-handling system, e-mail, standard PC software for word-processing, spreadsheets, graphical presentation and access to the Internet. The development of computer solutions to the SHE information system is a central task in the improvement process. It is natural to choose solutions that utilise existing computer resources at the company.

Finding and customising the right software is only one of the many concerns in the design of the SHE information system. From Section 1.1 we remember that a SHE information system is made up of the following parts:

- People with their knowledge, experience and motive structure.
- Organisation and routines.
- Instruments and tools, including, for example, reporting forms, analysis techniques and computer software.

We have to review all the different parts in parallel. The Deming circle offers a structured development process for this work. It is also important to involve all concerned interest groups at the company in the process in order to ensure adequate ownership to and quality of the results.

By applying computer support, our immediate aim is to make the feedback to decision-makers of information about accident risks quicker and more efficient. A further aim is to reduce man-hours needed for the administration of reports. Computer support also makes information retrieval much easier, and the accumulated experience from SHE practice will be available through a few keystrokes. Long-term effects include a reduced risk of accidents, improved production regularity, more efficient safety work, improved knowledge and motivation, improved goodwill, etc.

The recommendations presented here concerning computer support and accident statistics do not primarily apply to small companies with a low number of accident cases per year. Here it is vital to secure a thorough reporting and investigation of accidents and near accidents, and follow up on a case-by-case basis. Risk analysis is also an activity that is well suited to the small company. Systematic activities in these areas may generate the need for computer support in follow-up and in storing and retrieval of analysis results.

29.1 Evaluation of existing conditions

We ask three basic questions in the planning phase of the Deming circle: Where are we? Where do we want to go? How do we get there? Any plan to improve the SHE information system of a company should start by evaluating the present conditions. Such an evaluation may be carried out as an audit. Persons that do not have a vested interest in the present solutions should carry out the audit. There are a number of questions that we can ask in such an audit. The requirements for SHE information systems according to Chapter 11 give some important input. At a more general level, the following basic questions should be looked at:

- What is the scope of the SHE information system? Is the level of ambition limited to meeting regulatory requirements concerning the reporting and prevention of occupational accidents? Or is the SHE information system used in a larger context in order to prevent all types of deviations and incidents that may result in accidents and illnesses with injury to personnel, environmental pollution, material damage, reduced production quality, etc.?

- Are the feedback loops closed? Are the routines for the reporting and investigation of accidents and near accidents and for the execution of workplace inspections and SHE audits adequate? Are the results used in decisions to improve safety, and is there a follow-up of the implementation of the actions decided upon? Does the company meet the requirements in relation to reporting to the authorities? Have the routines been documented and are they known to the people concerned?

- At what feedback level does the company operate? Is the focus on short-term measures such as correction of deviation? Or are higher levels of learning from experience involved?
- Have goals been established for SHE performance? Are the goals followed up through comparison with actual performance? Are actions taken when goals are not met?
- Are the accumulated experiences from SHE practice available for use in SHE practice? Are experiences concerning severe and frequently recurring accidents allowed to be forgotten or are they available at a later stage for active use in the decision-making? Are they available to a small group of SHE experts only or are they generally available to the different decision-makers with responsibilities that affect the accident risk?
- Are routines in place for a systematic identification and assessment of accident risks? Are the results used in actions to improve safety and are the results followed up? Are the results of the risk analyses adequately documented and retrievable?
- Do the decision-makers have adequate knowledge about the SHE regulations that apply to their work? Do they have access to these governing documents in relation to SHE?

29.2 Establishing goals and defining user needs

The further development work preferably takes place in a problem-solving group, where typical users of the SHE information system are represented. It is important to ensure a shared understanding between systems designers and users on the different needs that the system has to meet. Audit results are important input to the group's work. The group should establish the level of ambition and goals for the development work that follows.

An important issue is to define the user needs. Table 29.1 shows a checklist of typical modules of a SHE information system that meet different needs. The checklist could be used to establish a long list of immediate and potential needs, and these have to be prioritised.

It is not possible or even desirable to try to meet all the different needs in one big leap. A stepwise process with the following intermediate goals or milestones is here recommended:

1 Access to steering documents on SHE (via Intranet and Internet).
2 Simplify and improve existing routines for recording, storage and distribution of reports, generation of standard statistics and official reporting to insurance companies and the authorities.
3 Simplify and improve routines for follow-up of actions.
4 Information-retrieval capabilities including free-text searches and application of standard statistics for manipulation of the database or subsets of it.
5 Tools for special applications (accident investigations, risk analyses, special statistical analyses, etc.).

Table 29.1 Checklist of uses of a SHE information system

Type of use:	Comments:
A: Output	
Standard reports	
• Internal periodic reports	See Chapters 13–14
• Reports to the authorities	
• Reports to insurance company	
Internal distribution of individual reports	See Chapter 13
• Accidents and near accidents	
• Unsafe conditions	
• Workplace inspections	
• Nonconformities	
• Risk analysis	
• SHE audits	
Retrieving individual reports	See Chapter 15
• Accidents and near accidents	
• Unsafe conditions	
• Workplace inspections	
• Nonconformities	
• Risk analysis	
• SHE audits	
Follow-up of status of actions from reports on	See Chapters 13–15
• Accidents and near accidents	
• Unsafe conditions	
• Workplace inspections	
• Nonconformities	
• Risk analysis	
• SHE audits	
Input to experience carriers	See Chapter 15
• Work instructions	
• Training programmes	
• Risk analyses	
• Technical specifications as to design	
• Other experience carriers	
Results on SHE performance measures	See Chapters 16–20
• Loss-based	
• Process based	
• Causal-factors based	
Statistical analyses of data on	See Chapter 15
• Accidents and near accidents	
• Unsafe conditions	
• Workplace inspections	
• Nonconformities	
• Accident risks identified in risk analyses	
• Remedial actions	
by applying	
• Uni- and bi-variate distribution analysis	
• Accident-concentration analysis	

Table 29.1 cont'd

Type of use:	Comments:
• Severity-distribution analysis • Extreme-value projection • Other methods	
Access to regulations, standards and internal procedures	Preferably through pointers to other databases and to authority web sites on the Internet
B: Input Recording of results of • Accident and near-accident investigation • Reporting of unsafe conditions • Workplace inspections • Filing of nonconformities • Risk analysis • SHE audits	See Chapter 13, Chapter 14
Support in planning and execution of • Accident and near-accident investigations • Workplace inspections • Risk analysis • SHE audits	See Chapters 13, 14 and 21–26

29.3 Developing solutions and following up results

The developments of people, organisation, routines, instruments and tools have to proceed in parallel. The selection of adequate computer software is a central issue. We will here recommend a strategy for the development of computer support that uses the company's standard PC software. There are also other possibilities. A common solution is to buy a proprietary SHE information system. Still another alternative is to develop new software to suit the company's specific needs by applying a software-engineering approach.

We here choose the first alternative in order to help the users by using a familiar computer environment, and to integrate the development of the SHE information with the general IT development in the company. The second alternative is cost-efficient in the short run. The key point in refraining from the selection of this solution is to make management feel ownership of the new system rather than seeing it as an external imposition. Experience shows that it is difficult to reach potential user groups beyond SHE experts with a proprietary system.

In a traditional software-engineering approach, the definition of user needs and software-design specifications come early in the development process. At this stage, the users may not have the necessary knowledge and experience to be able to define their needs. Only after using the system for a while do they know what they actually want. It may then be too expensive to change the system, since the costs of changes increase by a factor of ten with each step in the development process.

An important issue at an early stage is to decide on a data model for the storage of information, see next section. Although the group may decide to start by storing accident and near-accident reports, this data model has to be flexible enough to account for future expansions to include data from other SHE activities as well. By using the company's standard PC software for database handling and data analysis and presentation (e.g. spreadsheets), it is then possible to feed the database with available information on accidents, near accidents, etc. This solution will require access to a specialist in the relevant types of software, in the first instance the database management system. It will, however, through its flexibility support the users in gaining experience about what they actually need. A further advantage is the minimal investment costs that are required. At the next step, the group may decide to develop customised solutions for input and output, for example, around some of these standard PC packages.

There are economic, legal and organisational issues that have to be addressed in the implementation process. It may be difficult to prove that the benefits of the improvements in the SHE information system are merited from an economic point of view. Their costs are relatively easy to calculate. They include investments in hardware and software and operational expenditures for running the system, training of users, etc. There are immediate benefits from the improvements that also are relatively easy to assess. They include administrative rationalisation through more efficient administration of reports. Ideally, these immediate benefits should exceed the costs, since the more long-term benefits of a reduced risk of accidents are more difficult to assess.

Improvements in the SHE information system may trigger intra-organisational conflicts by threatening the autonomy of managers, or highlight conflicts between company goals and incentive structures. There is no straightforward remedy to this problem. It is necessary that development work has top management support and that the different stakeholders are part of the development process.

Each step in the development process should be followed up through adequate evaluations. It is here not only a question of collecting the evaluations and experiences of the users. Development work should also be followed up through the establishment and monitoring of specific criteria, see Chapter 11 for examples.

30 Design of the system

30.1 Database definition

A basic decision in the design of a SHE information system is to define the types of data about accident risks to be collected and how they are going to be stored (as defined by the data model). Experience shows that there is a tendency to over-estimate the actual information needs and their complexity. As a consequence, the data-collection routines and the accident-investigation form are made more complicated than necessary. A sub-optimal design will affect the quality of the data as well as the motivation of the concerned personnel.

It is recommended that the database should only contain relatively few types of observations. A minimum is defined by the mandatory information to the insurance company and/or the authorities. The application of the principle of 'a smallest efficient data set' displayed in Table 15.1 will allow the company to concentrate its data-collection efforts in order to ensure an acceptable data quality.

This data set applies to reports on accidents and near accidents. Decisions must be made at an early stage on the totality of types of reports that are to be stored, even when the development work proceeds in steps and the immediate focus is on accident and near-accident reports. A generalised data model should be applicable to the storage of reports on the following SHE activities:

- accident and near-accident investigation;
- reporting of unsafe conditions;
- workplace inspections;
- reporting of nonconformities;
- risk analysis.

The accident-analysis framework of Chapter 6 provides a basis for tying together information from these different activities, see Section 15.1.

A decentralised SHE information system is recommended. This means that data collection and input to the database is made at the workplace level. The workplace should also have access to the database for retrieval of information. This will help in creating ownership of the data at the workplace level. The database should be located at one common server, however. This solution makes it easier to maintain and update the database.

30.2 Organisation and routines

Supervisors and safety representatives usually carry out the first investigation. An investigation team or a problem-solving group should be mobilised and allocated time for in-depth investigations of accidents and near accidents with a high potential for serious consequences. It is the responsibility of the management at a level above the affected department to make decisions about in-depth investigations.

The SHE department should be responsible for checking the quality of the input to the database and for the generation of periodic reports. Other concerns regarding organisation and routines that need to be addressed include:

- Reporting criteria.
- Definition of responsibilities for reporting, investigation, implementation and follow-up.
- Follow-up and information feedback to reporters.
- Etc.

Similar routines have to be developed for the reporting of unsafe conditions, workplace inspections, risk analyses, SHE audits, etc. Responsibilities and routines have to be documented in procedures and work instructions.

30.3 Personnel

We are here concerned with the personnel responsible for the different SHE activities including the reporting and investigation of accidents and near accidents and with the users of the information. There is a need to ensure that these groups of personnel have the necessary basic knowledge and motivation, in particular about accident causes, about information needs in accident prevention and about the efficiency of different types of remedial actions. They also need to know the new routines and responsibilities and acquire know-how in accessing the new database. There are efficient means of affecting knowledge and motivation. Formal training is important, but know-how is best accomplished through hands-on training in SHE practice. Management has an important task in promoting the right attitudes and behaviour through its own example. The question of whether the company needs to establish a group of trained accident investigators has to be addressed.

30.4 Instruments and tools

Modern computer technology makes it feasible to use computer support also in cases where relatively small sets of data are handled. The selection of software is a key decision. This issue has already been discussed in Chapter 29.

Efforts should be made to simplify the form for the registration of accidents and near accidents as far as possible. It is an important principle that the design of the form shall not create a barrier towards the reporting and investigation of accidents and near accidents. Some of the data are fed directly into the computer and need not show up on the form for the supervisor's first report. No specific form is needed for the in-depth investigations. Checklists should be developed to support the investigations. There are also other tools to be considered such as checklists and record sheets for use in workplace inspections, risk analyses and SHE audits.

31 Epilogue

During recent years, there has been a breakthrough in the area of SHE information systems mainly within large companies in highly hazardous industries. Experience transfer between companies and peer pressure at the top management levels has driven this breakthrough. It has manifested through increased focus on SHE performance. We see intensified activities in these companies in the areas of accident and near-accident reporting, management inspections, audits, computer-supported feedback and follow-up, etc. Many companies report significant reductions in their accident frequency rates as a consequence of this development.

Safety research has played a subordinated role in this knowledge build-up and transfer. We have seen that SHE management concepts receiving high acceptance such as ISRS are based to a large extent on research from the first part of the twentieth century. One of the author's ambitions with this book has been to support knowledge diffusion from recent research to practice.

The positive SHE performance developments experienced by some large companies have not been paralleled by companies in the small and medium-sized range. The large companies in highly hazardous industries remain an exclusive group and the knowledge and motive structures of their management have not spread to the management of other industries. Obviously, the incentives to promote such a development have not been large enough.

Current research of relevance to SHE information systems proceeds along two tracks. On the one hand, it utilises the dramatic developments in information technology. This development opens new possibilities for information retrieval and distribution. Managers at all levels in the companies now have direct access to a variety of databases via corporate computer networks. SHE management has until now not fully benefited from this development. Line management's accessing of accident databases for use in day-to-day decision-making still remains a utopia in most instances. New network solutions and software techniques aim at remedying this problem. They provide solutions that facilitate feedback and reuse of experiences on accident risks. Here, success is dependent on the possibilities to model and index the information adequately by use of relevant accident theory.

A second track is the developments in research into learning organisations. This research generates new knowledge on necessary changes in organisation and management to accomplish learning processes. We have seen that risk-analysis techniques such as job-safety analysis are well suited for use as tools in this context.

Future research into SHE information systems will need to integrate these two tracks of development. Learning organisations are dependent on prompt and correct feedback of results and access to information that its members find interesting. We need simple and verifiable theory and methods to support this development. Future research in the area has to be output-driven in order to take advantage of the principles from natural selection in generating effective solutions. It remains to be seen whether the end results will attract the attention of middle-sized and small company management.

Appendix I
Definitions

Acceptance criterion for the risk of accidents   Criterion used to express an accepted level of risk in an activity.

Accident   A sequence of logically and chronologically related deviating events involving an incident that results in injury to personnel or damage to the environment or material assets.

Accident risks   Collective name for a condition at the workplace or in the management system that increases the risk of accidents. It is a deviation, contributing factor or root cause.

ALARP   As low as reasonably practicable.

Analysis object   The part of the workplace which defines the scope of the risk analysis (e.g. an area in coarse job-safety analysis and a specific job in detailed job-safety analysis).

Assumed risk   Accident risk that has been identified, evaluated and accepted at a proper management level.

Cause of accident   Contributing factor or root cause.

Contributing factor   More lasting risk-increasing condition at the workplace related to design, organisation or social system.

Danger zone   Any zone within the reach of a hazard.

Deviation   Event or condition that departs from the norm for the faultless or planned-systems processes.

Diagnosis   A decision cycle consisting of identification of symptoms, determining causes and prescription of remedy.

Experience carrier   A document, database or piece of hardware that represents the company's collective experiences and governs its activities and decisions.

Experience, qualifying   Experience that makes an individual or organisation able to understand real-world phenomena (know-why) and able to act efficiently (know-how).

Feedback control (negative)   A regulating mechanism that produces corrective action.

Hazard   A source of possible injury to personnel or damage to the environment or material assets.

Hazardous event   Loss of control of energies in the system or body movements, resulting in a potential for exposure of personnel (or the environment/material assets) to the energy flow.

Incident   See HAZARDOUS EVENT.

Job   A series of inter-related manual activities that are performed to achieve organisational objectives.

Job-safety analysis (JSA)   A series of logical steps to enable systematic examination of the hazards associated with jobs.

LTI-rate   Lost-time injury frequency rate, i.e. the number of lost-time injuries at work per one million hours of work.

Near accident   A sequence of logically and chronologically related deviating events involving an incident that under slightly different circumstances could have resulted in injury to personnel or damage to the environment or material assets.

Nonconformity   Non-fulfilment of specified requirement.

Precision, degree of   The proportion of the retrieved reports from a database that correspond with what was wanted.

Proprietary SHE information system   Copyright-protected SHE information system software and procedures.

Recordable injury   Fatality, lost-time injury, medical treatment injury or injury resulting in loss of consciousness, transfer to another job or restricted work.

Reliability in the reporting of incidents   The probability that a reportable incident is reported. It is expressed as the number of reported incidents divided by the 'true' number of incidents, as defined by the reporting criterion.

Retrieval, degree of   The share of the total number of wanted reports from a database that have been retrieved.

Risk analysis   A systematic identification and categorisation of risk to personnel, the environment and material assets.

Risk assessment   A comprehensive estimation of the probability and the degree of possible injury or damage in a hazardous situation, made in order to select appropriate safety measures.

Risk of accidents   A combined measure of the probability or frequency of accidents involving losses per unit of exposure to a specified activity, and the extent of the losses (consequences).

Root cause of accidents   Most basic cause of an accident/incident, i.e. a lack of adequate management control resulting in deviations and contributing factors.

SHE   Safety, health and environment.

SHE audit   A systematic and independent examination of a company's SHE management system. The aim is to determine that the elements within the system have been established and are effective and suitable for achieving stated SHE requirements and goals.

SHE information system   System that provides the information needed for decisions and signalling relating to safety, health and the environment.

SHE management system   Organisational structure, procedures, processes and resources needed to implement SHE management.

SHE performance measure/indicator   A measure or indicator of the management system's ability to control the risk of accidents.

Symptom   A deviation of the system's behaviour from what is considered to be 'normal'.

Taxonomy   A classification system made up of a complete set of mutually exclusive classes.

Validity as indicator of the risk of accidents   A measure of the degree to which the indicator represents the risk of accidents.

Variety   The total number of states of a system.

Appendix II
SMORT questionnaire

Level 1   Sequence of events/risk situation

1.1   Deviations in production

1.1.1   Personnel deviations

1   Did persons other than the usual ones carry out the work?
2   Temporary reductions in the work crew?
3   Personnel in an unsatisfactory condition (due to inadequate sleep, drug use, etc.)?

1.1.2   Inadequate information

1   Insufficient or incomplete work instructions (written or oral)?
2   Technical documentation (drawings, user instructions, certificates, etc.) incomplete or lacking?
3   Did the personnel lack the required driver's license, certificate etc.?

1.1.3   Disturbance in material flow

1   Deficiencies in the quality of the work material?
2   Wrong dimensions or insufficient quantity of delivered materials?
3   Delays in material delivery?
4   Inadequate packaging or storage of materials?

1.1.4   Human error

1   Wrong work method or equipment used?
2   Actions omitted, delayed or out of sequence?
3   Misunderstanding within work crew?
4   Improvisations, horseplay?
5   Inadequate error recovery?

1.1.5   *Technical failure*

1   Faulty machinery, equipment or tools?
2   Errors in technical control system?
3   Faulty plant containment of hazardous substances?
4   Machinery, equipment or tools missing?
5   Wrong location of machinery?

1.2   Disturbances from the environment

1.2.1   *Intersecting/parallel activities*

1   Disturbances from adjacent work crew?
2   Delays affecting downstream activities?

1.2.2   *Poor housekeeping*

1   Waste from work materials or packaging materials present?
2   Spillage from production process present?
3   Unattended tools and equipment present?

1.2.3   *Poor environmental conditions*

1   Too high noise level?
2   Too high vibration level?
3   Inadequate lighting, glare?
4   Too high wind speed?
5   Too high/low temperature?
6   Unusually heavy rainfall, snow?
7   Contamination of the working atmosphere?
8   Uneven or slippery floor?

1.3   Incident

1.3.1   *Loss of control*

1   Loss of control of own body movements?
2   Loss of control of energies in the system?
  a   Operator loses control of manually operated tools/machinery/equipment?
  b   Loss of control of machinery movements due to technical or control system failure?
  c   Accidental release of hazardous substances?
  d   Other?

1.3.2   Failure in active safety systems

1   Energy release not automatically detected?
2   Active safety system not released or not functioning?

1.3.3   Failure in fixed barriers

1   Fixed barriers not present?
2   Fixed barriers penetrated by energy flow?

1.3.4   Failure in personal protection

1   Personal protection not used?
2   Wrong personal protection in use?
3   Personal protection did not give adequate protection?

1.3.5   Person in danger zone

1   Unauthorised person present in danger zone?
2   Too many people present in danger zone?

1.4   Development of injury/damage

1.4.1   Failure in alarm and mobilisation of emergency response team

1   Alarm not received or delayed?
2   Delayed or incomplete mobilisation of emergency team?

1.4.2   Failure in limiting injury/damage

1   Inadequate or inefficient combat of energy flow?
2   Inadequate evacuation or rescue of victims?
3   Inadequate protection of emergency team?

1.4.3   Failure in management of information

1   Delayed or inadequate information to the authorities?
2   Delayed or inadequate information to the next-of-kin?
3   Delayed or inadequate information to higher management and/or the employees?
4   Delayed or inadequate information to the media and general public?

Level 2 Department and work system

2.1 *SHE culture and values*

2.1.1 *Line management commitment*

1 Do the supervisors promote SHE by their own vigour and example?
2 Do the supervisors take active part in SHE activities (accident and near-accident investigations, workplace inspections, job-safety analyses)?

2.1.2 *Adherence to safety rules*

1 Do the supervisors monitor the safety behaviour of the workers and take action in case of non-compliance?
2 Do the supervisors themselves show strict compliance with safety rules and procedures?

2.1.3 *Communication on SHE*

1 Is communication between supervisors and higher management and workers on SHE adequate?
2 Are there regular meetings between supervisors and workers where SHE issues are highlighted?

2.1.4 *Employee involvement*

1 Is it ensured that the employees are involved in decisions on SHE, where they possess relevant knowledge and experience?
2 Is ownership of SHE measures ensured though employee involvement?
3 Do employees participate in accident and near-accident investigations, workplace inspections, job-safety analyses?

2.2 *SHE management*

2.2.1 *SHE organisation*

1 Has an adequate SHE organisation been established with safety representatives and a working-environment committee?
2 Does this include contractors and hired labour?
3 Has the division of responsibility between line management and the SHE organisation been adequately defined?

2.2.2 SHE goals and action plans

1 Have adequate SHE goals been established?
2 Is it ensured that they are implemented in action plans with clear responsibilities and deadlines?
3 Are the plans adequately followed up?

2.2.3 SHE routines and procedures

1 Are there adequate routines for accident and near-accident reporting and investigations, workplace inspections, job-safety analyses?
2 Do the routines for workplace inspections ensure that they are carried out with an acceptable frequency and scope?
3 Do these routines function adequately in practice?
4 Are the actions from these different activities adequately documented and followed up?
5 Are there adequate routines for the identification, analysis and documentation of new jobs with respect to safety?
6 Is the physical working environment monitored at regular intervals?
7 Are key characteristics of the operation and activities that can have a significant impact on the environment monitored and measured regularly?
8 Is compliance with relevant environmental legislation and regulations periodically evaluated?
9 Are there regular health checks of the personnel?

2.2.4 Work and safety instructions

1 Are there adequate work and safety instructions for the different jobs?
2 Have persons with direct experience of the jobs participated in the development of the instructions?
3 Are the instructions based on a systematic evaluation of the hazards?
4 Are hazards adequately identified and necessary control actions described?
5 Do the instructions coincide with the way work is actually carried out?
6 Are the instructions checked and updated at regular intervals and also when the jobs are changed?
7 Do the instructions cover start-up, shut-down, normal operation, handling of disturbances, cleaning and repair?
8 Have potential conflicts between safe operation and requirements for productivity been adequately evaluated?
9 Are the responsibilities and routines for housekeeping adequately defined and documented?
10 Do they ensure regular housekeeping to avoid accumulation of waste and litter at the workplaces?

2.2.5   *Co-ordination of SHE work with contractors*

1   Have the responsibilities for co-ordination of SHE work been adequately defined?
2   Is it ensured that contractors working on the premises have the necessary qualifications?
3   Is it ensured that the contractors have SHE management systems of adequate quality? Are they co-ordinated with the SHE management system of the principal enterprise?
4   Are the different SHE activities adequately co-ordinated with contractors on the premises?
5   Is it ensured that contractors' employees get the necessary SHE education and training before they start work?
6   Are necessary actions taken to follow up on the safety behaviour of contractors' employees? Are the necessary steps taken to ensure compliance?

2.2.6   *Emergency preparedness*

1   Does an adequate emergency-preparedness plan exist that describes the responsibilities and routines?
2   Has the plan been established based on an analysis of the different possible accident scenarios with consequences to personnel, the environment and material assets?
3   Are there adequate organisational and technical resources for emergency operations in relation to the potential accident scenarios? Are they available at all hours when they are needed?
4   Is a single line of command ensured in emergency situations?
5   Are there adequate evacuation and counting routines?
6   Is the plan for emergency training and drills adequate?
   a   Of the emergency organisation?
   b   Of all concerned personnel?
   c   Does it ensure that the emergency crew is familiar with the facilities?
   d   That the actions of internal and external resources are co-ordinated?
7   Does the plan ensure an adequate notification of accidents, mobilisation of emergency organisation, rescue operations and combat of fires, etc.?

2.2.7   *Documentation*

1   Has the SHE management system been adequately documented?
2   Does the documentation include:
   a   Policy and goals?
   b   Organisation and responsibilities?
   c   Identification of applicable procedures and instructions?
   d   Action plans?

3   Is the SHE management system available within the organisation? Has it been ensured that it is known?
4   Is the access to statutory regulations adequate?

2.3   Management responsibilities

2.3.1   Assignment of responsibilities

1   Have the SHE responsibilities been adequately defined:
    a   Between first-line supervisors and middle management?
    b   Between line management and safety officers?
    c   Between operation and maintenance?
    d   In hand-over between shifts?
    e   Within plant areas that are used by more than one department?
    f   In work outside normal working hours?
    g   For housekeeping?
    h   For hired personnel?
2   Has authority to make decisions been delegated to the lowest possible level?
3   Have the operators been provided with adequate autonomy and well-defined operational goals and decision criteria to be able to handle critical disturbances in a satisfactory way?
4   Do the authority and accountability match the responsibility?

2.3.2   Resources

1   Do the supervisors receive adequate support from the line organisation and staff officers?
2   Do the supervisors receive the necessary information?
3   Is selection and training of supervisors adequate?
4   Have the supervisors received the required SHE education and training?
5   Do the supervisors have adequate time at their disposal to follow up on SHE issues?
6   Do the supervisors have an adequate budget for SHE measures?

2.3.3   Supervision

1   Do the supervisors have required qualifications to make the necessary decisions?
2   Do the supervisors monitor hazardous conditions at the workplaces and take the necessary decisions on remedial actions?
3   Do the supervisors make adequate priorities between SHE, quality and progress?
4   Do the supervisors adequately address SHE issues in management meetings?

5   Is the co-ordination between different supervisors in different departments and shifts adequate?

2.3.4   *Production and activity plans, control of work pace*

1   Are there adequate production plans?
2   Are new jobs adequately planned?
3   Is it possible for the workers to control their own work pace?
4   Have working hours, shift schedule and breaks been adequately planned?
5   Are the activity plans and schedule acceptable considering the age and skills of the workers and their required needs of recovery?
6   Do the activity plans allow for adequate variations in job execution and work pace?

2.3.5   *Co-ordination between activities*

1   Is there adequate co-ordination of different activities to avoid interference with potential unwanted SHE effects?

2.3.6   *Planning for disturbances*

1   Are there adequate provisions for the handling of deviations from activity plans due to disturbances?

2.4   Human Resources

2.4.1   *Resources, manning*

1   Have the requirements as to manning levels been adequately evaluated and defined?
2   Have the requirements as to competence and skills in the different positions in the organisation been adequately evaluated and defined?
3   Have the requirements as to manning and competence taken the needs of back-up from colleagues during disturbance-handling and error-recovery adequately into account?
4   Is there adequate slack in the organisation to allow for consultation in the handling of new situations and to facilitate learning from experience?
5   Has it been possible to meet the requirements as to manning and competence?
6   Are there adequate extra resources (manning, competence) to fill positions in case of absenteeism?
7   Have the desires of the individual employees been adequately considered in the allocation of personnel?

2.4.2 Education, training

1 Is there a satisfactory education and training plan for the department?
2 Does the plan include an adequate general safety and emergency-preparedness training?
3 Does the plan ensure that the requirements as to skill and competence in the individual positions are met?
4 Have personnel that are required to perform critical recovery and emergency response actions been adequately trained until correct behaviour is automatic?
5 Does the plan ensure an adequate knowledge of the hazards at the workplace?
6 Have all personnel received adequate training according to the plan? Has it been documented?
7 Is it controlled and verified that the training is efficient?
8 Is there adequate practice and new training when required by changes in job content?
9 Does the education and training ensure correct attitudes towards safe work practices?
10 Does the training plan include service personnel, contractors and temporary hired personnel?

2.4.3 Remuneration

1 Are there adequate remuneration systems that promote safe behaviour?
2 Are there adequate routines to punish violations of safety rules? Are these experienced as fair?

2.5 Fixed assets

2.5.1 Plant layout, access

1 Is the building of an adequate size and construction considering the needs for a safe and efficient operation?
2 Are the different areas of the plan (production, maintenance and storage areas, transportation routes, rooms for personnel and administration, etc.) of an adequate size, and are they adequately located and segregated in relation to each other, considering fire hazards, transportation, communication, etc.?
3 Have the transportation routes been adequately designed and marked, considering traffic intensity, the separation of different types of traffic, sight lines, etc.?
4 Have access and escapeways been adequately designed and marked?
5 Is machinery and equipment adequately located? Is the access for operation, inspection and maintenance adequate?

6   Are there adequate provisions for housekeeping and waste handling?
7   Are there adequate facilities for the storage of combustible and toxic materials?
8   Are there adequate work platforms and guard-rails?

2.5.2   *Physical barriers against fires and explosions, safety systems*

1   Have the different plant areas been adequately classified in relation to the risk of fires and explosions?
2   Have ignition sources been adequately isolated?
3   Are buildings adequately ventilated to avoid build-up of explosive gases?
4   Have fire and explosion partitions been adequately established and equipment adequately insulated against fire?
5   Are buildings adequately designed to withstand the most severe fires and explosions during a specified time period in order to meet the defined risk-acceptance criteria?
6   Are there adequate detection systems for releases of gas, fires, and explosions?
7   Are there adequate systems for emergency shut-down?
8   Are there adequate provisions for fire-fighting and first aid?
9   Are there adequate communication systems in case of an emergency?

2.5.3   *Physical working environment*

1   Is the ventilation and temperature regulation adequate?
2   Are there adequate provisions to avoid contamination of the working atmosphere?
3   Are the noise and vibration levels acceptably low?
4   Is lighting adequate?
5   Are alarm and warning signals audible?

2.5.4   *Design of machinery/equipment*

1   Have the hazards been adequately eliminated in design?
2   Are there adequate barriers (guarding, etc.) to protect against hazards?
3   Are areas for operation, inspection and maintenance adequately located outside the danger zone?
4   Is there adequate access for operation, inspection and maintenance?
5   Is there adequate access to production and emergency-stop functions, and for isolation of the machinery from the energy sources?
6   Are work postures and methods of work adequate to avoid the risk of strain injuries?
7   Is the process enclosure of adequate quality to ensure containment of hazardous substances within the process?

2.5.5 Man–machine interfaces

1 Have controls and displays been designed in accordance with recognised ergonomic design standards?
2 Have controls and displays been located for adequate view and access in a normal working posture?
3 Does the operator have access to the necessary process parameters to be able to watch process safety performance adequately?
4 Are there adequate provisions to avoid information overload such as alarm filtration?
5 Have displays and controls been arranged in a way that is consistent with both normal and emergency response tasks?
6 Are there adequate provisions for error recovery? Are the consequences of actions transparent, are they traceable and are errors reversible?
7 Are there adequate provisions for communication with other operators and supervisors?

2.5.6 Availability of machinery and equipment

1 Is the reliability of the machinery adequate?
2 Are machinery, equipment and lifting aids available in a sufficient number?
3 Is the access to containers for waste disposal adequate?

2.5.7 Documentation

1 Are the routines for storage, retrieval and updating of documentation on machinery and equipment (drawings, instructions, certificates, inspection protocols, etc.) adequate?
2 Are the hazards of the machinery during operation and maintenance adequately documented in manuals and work instructions?
3 Is the technical documentation adequate and kept up to date?
4 Is the documentation available to those who need it in language that they understand?

2.6 Maintenance of fixed assets

2.6.1 Human resources

1 Are there adequate human resources (manning, qualifications) for inspection, maintenance and scaffolding (see further item 2.1)?

2.6.2 Technical resources

1 Is the access to tools and equipment for inspection, maintenance, material handling (lifting) and scaffolding adequate?
2 Are spare parts adequately available?

2.6.3   *Routines, procedures*

1   Has a maintenance philosophy been established that defines the balance between inspection, preventive maintenance and repair?
2   Is it based on experiences with each type of equipment? On the criticality of the equipment from a safety point of view?
3   Are there adequate instructions and checklists for inspection and maintenance?
4   Is the inspection and testing frequency of safety barriers adequate?
5   Are the routines for inspection and testing of barriers adequate? Do they ensure that the total barrier efficiency is tested?
6   Are there adequate routines for evaluation of maintenance work from a SHE point of view before start of work?
7   Are there adequate routines to prioritise between different maintenance tasks?
8   Is the quality of the maintenance work ensured?

2.6.4   *Co-ordination with operations*

1   Are there adequate routines for the reporting of maintenance and repair needs?
2   Is the division of responsibility between operation and maintenance clear?
3   Are there adequate routines to co-ordinate operation and maintenance tasks to avoid conflicts? Are there adequate communication links between operation and maintenance personnel?
4   Is it ensured that the necessary safety precautions are taken before start of maintenance? Is it ensured that the equipment is secured and free from hazards? That the area is roped off?

2.6.5   *Permit-to-work system*

1   Have the needs of a formal work-permit system been adequately evaluated, considering the hazards in the process?
2   Do the routines for work permits ensure:
   a   Adequate definition of the types of work for which a permit is required?
   b   Responsibilities for application and approval?
   c   When and where the work permit is valid?
   d   Identification and implementation of the necessary safety precautions before the equipment is released for maintenance?
3   That the work is performed at the right place?
4   That the equipment is brought back to an adequate operational status before start-up?
5   Overview of active work permits by person responsible for operations?

2.6.6 *Modifications*

1 Have routines for all phases of the modification work been established?
2 Have responsibilities for SHE been adequately defined?
3 Have the SHE requirements as to design been adequately defined?
4 Do the routines ensure an adequate and timely definition of the safety scope of the modification work? Does it ensure that the area affected by the modification is brought up to today's SHE standard?
5 Have plans for risk analyses and SHE design reviews been adequately established?
   a Do they cover all relevant hazards?
   b Are the activities adequately scheduled in relation to the progress of design work?
   c Is access to competent personnel including operations personnel ensured?
   d Are the results adequately documented and followed up?
6 Have routines been established to verify that the modifications meet all relevant requirements?

2.6.7 *Documentation*

1 Are inspection, repair and maintenance activities adequately documented? Does an overview of the status of the equipment exist that identifies deviations from an acceptable technical standard?
2 Are they adequately used as a basis in determining equipment status and in the planning of maintenance activities?
3 Are modifications of machinery and equipment adequately evaluated from a SHE point of view and documented?

2.7 **Material supply**

2.7.1 *Composition*

1 Have the chemical hazards in connection with the handling of materials (raw materials, materials in the process and end product) been adequately evaluated and controlled?
2 Is the weight and shape of the materials adequate for safe handling?
3 Is the material adequately wrapped up?

2.7.2 *Material transportation, storage*

1 Are the provisions for material handling and storage (buildings, equipment, routines) adequate? Is manual handling avoided when this represents a hazard?
2 Is the access to materials adequate to avoid production disturbances?

2.7.3   *Waste handling*

1   Are the routines and facilities for waste removal and storage adequate?
2   Is waste disposed of in an adequate way?

2.7.4   *Documentation*

1   Are all products containing hazardous chemicals documented in approved safety data sheets? Are these available to the workers?
2   Are all materials and products adequately marked?

2.7.5   *Quality control*

1   Is the quality of the materials adequate to avoid production disturbances?
2   Are there adequate routines to control and document the quality of the materials in relation to the specifications?

Level 3   Project management

3.1   SHE culture and values

3.1.1   *Project management commitment*

1   Do the project managers demonstrate an adequate concern regarding SHE issues?
2   Does project management take an active part in relevant SHE activities?
3   Does project management demonstrate a proactive attitude by addressing potential SHE problem areas and making the necessary decisions in due time?
4   Does project management take the necessary actions when sub-standard SHE conditions are brought to their attention?
5   Is adequate priority given to SHE in relation to other project goals (budget, schedule, etc.)? Also during cost-cutting exercises?

3.1.2   *Communication on SHE*

1   Have adequate routines been established to ensure communication within the project on SHE issues?
2   Do the routines ensure adequate communication to all project members of the project's SHE policy, goals and requirements and the client's expectations?

3.2 SHE management

3.2.1 Organisation and responsibilities

1 Have the responsibilities for the management, execution and verification of work affecting SHE been adequately defined?
2 Does it include responsibilities for the identification, implementation and verification of requirements to SHE?
3 Has the distribution of responsibilities between the project and the client (operations) been adequately defined?
4 Has the distribution of responsibilities between the project and suppliers (contractors and vendors) been adequately defined?
5 Have the responsibilities for review and acceptance of changes and nonconformities been adequately defined?
6 Have the responsibilities been adequately documented in organisational charts and job descriptions?

3.2.2 Project's SHE programme

1 Is it ensured that the project's SHE programme is established at an early phase?
2 Does the programme adequately define:
  a The project's SHE objectives, goals and acceptance criteria?
  b Governing documents relating to SHE?
  c Responsibilities for the implementation and follow-up of SHE?
  d Plans for control and verification activities?
3 Does the programme adequately cover:
  a Major accident risks, working environment and environmental care in design?
  b SHE during fabrication and construction?
4 Do the routines ensure a timely update of the SHE programme in new project phases and when the conditions change?
5 Do the routines ensure an adequate verification of the SHE programme by qualified personnel, project management and the client?

3.2.3 Suppliers' SHE programmes

1 Are requirements as to a documented SHE management system adequately defined in contracts and purchase orders?
2 Do the requirements adequately cover aspects relating to SHE in design and fabrication/construction?
3 Do the requirements also apply to contractors' subcontractors?
4 Are the routines adequate to ensure a proper evaluation of the programme before start of design and fabrication/construction activities?

3.2.4   *Experience transfer*

1   Are the routines to ensure experience transfer relating to SHE from operations and other companies to the project adequate?

2   Does the experience transfer to the project include adequate information about accidents, incidents and production disturbances in similar types of production?

3   Does the experience transfer include information about previous SHE analyses and evaluations of similar plant concepts?

4   Does the experience transfer include adequate information about occupational health problems in similar types of production?

5   Does the experience transfer include adequate information about environmental pollution in similar types of production?

6   Is experience transfer adequately ensured through participation of personnel with operations experience in design work?

7   Does the experience transfer from the project to operations ensure adequate information on requirements to operational instructions and procedures to avoid hazards?

3.2.5   *Regulations, standards and specifications*

1   Have applicable regulatory requirements been identified, evaluated and implemented in the project's governing documents?

2   Have adequate and verifiable goals concerning SHE in design been established?

3   Have adequate and verifiable acceptance criteria for the risk of accidents with effects on personnel and environment been established?

4   Have adequate SHE design specifications been established?

5   Do the specifications cover all relevant SHE aspects concerning:

  a   The prevention of major accidents? Are they based on adequate intrinsic safety and defence-in-depth philosophies?

  b   The working environment, including requirements as to the prevention of occupational accidents and strain injuries, industrial hygiene, man–machine interfaces and psycho-social working environment?

  c   Environmental care including prevention of emissions to the air, discharges to the sea, contamination of land, waste handling, energy conservation and recycling?

6   Do the specifications adequately cover operation, maintenance and transportation?

7   Have the client's (/operation's) requirements and expectations in relatia to SHE been adequately implemented in the specifications?

8   Is it ensured that the SHE requirements are unambiguous? That the order of priority between SHE requirements has been adequately defined?

9   Are all relevant requirements available and made known to the personnel with responsibility for implementation and verification?

3.2.6 Analyses and verifications

1 Have the requirements and responsibilities as to analyses and verifications of SHE been adequately specified?
2 Do the activity plans ensure adequate analyses and verifications during concept selection and definition, engineering, procurement, construction and commissioning in relation to all relevant SHE requirements?
3 Have the work processes for analysis and verification been adequately documented?
4 Do the plans ensure that the analyses and verifications are performed in due time, with the necessary design documentation available, by competent personnel and with an acceptable quality?
5 Is it ensured that the analyses and verifications adequately cover:
   a Risk analyses of accident scenarios with the potential of major losses for comparison with the project's acceptance criteria?
   b Working-environment aspects including analyses of the risk of occupational accidents and strain injuries, man–machine interface evaluations, predictions of air quality, noise, vibration and illumination levels and evaluation of psycho-social conditions?
   c Environmental-impact assessment including analysis of the risk of accidental pollution and predictions of emissions to the air, discharges to the sea and land contamination, energy analysis, waste-handling evaluations and product-life-cycle analyses?
   d Emergency-preparedness analysis?
   e Analysis of design with respect to risk of accidents during construction (constructability analysis)?
6 Is it ensured that the assumptions and findings from the analyses and verification activities are adequately documented and followed up?

3.2.7 Audits and management reviews

1 Are the plans for audits and management reviews adequate to ensure that the project's and suppliers' SHE management systems satisfy the client's and project requirements and expectations and that they are adequately implemented?
2 Are audits and management reviews performed at adequately regular intervals?
3 Are the audits and management reviews adequately comprehensive to cover critical aspects of the SHE management systems?
4 Are the audits and management reviews conducted by personnel with adequate competence?

3.2.8   *Nonconformity handling*

1   Has a procedure for the handling of nonconformities been established? Does it ensure that nonconformities relating to SHE are adequately identified, recorded and reviewed?
2   Is it ensured that nonconformities relating to SHE are approved at the right level of the organisation and that adequate compensatory measures are taken?
3   Is it ensured that design not conforming to regulatory requirements is adequately evaluated and approved?

3.3   *Resource management*

3.3.1   *Human resources*

1   Is it ensured that positions of significance to SHE are identified and staffed with personnel with adequate qualifications?
2   Is it ensured that the project members have adequate knowledge of their own responsibilities for the implementation and follow up of SHE?
3   Is it ensured that the project members have adequate knowledge of the SHE requirements within their area of responsibility?
4   Is the project organisation adequately manned with personnel with expert knowledge of SHE?

3.3.2   *Budget*

1   Is it ensured that the control estimate is accurate enough to avoid cost overrun and subsequent cost-cuts? Has the SHE scope been adequately implemented in the estimate?
2   Is the budget adequate to ensure the procurement of buildings and machinery of an adequate SHE standard?
3   Are adequate resources allocated to the amelioration of SHE problems identified during project work?
4   Does the man–hour budget ensure adequate access to the necessary SHE expertise?
5   Have adequate working hours been allocated to SHE-related activities?

3.3.3   *Schedule*

1   Has the schedule been subdivided into phases and milestones with adequate goals concerning SHE activities and documentation?
2   Does the schedule enable the project to generate sufficient information about design and to make the right decisions related to SHE as early as feasible with minimal expenditure of time and money?

3 Is it ensured that critical decisions related to SHE are not postponed until later phases when a proper solution will have major cost or schedule impact?
4 Is the schedule realistic? Does it allow for adequate evaluations of SHE and implementation of results?
5 Is the schedule adequately flexible to absorb possible delays without undue SHE effects?

3.4 Relation to stakeholders

3.4.1 Co-ordination of stakeholders

1 Is it ensured that the project's stakeholders have been adequately identified, including:
   a The client?
   b Business parties?
   c The authorities?
   d Employees and their organisations?
   e Contractors and suppliers?
   f External interest groups?
2 Is it ensured that the stakeholders' needs and expectations relating to SHE are adequately identified and evaluated? That potential conflicts are adequately addressed?

3.4.2 Liaison with Operations

1 Is it ensured that the co-ordination between the project and Operations is adequate?
2 Does this co-ordination ensure that the project has adequate access to personnel with adequate operations and maintenance experience of significance to SHE?
3 Does this co-ordination ensure that operational experience related to SHE is adequately considered:
   a In the definition of scope?
   b In choosing technical solutions?
   c In the handling of changes and nonconformities?
   d In developing documentation for operation and maintenance?
4 Is it checked that the permit-to-work system is adequate for the needs during the tie-in of new systems to the existing plant and during commissioning and start-up of new systems?
5 Is it ensured that the take-over of the plant by Operations functions smoothly and with an acceptable level of safety?

3.4.3   Contact with the authorities

1   Have the responsibilities for contacts with the authorities been adequately defined? Is the responsibility-split between the client and the project clear?

2   Is it ensured that the required document deliveries to the authorities are identified and adequately scheduled?

3   Are there adequate routines to ensure timely delivery to the authorities of the required documents? The delivery of documentation of acceptable quality? Of documentation that is adequately co-ordinated within the company?

4   Is it ensured that the authorities are kept adequately updated on the progress of the project through status meetings, inspections, document reviews, etc.?

3.5   Study phases

3.5.1   Concept exploration and definition

1   Is it ensured that the main SHE issues relating to the concept for the plant are adequately identified and addressed?

2   Is it ensured that SHE aspects relating to plant location are adequately evaluated such as:

    a   Security zones?

    b   Distance to residential areas, other plants and transportation routes?

    c   Transportation of goods?

    d   Access to air and water?

    e   Cultural heritage?

    f   Environmental baseline data and proximity to sensitive environmental resources?

3   Is it ensured that the advantages and disadvantages of the different concept-development alternatives are adequately evaluated based on SHE criteria?

4   That SHE characteristics disqualifying any of the alternatives are adequately identified and evaluated?

5   Is it ensured that technological uncertainties relating to SHE are adequately identified and evaluated? That back-up solutions are provided if the new technologies are found unfeasible?

3.5.2   Demonstration and validation

1   Is it ensured that the evaluation of the economic and technical feasibility of the selected plant concept takes SHE acceptance criteria and requirements adequately into account?

2   Is it ensured that basic decisions regarding safety systems are addressed, such as:

   a Segregation into fire cells?
   b Plant shutdown philosophy?
   c Facilities for emergency response such as fire-fighting?
   d Escape?

3 Is it ensured that basic decisions regarding the working environment are addressed, such as:
   a Layout promoting safe and efficient operation, maintenance and transportation?
   b An acceptable working atmosphere (climate, pollution)?
   c Storage and handling of hazardous substances?
   d Location of major noise and vibration sources?
   e Solitary work?

4 Is it ensured that basic decisions regarding environmental care are addressed, such as:
   a Prevention of accidental pollution of the environment?
   b Control of emissions to the air, discharges to water recipients and contamination of land?
   c Waste disposal?
   d Sampling and monitoring of effluent streams and exhaust gases?

5 Is it ensured that the plant design has been defined and documented in adequate detail concerning SHE issues before making decisions to go ahead with realisation?

3.5.3   Definition of scope, change control

1 Is it ensured that the scope of the project work is adequately defined and documented before deciding on realisation and entering into agreements with external parties (contractors, vendors)?

2 Is it ensured that the SHE-related parts of the scope are adequately defined? Does the scope include an upgrading of existing facilities to today's SHE standard?

3 Has a procedure been established that defines the routines and responsibilities for the handling of changes in scope?

4 Does this ensure an identification and evaluation of changes that are critical with respect to SHE? Is the authority to approve changes with safety impact placed at the right level of the organisation (project/client)?

3.6   Project execution

3.6.1   Engineering

1 Are the plans for identification, documentation and review of design input (statutory requirements, client requirements, etc.) adequate?

2 Is it ensured that the design concept has been adequately defined and qualified from a SHE point of view before placement of major contracts and purchase orders?

3   Are the routines for implementation of SHE requirements in contracts and purchase orders adequate?
4   Are the plans for documentation of design output for verification and validation adequate?
5   Are the plans for design reviews adequate? Do they ensure reviews at regular intervals and with the participation of competent personnel?
6   Are the routines for control of design changes adequate to ensure proper identification, review and approval by authorised personnel?
7   Are the routines for documentation and follow-up of actions adequate?
8   Are the routines for document control adequate? Do they ensure that the documents are traceable and that they are complete, updated and accessible?
9   Are there adequate routines for documentation of design changes, nonconformities and ameliorative actions?

3.6.2   *Procurement*

1   Is it ensured that the suppliers (contractors and vendors) are evaluated and selected in pre-qualifications and bid evaluations on the basis of their ability to meet the project's SHE goals, requirements and acceptance criteria related to:
   a   SHE in design?
   b   SHE in fabrication and construction?
2   Is it ensured that the evaluation includes the tenders':
   a   Organisational resources and competence?
   b   Technical resources?
   c   Planned activities to meet the requirements?
   d   Documented procedures for control and verification?
   e   Experience data on the tenders' previous performance?
3   Is it ensured that the technical evaluation of the proposed design includes a qualification with respect to the project's SHE goals, acceptance criteria and requirements?
4   Is it ensured that the tenders provide the necessary documentation (risk analyses, verifications) of the feasibility of the proposed design?
5   Is it ensured that the contract adequately defines:
   a   The SHE goals, acceptance criteria and requirements?
   b   The scope of work related to SHE including activities to analyse, verify and document compliance with the contractual requirements to SHE?
   c   The required delivery of documents including:
      i   Suppliers' and sub-suppliers' SHE programmes for the project?
      ii   Reports on analyses and verifications?
      iii   Reports on SHE activities and results in fabrication and construction, including accident reports?
   d   The responsibilities for SHE including the split of responsibility between subcontractors and vendors?

6 Has the extent of control exercised over the suppliers been determined on the basis of the results of pre-qualifications and bid evaluations?

7 Is it ensured that the suppliers adequately understand the scope of work and SHE requirements and work according to these?

8 Is it ensured that the product's compliance with the contractual SHE requirements is adequately verified before delivery? Is the verification documented?

3.6.3 Mechanical completion and commissioning

1 Is it ensured that equipment and plant areas are adequately inspected to verify compliance with the contractual requirements to SHE, design specifications, drawings and other engineering documents?

2 Is it ensured that the dynamic testing adequately verifies compliance with the contractual requirements as to SHE, design specifications, drawings and other engineering documents?

3 Is it ensured that the findings from inspections and testing are adequately evaluated and followed up?

4 Is it ensured that the inspection and testing results during mechanical completion and commissioning and follow-up of nonconformities and deficiencies are adequately documented?

3.6.4 Follow-up of SHE in fabrication and construction

1 Is it ensured that the suppliers have established adequate SHE programmes for fabrication and construction before start-up of work?

2 Are suppliers adequately followed up through monitoring of SHE performance?

3 Are the routines for reviews and audits of suppliers' SHE activities adequate?

4 Are SHE issues in fabrication and construction adequately brought up in management meetings with suppliers?

5 Are the routines adequate to implement corrective measures when suppliers' performance is sub-standard or deteriorates?

3.7 Take-over by operations

3.7.1 Recruitment and training of operations personnel

1 Have the requirements as to knowledge and skills been adequately defined for the different positions in the operations organisation?

2 Is the recruitment of personnel adequate, and does it take place in due time to allow for satisfactory education and training?

3 Does the recruitment include experienced personnel?

4 Are personnel recruited in due time to participate in commissioning and start-up?

5 Has the training programme adequate contents and schedule to ensure that the personnel meet the requirements as to knowledge and skills?

3.7.2 Development of procedures and work instructions

1 Are the routines to ensure timely development of work and safety instruction of acceptable quality adequate?
2 Do the instructions adequately implement suppliers' instructions, experience from evaluations in design and commissioning and experience from the operation of similar facilities?
3 Are the routines to ensure evaluations of the instructions from a SHE and functionality point of view adequate?

3.7.3 Start-up

1 Are the SHE criteria for start-up adequate? Do they adequately cover:
   a Requirements as to the completeness of technical systems?
   b The availability of safety systems?
   c Requirements as to the availability and qualifications of personnel?
   d Requirements as to the availability and quality of procedures?
2 Is it ensured that the status of the plant is adequately evaluated and documented before the introduction of hazardous substances into the process?
3 Are any nonconformities and changes adequately evaluated and accepted by persons with satisfactory competence and authority before start-up?
4 Is it adequately ensured that the drawings and other design documentation with as-built status are adequately updated, verified and approved before hand-over to Operations?

Level 4 Higher management and management of SHE

4.1 SHE culture and values

4.1.1 Leadership and commitment

1 Does company management promote SHE by demonstrating adequate leadership and commitment?
2 Is adequate priority given to SHE in relation to other company goals?
3 Does management show adequate acceptance of its own responsibility and accountability for SHE activities, performance and results?
4 Does management take the necessary actions when sub-standard conditions and work practices are brought to its attention?
5 Does management demonstrate a concern by actively querying possible new hazards?

4.1.2   Adherence to safety rules

1   Has adequate company policy been established to ensure compliance at all levels with the SHE policy, procedures and rules?
2   Has the policy been adequately implemented and followed up?
3   Does management adequately promote safe behaviour through its own vigour and example?

4.1.3   Communication on SHE

1   Have adequate routines been established to ensure communication within the company on SHE issues?
2   Do the routines ensure an adequate communication of company SHE policy, goals and rules as well as hazards to the employees?
3   Do the routines ensure adequate communication on the employees' SHE experiences and concerns?

4.1.4   Employee involvement

1   Have adequate routines been established to promote employee involvement in SHE activities at all levels of the company?
2   Does management promote employee involvement at all levels in SHE related activities?
3   Do the routines ensure adequate use of employees' experience and their ownership of the results of the activities?

4.2   Policy, goals and action plans

4.2.1   Policy

1   Does a documented company SHE policy exist? Is it consistent with the company's other policies?
2   Does it include a management commitment to meet all relevant statutory requirements?
3   Does it commit management to continuous efforts to improve SHE performance and to reduce risks as low as reasonably practicable?

4.2.2   Goals and acceptance criteria

1   Has the company's SHE policy been translated into verifiable SHE goals and acceptance criteria for the risk of losses due to accidents?
2   Do the goals and acceptance criteria cover all relevant aspects of SHE?
3   Is it ensured that goals are established for all departments of the company?
4   Is it ensured that the line management and employees are involved in the development of goals and acceptance criteria?

5　Do the goals ensure continuous improvements in SHE performance?
6　Are the goals regularly reviewed and updated in line with new legislation and other changes?

4.2.3　Action plans

1　Are there adequate routines to ensure the development of action plans at all levels of the company to implement the SHE policy and goals?
2　Are adequate resources allocated to the execution of the plans?
3　Is progress adequately monitored and followed up?

4.3　Resource management

4.3.1　Responsibilities

1　Have the responsibilities for SHE been clearly defined at all levels of the company?
2　Does adequate authority and accountability accompany these responsibilities?
3　Is the division of responsibilities between line management, SHE staff officers and safety representatives of the workers on SHE made clear? Is it clearly stated that SHE is a line-management responsibility?
4　Is it ensured that the line management knows about and accepts its responsibilities?

4.3.2　Human resources

1　Have adequate resources been allocated to human-resource management including the education and training of personnel?
2　Have personnel with adequate SHE competence and management skills been allocated to the SHE management and specialist positions?
3　Does management have access to adequate SHE expertise for advice and support in the handling of hazards connected with production?
4　Have sufficient man-hours been allocated to SHE activities?
5　Have sufficient resources been allocated to the follow-up and implementation of results of SHE activities?
6　Do the resources ensure adequate expert support in the identification and evaluation of accident risks, SHE education and training, advice on SHE regulations and on solutions to SHE problems, contacts with the authorities, etc.?
7　Have adequate resources been allocated to support the line management?
8　Have adequate resources been established for the development and maintenance of procedures and work instructions?
9　Do they ensure that procedures and instructions are complete and kept updated?

10 Are measures to avoid accidents adequately identified and defined?
11 Do the instructions represent best practice?
12 Do the employees know about, accept and comply with the instructions?

4.3.3 *Fixed assets and materials*

1 Have adequate routines been established in the area of fixed-asset management? Do they ensure that building, machinery and equipment have an adequate SHE standard?
2 Are the routines adequate for the procurement of materials, equipment and contract work?
   a Do the routines ensure that suppliers with sub-standard SHE performance are avoided?
   b Do the routines ensure that SHE requirements to the supply are defined in the contract and that the supply meets the requirements?

4.3.4 *SHE organisation*

1 Have safety representatives of the workers been elected in all departments?
2 Have the safety representatives been given adequate duties and status?
3 Have working-environment committees been established?
4 Do the working environment committees function adequately in accordance with the intentions and requirements? Does management take an active part in the work?
5 Does the SHE organisation ensure adequate employee involvement and influence on decisions affecting SHE?

4.4 *Identification and evaluation of risks*

4.4.1 *Routines*

1 Are the routines for systematic identification and evaluation of risks to personnel, environment and assets adequate?
2 Do they cover all phases of the life cycle of industrial systems?
3 Do the routines ensure an adequate distribution, follow-up and implementation of results?
4 Do the routines ensure the implementation of corrective actions to prevent recurrence?
5 Do the routines ensure adequate experience transfer between departments and to projects for new plants and modifications?
6 Do the routines ensure involvement by the personnel concerned?
7 Are the routines adequately adapted to the different needs within the company?

8   Are the routines kept updated in line with changes in production, organisation, legislation, etc.?
9   Are the routines adequately documented?

4.4.2   *Data collection*

1   Do the routines ensure a systematic identification of all different types of hazard?
2   Do the routines ensure prompt detection of new hazards?
3   Do the routines ensure adequate reporting and investigation into accidents and near accidents?
4   Are incidents relating to risks of personal injury, fires and explosions and environmental releases reported?
5   Are the investigation resources adequately prioritised in relation to the degree of severity (actual/potential)? Have routines been established for investigations at different levels (e.g. supervisor's first investigation, problem-solving groups, and independent investigation commission)?
6   Do the routines ensure regular workplace inspections of a satisfactory quality?
7   Do the routines ensure an adequate identification of new hazards by means of risk analysis?
8   Are the routines for monitoring of the physical working environment adequate?
9   Are the routines for health checks of personnel adequate?
10 Is the quality of the data adequately checked?

4.4.3   *Storage, analysis, distribution and use of information*

1   Are the results analysed and summarised in an adequate way for decision-making?
2   Are the incidents evaluated concerning potential for severe loss as a basis for prioritising?
3   Are the results compared to established goals and acceptance criteria and are gaps used adequately as a basis for decisions?
4   Is the information stored in an adequate way for information retrieval and experience transfer?
5   Is it possible for decision-makers to access the information when it is needed?
6   Are the routines for a periodic summary and presentation of the information adequate?
7   Is relevant information adequately distributed to the line management, SHE specialists, working environment committees, safety representatives?
8   Is it ensured that management and the SHE organisation gets the necessary information to evaluate the hazards at the workplaces and to take necessary preventive measures?

9 Is it possible to get an overview of all identified hazards and non-conformities that have not been resolved and the status of actions?

4.5 Handling of governing SHE documents

4.5.1 Routines

1 Are the routines for the handling of governing documents in the area of SHE adequate?
2 Do the routines ensure that the documents are complete and available to the personnel needing them?
3 Is it ensured that the documentation is kept updated in line with changes in legislation, organisation or production?

4.5.2 Regulations, codes and standards

1 Are all relevant regulatory requirements, codes and standards available at the company?
2 Are changes adequately identified and implemented?

4.5.3 SHE management programme

1 Have the different elements of the SHE management programme been adequately documented?
2 Does the documentation cover established policy, goals and acceptance criteria, responsibilities and activities?
3 Does the documentation cover operation and project work?
4 Is the programme regularly reviewed and updated?

4.5.4 Procedures

1 Has the need for the documentation of SHE activities in procedures been adequately defined?
2 Do the procedures adequately cover daily operation, modifications and building of new plants?
3 Do the procedures adequately cover the procurement of new materials and equipment and contract work?
4 Is it ensured that the procedures reflect best practice?
5 Is it ensured that the procedures are stated simply, unambiguously and understandably?

4.6   *Performance monitoring and auditing*

4.6.1   *Performance indicators*

1   Have adequate performance indicators been established?
2   Do the indicators adequately cover the different areas of SHE?
3   Do the indicators adequately cover operation and project work?
4   Are the indicators adequately congruent with the SHE policy and goals?
5   Are the indicators adequately accepted within the organisation as fair measures of performance?

4.6.2   *Follow-up of results*

1   Are the performance indicators used adequately in performance monitoring and follow-up of SHE results?
2   Are adequate actions taken when performance indicators show substandard development?

4.6.3   *SHE auditing and management reviews*

1   Has an adequate plan been established for SHE audits and management reviews?
2   Has the plan been adequately implemented?
3   Does the plan ensure a systematic and independent examination of the SHE management system and activities?

Bibliography

AAAM, 1985. *Abbreviated Injury Scale*. American Association for Automotive Medicine, Committee on Injury Scale, Arlington Heights.

Aaltonen, M.V.P., Uusi-Rauva, E., Saari, J., Antti-Poika, M., Räsänen, T. and Vinni, K., 1996. The accident consequence tree and its application by real-time data collection in the Finnish furniture industry. *Safety Science*, 23: 11–26.

ACSNI, 1993. *Organising for Safety*. Advisory Committee on the Safety of Nuclear Installations, HSE Books, Suffolk.

Adams, N., Barlow, A. and Hiddlestone, J., 1981. Obtaining ergonomics information about industrial injuries – a five-year analysis. *Applied Ergonomics*, 12(2): 71–81.

Alteren, B., 1999. Implementation and evaluation of the Safety Element Method at four mining sites. *Safety Science*, 31: 231–264.

Andersson, R. and Lagerlöf, E., 1983. Accident data in the new Swedish information system on occupational injuries. *Ergonomics*, 26: 33–42.

Arbeidstilsynet, 1996. *Regulations relating to systematic health, environmental and safety activities in enterprises*. Regulations No. 1127, Oslo.

Arbetarskyddsstyrelsen, 1996. *Internkontroll av arbetsmiljön. Arbetarskyddsstyrelsen författningssamling*, Regulations No. AFS1996: 06, Stockholm.

Argyris, C., 1992. *On Organizational Learning*. Blackwell Publishers Inc., Cambridge, Massachusetts.

Avedal, C., 1998. *Accidents Involving Commercial Vehicles in Sweden*. Volvo Truck Corporation, Gothenburg.

Benner, L., 1975. Accident investigations. Multilinear events sequencing methods. *Journal of Safety Research*, 7(2): 67–73.

Bird, F.E. and Germain, G.L., 1985. *Practical Loss Control Leadership*. Institute Publishing, Division of International Loss Control Institute, Loganville, Georgia.

Blindheim, G. and Lindtvedt, J., 1996. *Erfaringsoverföring i sikkerhetsarbeidet ved hjelp av databasen Synergi*. Norwegian University of Science and Technology, Project report, Trondheim.

Boe, K., 1996. Evaluering av HMS-styring i byggeprosjekter. Norwegian University of Science and Technology, Masters thesis, Trondheim.

Bolman, L.G. and Deal, T.E., 1984. *Modern Approaches to Understanding and Managing Organizations*. Jossey-Bass, San Fransisco.

Booth, M.J., 1991. The incident potential matrix. Paper presented at the Society of Petroleum Engineers' First International Conference on Health, Safety and Environment, The Hague, 10–14 November 1991.

Briscoe, G.J., 1982. *Risk Management Guide*. Systems Safety Development Center, EG&G Idaho, Inc., Report No. SSDC-11, Idaho Falls, Idaho.

Briscoe, G.J., 1991. MORT-based risk management. Systems Safety Development Center, EG&G Idaho, Inc., Working paper No. 28, Idaho Falls, Idaho.

British Standards Institution, 1996. *Guide to Occupational Health and Safety Management Systems*. British Standard BS 8800:1996, London.

British Standards Institution, 1999. *Occupational Health and Safety Management Systems – Specification*. OHSAS 18001:1999, London.

Brown, R.L. and Holmes, H., 1986. The use of a factor-analysis procedure for assessing the validity of an employee safety climate model. *Accident Analysis and Prevention*, 18: 455–470.

Carter, N. and Menckel, E., 1985. Near-accident reporting: A review of Swedish research. *Journal of Occupational Accidents*, 7: 61–64.

CEN, 1991. *Safety of Machinery – Basic Concepts, General Principles for Design – Part 1: Basic Terminology, Methodology*. European Standard EN 292-1:1991, Brussels.

CEN, 1996. *Safety of Machinery – Principles for Risk Assessment*. European Standard EN 1050: 1996, Brussels.

Cohen, A., Smith, M. and Cohen, H., 1975. *Safety Program Practices in High vs. Low Accident Rate Companies – an interim report*. National Institute for Occupational Safety and Health, Cincinnati.

Cornelison, J.D., 1989. MORT based root cause analysis. Systems Safety Development Center, EG&G Idaho, Inc., Working paper No. 27, Idaho Falls, Idaho.

Cox, S. and Flin, R., 1998. Safety culture: philosopher's stone or man of straw? *Work & Stress*, 12: 189–201.

Dedobbeleer, N. and Béland, F., 1991. A safety climate measure for construction sites. *Journal of Safety Research*, 22: 97–103.

DeJoy, D.M., 1994. Managing safety in the workplace: An attribution theory analysis and model. *Journal of Safety Research*, 25: 3–17.

Det Norske Veritas, 1994. *ISRS Revisjonsprotokoll*. Høvik.

Döös. M., 1997. Den kvalificerande erfarenheten. Lärande vid störningar i automatiserad produktion. Thesis, Arbete och hälsa, 1997: 10.

E&P Forum, 1994. *Guidelines for the Development and Application of Health, Safety and Environmental Management Systems*. Report No. 6.36/210, London.

Eisner, H.S. and Leger, J.P., 1988. The international safety rating system in South African mining. *Journal of Occupational Accidents*, 10: 141–160.

Elmasri, R. and Navathe, S.B., 1994. *Fundamentals of Database Systems (2nd ed.)*. The Benjamin/Cummings Publishing Company Inc., Redwood City, CA.

Englund, A., Gregersen, N.P., Hydén, C., Lövsund, P. and Åberg, L., 1998. *Trafiksäkerhet – En kunskapsöversikt*. Studentlitteratur, Lund.

EPA, 1990. *The Clean Air Act Amendments of 1990*. U.S. Environmental Protection Agency, Washington, DC.

European Council, 1989. *The Introduction of Measures to Encourage Improvements in the Safety and Health of Workers at Work*. Council Directive 89/391/EEC, Brussels.

European Council, 1989/98. *Machinery*. Council Directives 89/392/EEC, 91/68/EEC and 98/37/EC, Brussels.

European Council, 1993. *The EC Eco-management and Audit Scheme*. EC Regulation 1836/93, Brussels.

European Council, 1996. *The Control of Major-Accident Hazards Involving Dangerous Substances*. Council Directive 96/82/EC, Brussels.

European Council, 1998. *The Protection of the Health and Safety of Workers from the Risks Related to Chemical Agents at Work*. Council Directive 98/24/EC, Brussels.

Eurostat, 1998. *European Statistics on Accidents at Work – Specification for Case-by-Case Data*. Eurostat E-3, Luxembourg.

FERSI, 1994. *In-Depth Investigation*. Forum of European Road Safety Research Institutes (FERSI), Lyon.

Feyer, A.-M. and Williamson, A. (eds), 1998. *Occupational Injury: Risk, Prevention and Intervention*. Taylor & Francis, London.

Folkard, S., 1997. Black times: temporal determinants of transport safety. *Accident Analysis and Prevention*, 29: 417–430.

Gibson, J., 1961. The contribution of experimental psychology to the formulation of the problem of safety. In: *Behavioral Approaches to Accident Research*. Association for the Aid of Crippled Children, New York.

Grimaldi, J.V., 1970. The measurement of safety engineering performance. *Journal of Safety Research*, 2: 137–159.

Grimaldi, J.V. and Simonds, R.H., 1975. *Safety Management*. Richard D. Irwing, Homewood, Illinois.

Guastello, S.J., 1993. Do we really know how well our occupational accident prevention programs work? *Safety Science*, 16: 445–464.

Haddon, W., 1968. The changing approach to epidemiology, prevention and amelioration of trauma. *American Journal of Public Health*, 58(8): 1431–1438.

Haddon, W., 1980. The basic strategies for reducing damage from hazards of all kinds. *Hazard Prevention*, September/October: 8–12.

Hale, A.R. and Glendon, A.I., 1987. *Individual Behaviour in the Control of Danger*. Elsevier, Amsterdam.

Hale, A., Wilpert, B. and Freitag, M. (eds), 1997. *After the Event – From Accident to Organisational Learning*. Elsevier Science Ltd, Oxford.

Harms-Ringdahl, L., 1993. *Risk Analysis – Principles and Practice in Occupational Safety*. Elsevier, London.

Harrison, F.E., 1987. *The Managerial Decision-Making Process*. (3rd ed.). Houghton Mifflin, Boston.

Hatakka, M., Keskinen, E., Katila, A. and Laapotti, S., 1997. Self-reported driving habits are valid predictors of violations and accidents. In: Rothengatter, T. and Vaya, E.C. (eds), *Traffic and Transport Psychology*. Pergamon, Amsterdam.

Heinrich, H.W., 1959. *Industrial Accident Prevention – A Scientific Approach (4th ed.)*. McGraw-Hill, New York.

Hendrick, K. and Benner, L., 1987. *Investigating Accidents with STEP*. Marcel Dekker, Inc., New York.

Hockey, G.R.J. and Maule, A.J., 1995. Unscheduled manual intervention in automated process control. *Ergonomics*, 38: 2504–2524.

Hovden, J. and Larsson, T.J., 1987. Risk: culture and concepts. In: Singleton, W.T. and Hovden, J. (eds), *Risk and Decisions*. Wiley, New York.

HSE, 1996. *A Guide to the Reporting of Injuries, Diseases and Dangerous Occurrences Regulations 1995*. HSE Books, Guidance No. L73, Her Majesty's Stationery Office, Sheffield.

HSE, 1997a. *Successful Health and Safety Management. HSE Books, Guidance No. HS(G) 65*, Her Majesty's Stationery Office, Sheffield.

HSE, 1997b. *The Application of Risk Assessment to Machinery.* Report SME/484/ 94/95, Health and Safety Laboratory, Health and Safety Executive, Sheffield.

ILO, 1998. *Meeting of Experts on Labour Statistics.* Report MELSOI/1998/1, Geneva.

Ingstad, O. and Bodsberg, L., 1989. CRIOP: *A Scenario-Method for Evaluation of the Offshore Control Centre.* SINTEF, Report No. STF75 A89028, Trondheim.

ISO, 1990/91/92. *Guidelines for Auditing Quality Systems. International Standards ISO 10011-1-3.* International Organization for Standardization, Geneva.

ISO, 1994. *Quality Management and Quality Assurance – Vocabulary (International Standard ISO 8402: 1994), Quality Systems – Model for Quality Assurance in Design, Development, Production, Installation and Servicing (International Standard ISO 9001: 1994,* International Organization for Standardization, Geneva.

ISO, 1996. *Environmental Management System – Specification with Guidance for Use. International Standard ISO 14001: 1996,* International Organization for Standardization, Geneva.

ISO, 1999. *Petroleum and Natural Gas Industries – Offshore Production Installations – Guidelines on Tools and Techniques for Identification and Assessment of Hazardous Events. Draft International Standard ISO/DIS 17776,* International Organization for Standardization, Geneva.

Johnson, W.G., 1980. *MORT Safety Assurance System.* Marcel Dekker, New York.

Juran, J.M., 1989. *Juran on Leadership for Quality – An Executive Handbook.* The Free Press, New York.

Kjellén, U., 1982. An evaluation of safety information systems at six medium-sized and large firms. *Journal of Occupational Accidents,* 3: 273–288.

Kjellén, U., 1983. *Analysis and Development of Corporate Practices for Accident Control.* Royal Institute of Technology, Thesis, Report No. Trita AVE-0001, Stockholm.

Kjellén, U., 1984. The role of deviations in accident causation and control. *Journal of Occupational Accidents,* 6: 117–126.

Kjellén, U., 1987. Simulating the use of a computerized injury and near accident information system in decision making. *Journal of Occupational Accidents,* 9: 87–105.

Kjellén, U., 1990. Safety control in design – Experience from an offshore project. *Journal of Occupational Accidents,* 12: 49–61.

Kjellén, U., 1992. Arbeidsulykker. In: *Grunnbok i arbeidsmiljøopplæring.* Tiden Norsk Forlag, Oslo.

Kjellén, U., 1993. *Skade- og hendelsesrapportering på Oseberg feltsenter og Oseberg C – Resultat av evaluering.* Norsk Hydro, Report No. NHT-F15-00026, Oslo.

Kjellén, U., 1995. Integrating analyses of the risk of accidents into the design process – Part II: Method for prediction of the LTI-rate. *Safety Science,* 19: 3–18.

Kjellén, U., 1997. Feedback control of accidents. In: Brune, D., Gerhardsson, G., Crockford, G.W. and D'Auria, D. (eds), *The Workplace,* Volume 1, Fundamentals of Health, Safety and Welfare. Scandinavian Science Publisher, Oslo.

Kjellén, U., 1998. Adapting the application of risk analysis in offshore platform design to new framework conditions. *Reliability Engineering and System Safety,* 60: 143–151.

Kjellén, U. and Hovden, J., 1993. Reducing risks by deviation control – a retrospection into a research strategy. *Safety Science,* 16: 417–438.

Kjellén, U. and Larsson, T.J., 1981. Investigating accidents and reducing risks – a dynamic approach. *Journal of Occupational Accidents*, 3: 129–140.

Kjellén, U., Menckel, E., Lauritzen, J. and Maijala, P., 1986. Utredning av arbetsolycksfall och tillbud som del i ett lokalt skyddsinformationssystem. *Arbete och Hälsa*, No. 3.

Kjellén, U., Tinmannsvik, R.K., Ulleberg, T., Olsen, P.E. and Saxvik, B., 1987. *SMORT – Sikkerhetsanalyse av industriell organisasjon*, offshore versjon (in Norwegian). Yrkeslitteratur, Oslo.

Kjellén, U., Boe, K. and Løge Hagen, H., 1997. Economic effects of implementing internal control of health, safety and environment: A retrospective case study of an aluminium plant. *Safety Science*, 27: 99–114.

Kletz, T., 1994. *Learning from Accidents*. Butterworth-Heinemann Ltd, Oxford.

Knox, N.W. and Eicher, R.W., 1992. *MORT User's Manual*. Systems Safety Development Centre, EG&G Idaho, Inc., Report No. SSDC-4, Rev. 3, Idaho Falls.

Kolb, D., 1984. *Experiential Learning. Experience as the Source of Learning and Development*. Prentice-Hall, Englewood Cliffs, NJ.

Komaki, J., Barwick, K.D. and Scott, L.R., 1978. A behavioral approach to occupational safety: Pinpointing and reinforcing safe performance in a food manufacturing plant. *Journal of Applied Psychology*, 63: 434–445.

Kontogiannis, T., 1999. User strategies in recovering from errors in man–machine systems. *Safety Science*, 32: 49–68.

Krause, T.R., Seymor, K.J. amd Sloat, K.C.M., 1999. Long-term evaluation of a behavior-based method for improving safety performance: a meta-analysis of 73 interrupted time-series replications. *Safety Science*, 32: 1–18.

Laflamme, L., 1996. Age-related accident ratios in assembly work: A study of female assembly workers in the Swedish automobile industry. *Safety Science*, 23: 27–37.

LaPorte, T.R. and Consolini, P.M., 1991. Working in practice but not in theory: Theoretical challenges of 'High-reliable organizations'. *Journal of Public Administration Research and Theory*, 1(1): 19–47.

Laufer, A., 1987. Construction accident costs and management motivation. *Journal of Occupational Accidents*, 8: 295–315.

Lazarus, R.S. and Folkman, S., 1984. *Stress, Appraisal and Coping*. The Free Press, New York.

Leplat, J., 1978. Accident analyses and work analyses. *Journal of Occupational Accidents*, 1: 331–340.

Levitt, R.E. and Samelson, N.M., 1993. *Construction Safety Management* (2nd ed.). John Wiley & Sons, Inc., New York.

Loeb, P.D., Talley, W.K. and Zlatoper, T.J., 1994. *Causes and Deterrents of Transportation Accidents – An Analysis by Mode*. Quorum Books, Westport, Connecticut.

March, J.G. and Simon, H.A., 1958. *Organizations*. John Wiley & Sons, New York.

Matson, E., 1988. *Beregningsprinsipper for kostnader ved yrkesulykker* (in Norwegian). SINTEF Report STF83 A88007, Trondheim.

Maycock, G., 1997. Accident liability – the human perspective. In: Rothengatter, T. and Vaya, E.C. (eds), *Traffic and Transport Psychology*. Pergamon, Amsterdam.

McKennan, F., 1983. Accident proneness – a conceptual analysis. *Accident Analysis and Prevention*, 15: 65–71.

Menckel, E., 1990. Intervention and co-operation – occupational health services and prevention of occupational injuries in Sweden. Doctoral thesis, Arbete och hälsa, 1990: 31.

Mintzberg, H., Raisinghani, D. and Thèorèt, A., 1976. The structure of 'unstructured' decision processes. *Administrative Science Quarterly*, 21: 246–275.

Niskanen, T., 1994. Safety climate in the road administration. *Safety Science*, 17: 237–255.

Nonaka, I. and Takeuchi, H., 1995. *The Knowledge Creating Company*. Oxford University Press, New York.

Norsk Standard, 1996/97/98/99. *Technical Safety* (Norsok Standard S-001). *Working Environment* (Norsok Standard S-002). *Environmental Care* (Norsok Standard S-003). *Machinery – Working Environment Assessment and Documentation* (Norsok Standard S-005). *Health, Safety and Environment During Construction* (Norsok Standard S-CR-002), *Risk and Emergency Preparedness Analysis* (Norsok Standard Z-013). Norsk Standard, Oslo.

Norwegian Petroleum Directorate, 1990. *Regulations relating to the implementation and use of risk analysis in the petroleum activities*. Stavanger.

Norwegian Petroleum Directorate, 1995. *Regulations relating to systematic follow-up of the working environment in the petroleum activities*. Stavanger.

OSHA, 1971/97. *Recording and reporting occupational injuries and illness. Regulations*, Part 1904, Occupational Safety & Health Administration, U.S. Department of Labor, Washington, D.C.

OSHA, 1989. *Safety and Health Program Management Guidelines; Issuance of Voluntary Guidelines*. Federal Register No. 54: 3904–3916, Occupational Safety & Health Administration, U.S. Department of Labor, Washington, D.C.

OSHA, 1994. *Process Safety Management*. Occupational Safety & Health Administration, Report OSHA-3132, U.S. Department of Labor, Washington, D.C.

Persson, L.-O. and Sjöberg, L., 1978. The influence of emotions on information processing. University of Gothenburg, *Psychology Reports*, Vol. 8, No. 7, Gothenburg, Sweden.

Perrow, C., 1984. *Normal Accidents*. Basic Books, New York.

Rasmussen, J., 1993. Learning from experience? How? Some research issues in industrial risk management. In: Wilpert, B. and Qvale, T. (eds), *Reliability and Safety in Hazardous Work Systems*. Lawrence Erlbaum Associates, Hove, East Sussex.

Rasmussen, J., Duncan, K. and Leplat, J. (eds), 1987. *New Technology and Human Error*. John Wiley & Sons, Chichester.

Reason, J., 1990. *Human Error*. Cambridge University Press, New York.

Reason, J., 1991. Too little and too late: A commentary on accident and incident reporting systems. In: Van der Schaaf, T.W., Lucas, D.A. and Hale, A.R. (eds): *Near-miss Reporting as a Safety Tool*. Butterworth-Heinemann, Oxford.

Reason, J., 1997. *Managing the Risks of Organizational Accidents*. Ashgate, Hampshire.

Reason, J., 1998. Achieving a safe culture: theory and practice. *Work & Stress*, 12: 293–306.

Reynard, W.D., 1986. The development of the NASA Aviation Safety Reporting System. *NASA*, Reference Publication No. 1114, Mountain View, CA.

Rijpma, J.A., 1997. Complexity, tight-coupling and reliability: Connecting normal accident theory and high reliability theory. *Journal of Contingencies and Crisis Management*, 5(1): 15–23.

Rockwell, T.H., 1959. Safety performance measurement. *Journal of Industrial Engineering*, 10(1): 12–16.

Rognstad, K., 1993. Costs of occupational accidents and diseases in Norway. *European Journal of Operational Research*, 75: 553–566.

Rosness, R., 1995. *Kostnadseffektiv prioritering av HMS-tiltak* (in Norwegian). SINTEF Report STF75 A95031, Trondheim.

Rothengatter, T. and Vaya, E.C. (eds), 1997. *Traffic and Transport Psychology. Theory and Application.* Pergamon, Amsterdam.

Ruuhilehto, K., 1993. The management oversight and risk tree (MORT). In: Suokas, J. and Rouhiainen, V.: *Quality Management of Safety and Risk Analysis.* Elsevier, Amsterdam.

Saari, J. 1998. Participatory workplace improvement process. *ILO Encyclopaedia of Occupational Health and Safety*, Geneva.

Saari, J. and Näsänen, M., 1989. The effect of positive feedback on industrial housekeeping and accidents; a long-term study at a shipyard. *International Journal of Industrial Ergonomics*, 4: 201–211.

Salminen, S., Saari, J., Saarela, K.L. and Räsänen, T., 1992. Fatal and non-fatal occupational accidents: identical versus differential causation. *Safety Science*, 15: 109–118.

Salvendy, G. (ed.), 1987. *Handbook of Human Factors.* John Wiley & Sons, New York.

Sanders, M.S. and McCormick, E.J., 1992. *Human Factors in Engineering and Design.* McGraw-Hill, Inc., New York.

Senneck, C.R., 1973. Over 3-day absence and safety. *Applied Ergonomics*, 6(3): 147–153.

Shannon, H.S. and Davies, J., 1998. MAIM: The Merseyside accident information model. *ILO Encyclopaedia of Occupational Health and Safety*, International Labour Office, Geneva.

Shannon, H.S. and Manning, O.P., 1980. Differences between lost-time and non-lost time industrial accidents. *Journal of Occupational Accidents*, 2: 265–272.

Simonds, R. and Shafai-Sahrai, Y., 1977. Factors apparently affecting injury frequency in eleven matched pairs of companies. *Journal of Safety Research*, 9(3): 120–127.

Skaar, S., 1994. *Internkontroll – ørkenvandring eller veien til det forjettede land?* SINTEF Report STF 82 A94002, Trondheim.

Sklet, S. and Mostue, B.A., 1993. *Kostnader ved arbeidsulykker i prosess- og verkstedsindustrien* (in Norwegian). SINTEF Report STF75 A92032, Trondheim.

Smith, M., Cohen, H., Cohen, A. and Cleveland, R., 1978. Characteristics of successful safety programs. *Journal of Safety Research*, 10: 5–15.

Stanton, N. and Glendon, I., 1996. Risk homeostasis and risk assessment. *Safety Science*, 22: 1–13.

Statoil, 1997. Contractor HES qualification and evaluation. *Document No. ANS 0121E*, Stavanger.

Stellman, J.M. (ed.), 1998. *Encyclopaedia of Occupational Health and Safety. 4th Edition.* International Labour Office, Geneva.

Sulzer-Azaroff, B., Haris, T.C. and McCann, K.B., 1994. Beyond training; Organizational performance management techniques. *Occupational Medical: State Art Review* 9(2): 321–339.

Suokas, J., 1985. On the reliability and validity of safety analysis. Doctoral dissertation. Technical Research Centre of Finland, Publication No.25, Espoo.

Suokas, J. and Rouhiainen, V. (eds), 1993. *Quality Management of Safety and Risk Analysis*, Elsevier, Amsterdam.

Surry, J., 1974. *Industrial Accident Research. A Human Engineering Appraisal.* Labour Safety Council, Ontario Ministry of Labour, Toronto.

Svenson, L. and Vidén, S., 1994. *Accidents involving Volvo trucks resulting in driver injury, and the estimated effect of the SRS airbag*, Volvo Truck Corporation, Accident Investigation Report 4, Gothenburg.

Swain, A.D., 1974. *The Human Element in Systems Safety.* Industrial and Commercial Techniques Ltd., Surrey.

Swedish National Road Administration, 1999. *Road Traffic Safety Report 1998.* Publication 1999: 35E, Vägverket, Borlänge, Sweden.

Tarrants, W.E., 1980. *The Measurement of Safety Performance.* Garland STPM Press, New York.

Tinmannsvik, R.K., 1991. *Bruk av diagnoseverktøy i sikkerhetsstyring. Norges Tekniske Høgskole*, Doctoral dissertation 1991: 32, Trondheim.

Tuominen, R. and Saari, J., 1982. A model for analysis of accidents and its applications. *Journal of Occupational Accidents*, 4: 263–273.

Van Court Hare, 1967. *System Analysis: A Diagnostic Approach.* Harcourt Brace & World, New York.

Van der Schaaf, T.W., Lucas, D.A. and Hale, A.R. (eds), 1991. *Near-miss Reporting as a Safety Tool.* Butterworth-Heinemann, Oxford.

Van der Want, P.D.G., 1997. Tripod incident analysis methodology. In: J. van Steen (ed.), *Safety Performance Measurement.* Institution of Chemical Engineers, Warwickshire, UK.

Vinnem, J.E., 1999. *Risk Levels on the Norwegian Continental Shelf.* Preventor, Bryne, Norway.

Watson, I.D., 1997. *Applying Case-Based Reasoning: Techniques for Enterprise Systems.* Morgan Kaufmann Publishers, San Francisco, CA.

Weddle, M.G., 1996. Reporting occupational injuries: The first step. *Journal of Safety Research*, 27: 217–223.

Wendel, E., 1998. SHE experience transfer from operations to projects (in Norwegian). Norwegian University of Science and Technology, Masters thesis, Trondheim.

Wig, B.B., 1996. *Quality Improvement as a Craft.* Norwegian Association for Quality and Leadership, Stavanger.

Wilde, G.J.S., 1982. The theory of risk homeostasis: implications for safety and health. *Risk Analysis*, 2: 209–225.

Wilpert, B. and Quale, T. (eds), 1993. *Reliability and Safety in Hazardous Work Systems. Approaches to Analysis and Design.* Lawrence Erlbaum Associates Ltd., East Sussex.

Zohar, D., 1980. Safety climate in industrial organizations: Theoretical and applied implications. *Journal of Applied Psychology*, 65: 96–102.

Name index

Aaltonen, M.V.P. 63
Adams, N. 8
Alteren, B. 251
Andersson, R. 38–9
Argyris, C. 129, 258
Ashby, W.R. 124–5, 137, 144, 160, 190, 318
Avedal, C. 359

Béland, F. 255
Benner, L. 38–9, 166
Bird, F.E. 34, 39, 67–8, 70–1, 77, 207
Blindheim, G. 257
Bodsberg, L. 302
Boe, K. 257
Bolman, L.G. 13, 21
Booth, M.J. 61, 64
Briscoe, G.J. 76, 216, 218
Brown, R.L. 255

Carter, N. 155, 168
Cohen, A. 8
Consolini, P.M. 110
Cornelison, J.D. 70–1
Cox, S. 255

Davies, J. 165
Deal, T.E. 13, 21
Dedobbeleer, N. 255
DeJoy, D.M. 80
Deming 122, 159, 366
Döös, M. 101, 108

Eicher, R.W. 45
Eisner, H.S. 249
Elmasri, R. 199
Englund, A. 344, 346, 349, 351–2, 355–6

Flin, R. 255
Folkard, S. 349
Folkman, S. 104

Germain, G.L. 34, 39, 67–8, 70–1, 77, 207
Gibson, J. 39
Glendon, A.L. 97–9
Grimaldi, J.V. 37, 57, 62–3, 330
Guastello, S.J. 249

Haddon, W. 20, 38–41, 46, 82–4, 346, 354
Hale, A.R. 13, 79, 98–9, 131, 170, 185
Harms-Ringdahl, L. 267
Harrison, F.E. 120, 137
Hatakka, M. 355
Heinrich, H.W. 8, 13, 32–3, 61–2, 67–8, 95, 114, 155, 215
Hendrick, K. 166
Hockey, G.R.J. 107–8
Holmes, H. 255
Hovden, J. 13, 31, 55

Ingstad, O. 302
Ishikawa, K. 165
Ives, G. 157

Johnson, W.G. 8, 45, 48, 178
Juran, J.M. 21, 123

Kjellén, U. 4–5, 8, 36–7, 39, 48, 55–6, 63, 67–70, 71, 73, 79–80, 100, 125, 127, 144–5, 151–2, 160–1, 168, 170, 181, 198, 213, 294–5, 328, 330
Kletz, T. 146
Knox, N.W. 45
Kolb, D. 6

Komaki, J. 8, 244
Kontogiannis, T. 102–3, 106
Krause, T.R. 245

Laflamme, L. 214
Lagerlöf, E. 38–9
LaPorte, T.R. 110
Larsson, T.J. 31, 36–7, 39, 68–70, 73
Laufer, A. 62
Lazarus, R.S. 104
Leger, J.P. 249
Leplat, J. 43, 68
Levitt, R.E. 236, 330
Lindtvedt, J. 257
Loeb, P.D. 344, 348–51, 357
Lucas, D.A. 51–2, 77

McCormick, E.J. 44, 105–6, 119
McKennan, F. 32, 348
Manning, O.P. 154
March, J.G. 118
Matson, E. 62
Maule, A.J. 107–8
Maycock, G. 348
Menckel, E. 8, 155, 170
Miller, D.P. 100
Mintzberg, H. 118
Mostue, B.A. 62–3, 169

Navathe, S.B. 199
Niskanen, T. 255
Nonaka, I. 23
Näsänen, M. 245

Perrow, C. 109
Persson, L.-O. 104

Rasmussen, J. 55, 98, 101–2, 109, 109, 327, 347
Reason, J. 21, 35, 39, 51, 70–1, 77, 101–2, 104, 110, 162, 165, 256
Reynard, W.D. 157
Rijpma, J.A. 110
Rockwell, T.H. 233, 243
Rognstad, K. 62
Rosness, R. 63, 101
Rothengatter, T. 346

Rouhiainen, V. 43, 267, 323, 326
Ruuhilehto, K. 48

Saari, J. 152, 165, 245
Salminen, S. 154
Salvendy, G. 105–6
Samelson, N.M. 236, 330
Sanders, M.S. 44, 105–6, 119
Senneck, C.R. 151–2
Shafai-Sahrai, Y. 8
Shannon, H.S. 154, 165
Simon, H.A. 118
Simonds, R.H. 8, 62
Sjöberg, L. 104
Sklet, S. 62–3, 169
Skaar, S. 12
Smith, M. 8
Stanton, N. 97
Stellman, J.M. 17
Stout, N. 209
Sulzer-Azaroff, B. 245
Suokas, J. 43, 136, 265, 267, 323, 326
Surry, J. 95–6, 101
Svenson, L. 360
Swain, A.D. 72–3, 100–1

Takeuchi, H. 23
Tarrants, W.E. 135, 154, 243
Taylor, F.W. 133
Tinmannsvik, R.K. 75, 250
Tuominen, R. 165

Van Court Hare 124, 126, 189, 246, 256
Van Der Schaaf, T.W. 51, 77, 121, 155, 167
Van der Vant, P.D.G. 254–5
Vaya, E.C. 346
Vidén, S. 360
Vinnem, J.E. 311

Wagenaar, W.A. 74
Watson, I.D. 210
Weddle, M.G. 151
Wendel, E. 319
Wig, B.B. 78, 122, 165, 170, 272
Wilde, G.J.S. 24, 97

Zohar, D. 255

Subject index

AAAM 60
Abbreviated Injury Scale: see AIS
acceptance criterion 266, 294, 317–18, 376
accident analysis framework 53–8
accident concentration analysis 211–14, 328
Accident Consequence Tree method 62–3
accident costs 61–3, 168–9; direct and indirect 62
accident counter-measures 82–94
accident investigation 56–7; at three levels 147–9; bi-level 170; computer-supported 167–8; form 164; immediate 260–8; in-depth 173–86, 353–5, 359–60; legal aspects 186; procedure 187–8
accident investigation stairs 78, 147
accident models 31–52
accident proneness theory 32, 348
accident repeaters 209
accident reporting 146–54; procedure 187–8; to authorities 150
accident risk: definition 376
accident sequence 53–7; display of 163–7; start-end 55
accident statistics: analysis of 209–21; use in SHE management 334, 340
accident type classification 66
accounting model 62–3
ACSNI 51
action error analysis 268–9
action research 8
active barriers 86
active safety 349–50
acts of God 32
after-the-fact accident investigation 174
AIS (Abbreviated Injury Scale) 60

ALARP (as low as reasonably practicable) 64, 266, 277, 301, 325, 376
alcohol 349
Alexander L. Kielland 311
American National Standards Institute (ANSI) 33
analysis object 265, 268–9, 272, 280; definition 376
analytic methods 78
ANSI Z16.2 33
Arbeidstilsynet 15, 195
Arbetarskyddsstyrelsen 15
as low as reasonably practicable: see ALARP
Ashby's law of requisite variety 124–5, 137, 144, 160, 190, 318
ASRS 156
assembly of machinery 285
assumed risk 47, 376
attribution theory 80–1
Aviation Safety Reporting System: see ASRS

barrier 20–1, 82–94, 95, 109, 132; active and passive 86; availability of 260–1, 341–2; inspection of 193–4
behavioural theory 152
behaviour in the face of danger: model of 98–9
behavioural sampling 243–7
bi-level accident investigation 170
black-spot analysis 211; see also accident concentration analysis
boundary conditions 11–19
Bravo 311
British Airway Safety Services 162
British Standards Institution 18, 196
BS 8800 196

cake bake analogy 204
case-based reasoning 210
causal factor: *see* cause of accident
causal-sequence models 32–6
cause–effect diagram 166
cause of accident: analysis of 77–81, 215–16; definition 376
CE sign 285–6
CEN 39, 89, 265, 285
chain of multiple events theory 32
change analysis 178
checklist: on contributing factors 74; on deviations 69; on hazards 43, 274; on measures to grant hot-work permit 94; on inspection themes 191; on performance-shaping factors 308; on typical jobs 274; on uses of a SHE information system 368; SMORT 180–1
chemical hazards: analysis of exposure to 282–3; EC directive on 83–4, 266
Chemical Manufacturers Association 19
chi-square analysis 216
classification 33, 205–9; accident causes 207; accident type 66; consequences 58–61; deviations 68–9; human errors 101–2; injury 59; injury agency 66; part of body 59
Clean Air Act Amendment 15, 266
closed loop 5, 366
Coarse analysis 267–79, 328–9
coding: *see* classification
cognitive stages 101
comparison analysis 267–9, 294–301, 330
computer support: development of 365–73; in accident investigations 167–8; in distribution of information 186–7; in retrieval and analysis of accident data 198–209
concept study phase 314, 322–5, 327–9
consequences of accidents 40–2, 58–65; categories 60–1; measures of 59–61
construction-site safety 330–5
contributing factors 55, 70–5, 177; definition 376
control chart 228–36, 259–60, 335, 340
cost efficiency 139
costs of accidents: *see* accident costs

coverage 137
crash-involvement rate 345
crashworthiness 350
circadian rhythm 349
CRIOP analysis 267–70, 302–8, 326
culpability 131–2

danger zone 54, 95; definition 376
data model 202
database environment 202
database: accessing 200–5; definition 199–200, 371–2; on accidents and near accidents 198–209, 339; on potential accidents 278–9
decision gate 313–15, 321–2
decision-making process 118–22
declaration of conformity 285
deductive risk analysis methods 267
defences 35
defences in depth philosophy 84, 107, 109–10, 125
Deming's circle 122, 159, 366
Det Norske Veritas 18, 250
determining factors 37
deviation 36–8, 55–8, 67–9, 171; definition 376; checklist on 69
diagnosis 114–15, 117–22, 147; definition 376
diagnostic diagram 185
dimensioning accidental event 323
Domino theory 32–5, 95
double-loop learning 130
driver 348–9

E&P Forum's SHE management model 19, 49–50, 249
education and training 105
EMAS 195
emotions: influence of 104–5
employer's responsibilities 14–16
EN 1050 285–6, 290–3
EN 292 89–92, 285–6, 290–3
energy analysis 267–9
energy model 39–43, 172, 271
EPA 15, 266
ergonomics systems view 69
error recovery 102–4, 106
European Council 15–16, 83, 89, 193, 195, 266, 282–3, 285
Eurostat 58, 60
evaluation research 8
event tree analysis 268–9
experience carrier 221–3, 319; definition 376

experiential learning 6
expert judgements 78, 81
explicit effects 9, 23
ex-post facto analysis 8, 215
externalisation 23
extreme-value projection 218–21

FAA 156
factorial analysis of correspondence
  (FAC) 214
failure mode and effect analysis 268–9
failure state profile 254–5
failure types 35
fatal accident rate (FAR) 238
fatigue 348–9
fault tree analysis 43, 268–9
feasibility study phase 313
feed forward 123
feedback control 21, 123, 134;
  definition 376
feedback mechanisms in road
  transportation safety 356–7
FERSI 354
filters: in accident reporting 151; in
  data collection 144
Finnish model for accident analysis
  165
fires and explosions: protection against
  83–6
fishbone diagram 166
free-text: description 205–6; search 209

general failure types 74, 254
goal-oriented requirements 266, 316
group problem solving 168–73, 338,
  340, 367

Haddon's accident prevention
  strategies 39–41, 82–6, 126
Haddon's phase model 38, 346, 354
Hale's problem-solving cycle 120–1
harmonised standards 92
hazard: checklist on 43, 274; definition
  376
hazardous chemicals: *see* chemical
  hazards
hazardous event: definition 377
hazardous machines 91–2
HAZOP 268–9, 326
Health and Safety Executive: *see* HSE
Heinrich's model of decision making
  114, 118
hierarchical ascendant classification
  (HAC) 214

high potential incident rate 242–3
high reliability organisation 110
high-risk industry 11–12, 15
HSE 15–16, 286
human errors 32, 100–6, 307;
  definition 100; models 51–2;
  prevention 105; reporting 157;
  taxonomies 101–2
human factors 32
human information processing 6–7,
  44–5, 95–100, 119–20, 134, 308,
  346–7
human performance-shaping factors
  72–3, 306, 308
human reliability 101
human resource perspective 13

iceberg theory 155, 352–3
ILCI model 34–5, 39, 49, 56, 67,
  70–1, 76–7, 73, 79–80, 145, 161,
  206–7, 249
ILO 58, 60, 65–6, 68–9
in-depth accident investigation 173–86,
  353–5, 359–60
incidents 54, 65–6; definition 377; rate
  242–3
indirect costs of accidents 62
inductive risk analysis methods 267
industrial engineering systems view 69,
  132–3
information technology 6, 365, 374
injury agency classification 66
Injury Potential Matrix 61
injury: classification 58–9
input-driven development process 7
INRS model 43–4
internal control legislation 15, 189–90,
  195
internalisation 23
International Loss Control Institute
  (ILCI) 34
International Safety Rating System: *see*
  ISRS
international standards 18–19
interval scale 135
investigation commission 173, 175–9
ISA model 38–9, 206
ISA (Intelligent Safety Assistant)
  168
ISO 18, 67, 175, 178, 182, 194–6,
  265, 267, 320
ISO 10011 175, 178, 182, 195
ISO 14000 18, 195–6, 358
ISRS 34, 249–51, 258, 343, 374

job-safety analysis 267–70, 280–4, 329, 343

key performance indicator 137
knowledge-based behaviour 98, 102

latent errors 70, 109
learning from incidents: degree of 256–7
liability 131–2
Lloyds Register 18
logical tree 43, 166
losses: types of 53–4
lost-time injury (LTI): definition 59–60
lost-time injury frequency rate: *see* LTI-rate
LTI-rate 22, 135, 228–37, 377
Lucas' framework of organisational cultures 51–2, 77

machinery directives 89–92
machinery safety 39, 43, 89–92, 266, 285–93
machinery: definition 285
MAIM 165, 206
main study phase 314
major accidents 15, 83–6, 107; prevention of 322–7
Management of Health and Safety at Work Regulations 15
Management Oversight and Risk Tree: *see* MORT
man–vehicle–road–environment model 346
market-pricing model 61–3
Mintzberg's model of the decision-making process 118
MORT 45–9, 70–1, 76–7, 168
Multi-linear Events Chartering Method 38, 165

NASA 156
near accidents: definition 55, 377; reporting 146–7, 154–60
negative feedback 123
Nelson Aalen plot 240–1
nominal scale 135
nonconformity: definition 377
non-governmental organisations 19
normal accidents 109
normal distribution 112–13
norms: types of 36–7, 67–8
Norsk Standard 267, 286, 316, 323, 329, 331

NORSOK S-002 320
NORSOK S-005 286, 329
NORSOK S-CR-002 331, 334
NORSOK Z-013 267, 323
Norwegian Directorate of Labour Inspection 40–2
Norwegian Petroleum Directorate 294, 320

OARU accident-investigation method 170
OARU model 36, 39, 73
Occupational Accident Research Unit (OARU) 36
occupational accidents: reporting 16, 146; prevention of 327–30
Occupational Health and Safety Administration: *see* OSHA
OHSAS 18001 18
oil and gas industry 311–43
one-on-one inspection 192, 341
on-the-scene accident investigation 174
order of feedback: *see* Van Court Hare
ordinal scale 135
organisational defences 129–30
organisational learning 7, 22–4, 134, 375
organisational perspectives 13
OSHA 15–16, 65–6, 193
OSHA recordable accidents 16
outer context 14–19
output-driven development process 7

passive barrier 86
performance feedback 106
performance indicators for traffic safety 361–2
permit to work system 93–4, 197, 335
persistent feedback control 114–15, 123
phase model of investment projects 313–15
Piper Alpha 302, 311
Poission distribution 112–13
police investigation 132, 352
political model of decision-making 120
political perspective 13
potential losses of accidents 63–5
precision: degree of 203, 377
prescriptive requirements 266, 316
problem solving group 168–73, 338, 340, 367
procedure: accident reporting and investigation 187–8; workplace

inspections 190; permit-to-work system 93–4
process models of accidents 36–9
project execution phase 315, 325–7, 329–30
proprietary software 369, 377

qualifying experience 7, 22; definition 376
qualitative data 135
quality assurance management 76–7
quality improvement project 170, 246
quantitative data 135

Rasmussen's skill–rule–knowledge framework 98, 347
ratio scale 135
rational-choice model 118
recordable injury 60, 377
regression-to-the-mean effect 214
regulations on record keeping and reporting 16–17
reliability 137, 150, 377
remedial actions: evaluation of 172
REPA 314, 323–6, 343
restricted work injury (RWI) 60
retrieval: degree of 203, 377
risk analysis 263–308, 342–3; activities 265; definition 377; team 272–3
risk and emergency preparedness analysis: *see* REPA
risk assessment 15; definition 377; machinery 285–93, 329; model 295–6
risk estimation 275–7; matrix 276, 283
risk homeostasis theory 24, 97–8
risk management 3
risk of accidents: definition 237
risk taking 24
roads administration 360–2
root cause 34, 55, 70–2, 76–7; analysis of 76; definition 377
rule-based behaviour 98, 102

Safety and Health Programme Rule 15
safety campaign 106
safety climate 255–6
safety culture 77
safety data sheet 282
Safety Element Method: *see* SEM
safety information system 8
Safety Management and Organisation Review Technique: *see* SMORT
safety research 374–5

safety, health and environment: *see* SHE
scale of measurement: types of 135
scenario analysis 303–6
scientific management 133
selection of personnel 105
self reporting 355
SEM 251–3
semi-submersible platform 311–12
severity distribution analysis 216–18
severity rate 238
Seveso directives 15, 266
SHE audit 194–7, 331, 334, 343; definition 377
SHE culture 51–2, 132–3
SHE information system 4–10, 115–17, 131; definition 377; design of 371–3; improvement of 365–70; requirements for 134, 136–9; uses 368–9
SHE inspection 189–94; checklist 191
SHE management 4, 132–3, 248, 331, 336–7; definition 377; in project work 316; models 45–51; of contractor work 331–5; standards 18
SHE performance indicators 225–7; causal-factor based 248–57; definition 377; loss-based 228–41; process-based 242–7; requirements for 134–6; selection of 258–61
SHE performance measurement 225–61; problems of 233–6
SHE programme 320–2, 333
significant change 113
single-loop learning 130
skill-based behaviour 98, 102
small companies 12
smallest efficient data set 200–2, 371
SMORT 48–9, 70–1, 76–7; analysis 178–86, 335; in audits 196–7; questionnaire 379–408
socialisation 23
socio-technical research tradition 13
software-engineering approach 369
stakeholders 12
statistical methods 78
Statoil 249
STEP diagram 166, 172, 179, 305
stop rules 31
strategy for the selection of safety measures 89–92
structural perspective 13, 45
structured decision-making process 22

Surry's decision model 95–7
Swedish Information System on
    Occupational Accidents (ISA) 38,
    161
Swedish National Roads
    Administration 360–3
symptoms 77–8, 117, 377
systems ergonomics 69

tacit effects 9, 23
taxonomy 66, 135, 207, 377
total recordable injury rate: *see* TRI
    rate
traffic accident risk: measures of 345
traffic accidents 344–62; reporting of
    352–3
traffic-conflict technique 355–6
traffic environment 351–2
traffic-accident data system 360–1
trend 232
TRI rate 239–40, 258–9
trial and error learning 3
Tripod Beta 165
Tripod Delta 254
Tripod model 35–6, 39, 51, 70–1, 74,
    77
truck manufacturer 359–60
trucking company 357–9
trucking industry 344–62
tunnel vision 104

Tuttava 245–6
type I and II errors 203

under-reporting 151
uni- and bi-variate distribution analysis
    210–11
unsafe act 32, 35, 67–8
unsafe condition 32, 67–8; reporting of
    159
unscheduled manual intervention
    107–9

validity 136, 377
Van Court Hare's hierarchy of order of
    feedback 126–9, 148, 160, 172
variety 377
vehicle 349–50

Walk–Observe–Communicate (WOC)
    concept 192
well-defended system 109–10
why-why diagram 166
workers' compensation system 17, 151
Working Environment Committee 175,
    184, 341
workplace design 106
workplace inspection 190–3, 334,
    341

zero-goal mindset 236–7, 360